ALTERNATIVE DEVELOPMENT

T0304125

For Ragnhild Lund

Alternative Development
Unravelling Marginalization, Voicing Change

Edited by

CATHRINE BRUN
Norwegian University of Science and Technology, Norway

PIERS BLAIKIE
University of East Anglia, UK

MICHAEL JONES
Norwegian University of Science and Technology, Norway

LONDON AND NEW YORK

First published 2014 by Ashgate Publishing

2 Park Square, Milton Park, Abingdon, Oxon OX14 4RN
711 Third Avenue, New York, NY 10017, USA

Routledge is an imprint of the Taylor & Francis Group, an informa business

First issued in paperback 2016

British Library Cataloguing in Publication Data
A catalogue record for this book is available from the British Library

The Library of Congress has cataloged the printed edition as follows:
Alternative development : unravelling marginalization, voicing change / by Cathrine Brun, Piers Blaikie and Michael Jones.
     pages cm
 Includes bibliographical references and index.
  ISBN 978-1-4724-0934-8 (hardback)
1.   Developing countries–Economic conditions–Case studies. 2.   Developing countries–Social conditions–Case studies. 3.   Economic development–Developing countries–Case studies. 4.   Women in development–Developing countries– Case studies. I. Brun, Cathrine, author, editor of compilation. II. Blaikie, Piers M., 1942- author, editor of compilation. III. Jones, Michael, 1944 December 16- author, editor of compilation.
  HC59.7.A7758 2013
  338.9009172'4–dc23

                                                                      2013020839

ISBN 978-1-4724-0934-8 (hbk)
ISBN 978-1-138-25704-7 (pbk)

Contents

List of Figures and Tables

Figures

Tables

List of Contributors

Samuel Agyei-Mensah is Associate Professor in the Department of Geography and Resource Development, and also Dean of the Faculty of Social Studies, University of Ghana. He has worked at the University of Ghana since 1997. He has a PhD in Geography from the Norwegian University of Science and Technology (NTNU), Trondheim, Norway. His publications and research deal largely with issues relating to medical geography, demography and development issues.

Stuart C. Aitken is Professor and Chair of Geography at San Diego State University, and Director of the Center for Interdisciplinary Studies of Young People and Space (ISYS). He has also held a position as Adjunct Professor at the Norwegian University of Science and Technology (NTNU). His research interests include film and media, critical social theory, qualitative methods, children, families and communities, and he has published widely on these themes. He has worked for the United Nations on issues of children's rights, migration and dislocation.

Chamila T. Attanapola has a PhD in Geography from the Faculty of Social Science and Technology Management, Norwegian University of Science and Technology (NTNU). She currently works as coordinator of NTNU's research programme on globalization. She has published articles in international journals such as *Social Science & Medicine*; *Gender, Technology and Development*; *Norwegian Journal of Geography* and *Women & Environments*. Her present research interests include globalization, women's health, international migration, and gender and empowerment.

Fazeeha Azmi is Senior Lecturer in the Department of Geography, University of Peradeniya, Sri Lanka. She has an MPhil in Social Change and PhD in Geography from the Department of Geography at the Norwegian University of Science and Technology. Her PhD thesis focused on poverty and changing livelihoods in a large-scale state-sponsored development project in Sri Lanka. She is currently engaged in a research project focusing on youth in post war Sri Lanka in the Eastern Province.

Jonathan R. Barton is Director of the Centre for Sustainable Urban Development (CEDEUS) and Associate Professor in the Institute of Urban and Regional Studies, Pontificia Universidad Católica de Chile. He received his MA and PhD degrees from the University of Liverpool, and his current work focuses on the politics and planning of urban and regional sustainable development processes in Latin America.

Piers Blaikie is Professor Emeritus at the School of International Development, University of East Anglia, Norwich, UK, having served there since its inception in 1972. He took early retirement in 2002 and has remained active in the School. He has been Adjunct Professor at the Norwegian University of Science and Technology (NTNU). He is widely acknowledged to have been one of the founders of 'political ecology' and his work has been recognized by the Royal Geographical Society Edward Heath Award for published work on Africa and Asia (1994). He has been elected Life Member of the Norwegian Academy for Science and Letters, Oslo (2000). He was selected as one of 50 authors in *50 Key Development Thinkers*, edited by David Simon (2006), and as subject of a special issue of *Geoforum* 'In Honor of the Life Work of Piers Blaikie in Political Ecology and Development Studies' (2008).

Cathrine Brun is Associate Professor in Development Geography at the Norwegian University of Science and Technology (NTNU). Her teaching and research are in development geography with a particular focus on conflict, disaster, displacement and recovery. Her main geographical area of study is Sri Lanka, where she collaborates with universities, aid agencies and citizen groups.

Ingrid Eide is a sociologist from the University of Oslo. She was a cofounder of the Peace Research Institute (PRIO), Oslo, and was responsible for several studies on development cooperation practices. She has been active in politics (state secretary, member of parliament (Stortinget) and city councillor) as well as in research policy formulation, and has been a member of boards and committees in the United Nations system, including the United Nations International Research and Training Institute for the Advancement of Women (UN-INSTRAW) and the United Nations Educational, Scientific and Cultural Organization (UNESCO). She was Director of the United Nations Development Programme UNDP's Women in Development Programme in the late 1980s.

Arnt Fløysand is Professor at the Department of Geography, University of Bergen, Norway. He has a PhD in Human Geography from the University of Bergen. He is experienced in studies of capital transfer and economic restructuring at community and regional level (research area mainly in Latin America) and in studies of innovation and competitiveness at firm and clusters level (research area mainly in Europe). He has coordinated projects on the spatial dynamics of foreign direct investments (FDI) in Chile, Bolivia and Peru, and a project titled 'Negotiating New Political Spaces', following up findings from the FDI project.

Jennifer Hyndman is Professor and Director of the Centre for Refugee Studies at York University in Toronto, Canada, and formerly Adjunct Professor at the Norwegian University of Science and Technology (NTNU). Her research spans the continuum of forced migration from conflict zones to refugee resettlement in North America. Her work adopts a feminist approach that attends to issues of displacement, security and the geopolitics of asylum. She is the author of the books *Managing Displacement: Refugees and the Politics of Humanitarianism* (University of Minnesota Press, 2000) and *Dual Disasters* (Kumarian Press, 2011), and coeditor of *Sites of Violence: Gender and Conflict Zones* (University of California Press, 2004).

Michael Jones has been Professor of Geography at the Norwegian University of Science and Technology in Trondheim since 1985, having been a founding member of the Department of Geography in 1975. His main research fields are historical geography, landscape geography, legal geography and history of cartography, and he has published extensively in these fields. He has edited or coedited 10 books. He has contributed to the editing of two special issues on Sri Lanka for the *Norwegian Journal of Geography*, and is currently editor-in-chief of the journal. The first PhD completed under his supervision was that by Ragnhild Lund on gender and place in Sri Lanka.

Kyoko Kusakabe is Associate Professor of Gender and Development Studies at the School of Environment, Resources and Development, Asian Institute of Technology in Thailand. Her research focuses on gender issues in the informal economy and cross-border mobility. Her recent research involves citizenship of Burmese migrant workers in Thailand. She is an associate editor of the journal *Gender, Technology and Development*. She is currently engaged in a research project 'Mobile Livelihoods and Gendered Citizenship: The Counter-Geographies of Indigenous People in India, China and Laos', headed by Ragnhild Lund.

Haakon Lein is Professor in Geography at the Norwegian University of Science and Technology (NTNU). His main research interests lie within the field of development and environment in the Global South, with a particular focus on water resource management and water reforms, and the links between climate change, environmental hazards and social vulnerability. He has extensive fieldwork experience from Bangladesh, Tanzania and China (Xinjiang).

Li An is Associate Professor at the Department of Geography and Adjunct Professor at Research Center of Eco-Environmental Sciences (RCEES), Chinese Academy of Sciences. He has an MSc in Systems Ecology from the Chinese Academy of Sciences (1992) and in Statistics from Michigan State University (2002). He completed a PhD at Michigan State University in 2003. His research interests are in spatial analysis, geocomputation, landscape ecology and complexity theory. He was the recipient of the 2006 Outstanding Paper in Landscape Ecology from the USA Chapter of the International Association of Landscape Ecologists.

Ragnhild Lund is Professor in Geography, specializing in Development Studies, at the Norwegian University of Science and Technology (NTNU). Her research interests are in theories of development and geography, gender and place, development-induced displacement, transnational feminism, and women's activism. She has led externally funded projects on gender and development, forced migration, post-crisis recovery, and crisis communication, themes on which she has published widely. She is currently working on projects concerned with mobile livelihoods, rethinking gender, domestic violence, and cultural encounters. She has undertaken research especially in Asia and also in Africa.

George Owusu holds a PhD in Geography with a focus on urban and regional development and an MPhil in Social Change, both from the Norwegian University of Science and Technology (NTNU). He is Senior Research Fellow in the Institute of Statistical, Social and Economic Research (ISSER), University of Ghana, Legon. His main areas of research include urbanization and regional development, land tenure, decentralization and participatory approaches to development. He is a member of a core team of experts currently drafting a National Urban Policy (NUP) for Ghana. He has several publications on urban development and decentralization in national and international journals.

Smita Mishra Panda is Professor and Director of Research at the Centurion University of Technology and Management, Odisha, India. She has several years of experience of research and teaching in the area of gender and development and natural resource management (with a focus on water). She has been a guest researcher at the Department of Geography, Norwegian University of Science and Technology (NTNU). She has published extensively on gender and natural resource management issues, and her current research interests are in critiquing gender mainstreaming efforts in water and sanitation programmes, tribal women's rights, and mobility and gendered citizenship issues in the wake of neoliberalist policies in India.

Bernadette P. Resurrección is Senior Research Fellow at the Stockholm Environment Institute. She carries out research on livelihoods, migration and natural resource management. Her research has focused on the Philippines, Thailand, Vietnam and Cambodia. She has coedited the books *Gender and Natural Resource Management: Livelihoods, Mobility and Interventions* (Earthscan and IDRC, 2008) and *Water Rights and Social Justice in the Mekong Region* (Earthscan, 2011).

Hans Skotte is Professor in International Planning Studies at the Department of Urban Design and Planning, Faculty of Architecture and Fine Art, Norwegian University of Science and Technology (NTNU). After years of practice in Norway and abroad as architect and project manager in private and public service, he returned to academia to investigate how international nongovernmental organizations (INGOs) and governmental organizations (GOs) have contributed towards recovery of poor societies ravaged by war or disaster. He is engaged in a large research project in Bosnia on how private land claims justified by the 'sacrifice of war' affects the urban development of Sarajevo.

Vibeke Vågenes is Associate Professor at Bergen University College, where she is responsible for global studies and geography in teacher education. She completed her PhD in 1999 at the University of Bergen and has worked as Associate Professor at the University of Bergen and the University of Tromsø. Through years of working in Sudan she studied the gender system of Hadendowa Beja, and how this was changing in response to changes in the nation, such as Islamization and dramatic environmental instability with several periods of drought. Currently she is engaged in Tanzania, where a project is being set up concerning Koran schools in a developing society.

Berit Helene Vandsemb is Associate Professor of Development Studies, Faculty of Education and International Studies, Oslo University College. She received her MPhil and PhD in Geography from the Norwegian University of Science and Technology (NTNU). Her research interests include livelihoods and migration in South Asia.

Sarah M. Wandersee is a doctoral student in the Department of Geography at San Diego State University. Her research interests include landscape ecology, livelihoods, environmental security, remote sensing, agent-based and multilevel modelling, and survey methods. She graduated in geography in 2006 and earned her Master's degree in Costa Rica in Environmental Security and Peace. Her current dissertation and field research focuses on human–environment interactions in the Guizhou golden monkey habitat in China.

Charlotte Wrigley-Asante is Lecturer at the Department of Geography and Resource Development, University of Ghana, Legon. She has a PhD in Geography and Resource Development from the University of Ghana, Legon, and an MPhil in Human Geography from the University of Oslo, Norway. Her research areas include gender, poverty and empowerment issues of rural women and urban migrants.

Yeqin Yang is the Director of Guizhou Fanjingshan National Nature Reserve, China. He is Professor in plant ecology and animal behavior research with many years' experience in these areas. Some of his many publications include: *Research on the Fanjing Mountain* (Guizhou People's Publishing House, 1990), *Insects from Fanjingshan Landscape* (Guizhou Science and Technology Publishing House, 2006) and *Ecology of the Wild Guizhou Snub-nosed Monkey* (Guizhou Science and Technology Publishing House, 2002).

Preface

New development issues are emerging as people across the world face and negotiate the challenges arising from global economic recession, climate change and never-ending conflicts. Authors with a common interest in alternative visions and paths of development have come together in this book to show ways in which scholarship can contribute to unpack, understand and place under scrutiny processes of marginalization and how responses to change may be voiced. It is a book that speaks to current scholarship in development studies but at the same time questions and moves beyond such scholarship. There are challenges and possibilities inherent in the contexts in which alternative development is discussed today. The contributions to this book show that, despite shifting contexts, changing geopolitics and new development agendas, alternative development retains its relevance for securing continued focus on justice. We bring into the discussion the recent influences of feminism, postcolonial scholarship, studies of environment and society, conflict studies and discussions on the relationship between academic scholarship and practice.

In thinking critically about alternative development, many people have contributed with ideas, perspectives and words to this book. First, we would like to thank Professor Ragnhild Lund for the inspiration she has given the editors and contributors alike. In Ragnhild Lund's own words (from an unpublished draft):

> ... my research has reflected my empathy for the poor and marginalised, particularly women, and indigenous people, and their lived realities. During my research and teaching I have preached the inclusion of voices, focused on marginalisation and the 'subaltern' and how their agency should be activated.

The spirit of alternative development that we find in her words has been the starting point for this book and the underpinning of the individual contributions. We thank the contributors for their important texts and their endurance in developing the individual chapters and contributing to the joint book project. We thank the Department of Geography at the Norwegian University of Science and Technology (NTNU) for support and for initiating the project. Many thanks are due to Bodil Wold for administrative support and to Radmil Popovic for technical support in producing the map in Chapter 7.

I would like to thank the wonderful team of co-editors, Professor Piers Blaikie and Professor Michael Jones, for the ways in which they have engaged in this work: thanks for the many productive discussions, all the mutual support that has flowed through multiple emails and conversations, and not least the stamina exposed and work hours put into the project to enable the book to materialize.

Finally, many thanks to Valerie Rose and the team at Ashgate for their interest in the project and their work in finalizing the book. We also thank the anonymous reviewers for their generous and constructive feedback.

Cathrine Brun

Norwegian University Of Science And Technology, Norway

Chapter 1

Introduction

Alternative Development: Unravelling Marginalization, Voicing Change

Cathrine Brun and Piers Blaikie

Introduction – Alternative Visions and Paths of 'Development'

This book brings together a collection of essays that discuss alternatives to mainstream development thinking and practice, and how these alternatives may affect local and global processes of marginalization and change in the Global South. Alternative development is concerned with identifying and promoting alternative practices and redefining the goals of development. The book takes as its starting point the history of alternative ideas of development by engaging with the work of Professor Ragnhild Lund from the Norwegian University of Science and Technology, whose involvement in development geography spans four decades. Ragnhild Lund's career has balanced academic life with activism and policy work. The essays in the book honour her work by engaging with founding themes of alternative development such as local knowledges and practices, poverty, gender, environment and sustainable development, and by addressing recent debates such as forced migration, conflict and climate change. The themes of the book speak to academics, students of development studies, policy makers and activists in the Global South and North.

Development is a dynamic and fast moving concept and reality, and attaching the word 'alternative' to development must recognize this. Constant reappraisals occur of ideas, debates and policy. Alternative development questions who the producers of development knowledges and practices are, and aims at decentring development and geographical knowledge from the Anglo-American centre and the Global North. It involves resistance to dominant political-economic processes in order to understand the possibilities for nonexploitative and just forms of development (Watts 2003). 'Alternative' is frequently applied when we represent the marginal in a range of arenas – political, spatial, economic and environmental – and when we talk to and about the marginalized, i.e. individuals and groups whose voices are not heard, whose lives are deprived of basic needs and basic rights, and whose access to decision-makers is often restricted. In this book, we understand 'alternative' as writing and working with alternative agents, applying alternative methods and formulating alternative objectives to mainstream development.

However, rather than understanding 'alternative' in a binary relationship to mainstream, we find that alternative approaches should centre on a moral and political purpose that challenges the status quo leading to marginalization.

Alternative development has justice as one of its central themes. Distributive justice (the distribution of 'goods' and 'bads') is a principal theme together with procedural justice, whereby people have the right to be included in knowledge creation and decision-making concerning the social and physical environment in which they live. Participatory forums, citizens' associations, freedom of association and of the press, and transparency in decision-making are some of the enabling means by which procedural justice is served (Agyeman et al. 2003, Leach et al. 2010). Procedural justice also encompasses the research process and demands continuous introspection and discussion with research partners. Distributive justice and procedural justice, although analytically distinct, are inextricably linked since the former can only be achieved by the latter. Sen (2000) highlights how society may grant to individuals the capacity for taking part in creating their own livelihoods and securing basic needs of good health and physical security (all aspects of distributive justice) by means of governing their own affairs and participation in these processes (procedural justice). Two crosscutting approaches to distributional and procedural justice can be identified in the contributions to this volume. These are, first, a rights-based approach (for example, rights to human dignity, choice of identity and reasonable standard of living) and, second, the Benthamite principle of 'the greatest good for the greatest number' (see Agyeman et al. 2003 and Schofield 2006 for an extended discussion).

Marginalization means exclusion from resources, decision-making and rights. Justice and moral claims of the disempowered have been at the centre of alternative development throughout its shifting themes and agendas, and are means of addressing marginalization and voicing change. To highlight this focus, the book addresses the foundational and early themes and objectives of alternative development as well as identifying more recent themes. Researchers pursuing alternative development have been instrumental in theorizing and analysing the growing importance and role of civil society. They have also been instrumental in providing an understanding of the importance of environmental concerns and sustainability as opposed to narrowly defined 'economistic' objectives of economic growth, and as opposed to some of the mainstream development thinking of the Bretton Woods institutions. Our starting point is to understand development processes from the perspective of people's everyday lives by way of documenting and discussing how people and local communities help shape development themselves.

Mainstreaming Alternative Development?

The key notions of alternative development are 'social practice', 'participation' and 'empowerment' within the overarching concept of justice. According to Friedmann (1992) in his influential book, *Empowerment: The Politics of Alternative Development*, social practice can be seen to take place in four overlapping domains: the state, civil society, the corporate economy and the political community. Friedmann identifies a historical process of systematic disempowerment through the exclusion of a large part of the world's population from economic and political power. In opposition to mainstream development, critical development studies have focused on postcolonial and other marginal societies. However, marginalization is constantly constructed and reconstructed through time and takes different paths in different cultural contexts, as Lund's work exemplifies. It is a truism that the experience of marginalization varies across time and space, and that the Global South (and the Global North, for that matter) is not a homogenous entity, implying that we in this book must introduce the themes in their full historical and cultural context. The format of this book – 17 accounts of different situations of marginalization, agency and struggle – is particularly well suited to illustrate the great variety of culturally and historically constituted circumstances in which marginalized people find themselves. Agyei-Mensah and Wrigley-Asante (2014), for example, explore in the present volume the relationship between different political ideologies and how these have had an impact on different facets of development over time. In this context, we show how alternative development implies a striving for social justice and involves empowerment through the exercise of agency by subjects across different geographical scales and in different locations. The book's contributors come from geography and cognate disciplines and include academics from institutions in the Global South and Global North – although the main body of their research discussed in this book is situated in the Global South.

Alternative development might be claimed to be a success as some of its defining principles and discussions have become mainstream – at least on paper and in many cases as an explicit political aim. For example, key concepts and themes such as participation, empowerment and gender have all been adopted as part of the vocabulary of development actors in the World Bank and the International Monetary Fund, as well as in national development plans such as the *Poverty Reduction Strategy Papers* (IMF 2012). Indeed, so well-entrenched has participation become in development discourse and practice that a book titled *Participation: the New Tyranny?* (Cooke and Kothari 2001a) more than ten years ago reflected upon the unwarranted exercise of power by policy-makers, multilateral and bilateral aid agencies and nongovernmental organizations (NGOs) in the name of 'participation' and on the ways in which participation has been co-opted by special interest groups. One of the contributors writes (Francis 2001: 87):

> In a hall of mirrors, anything may seem true, even the slur that the price of admission for a new profession has been its collusion in the manufacture of a collective dream of participation and community, behind the screen of which the levers of business remain quite intact.

Nederveen Pieterse (2010) has acknowledged the well-established arrival of alternative development in terms of ideas and rhetoric, and has labelled it 'mainstream alternative development' (MAD). An implication of MAD is a shared language and understanding of a changeable, governable world in which actors at all levels are included. However, as the quotation above suggests, such shared language can often be a fig leaf for business as usual, whereby participation, empowerment and gender are tokens rather than realizable concepts and processes for achieving justice. Despite MAD, counterrevolutions continue to take place within development studies and global politics, and indicate the continuing necessity for alternative development to address new issues and face new challenges. Reactions against the mainstreaming of key (principally economic) concepts have, for example, resulted in a call for putting politics and political aims back into development through more radical forms of participation, feminism and environmental movements (Blaikie 2000, Hickey and Mohan 2004, Cornwall et al. 2007). As a result, core concepts in alternative development are proving to be dynamic, constantly critical and reforming from within.

Thinking Critically About Development and Marginalization

To honour the work of Ragnhild Lund, the contributions to this book all speak to the book's theme of 'unravelling marginalization, voicing change'. The essays address and contest the shifting mainstream development thinking within their respective themes. The book aims to make voices heard in various ways. First, the chapters analyse and discuss ways in which people act upon marginalization through activism, and more informal and invisible resistance, and how they become aware of and articulate the injustice of their circumstances. Second, the chapters analyse the realities of local communities and to what extent people's voices may be heard in various development projects and situations. Third, the book discusses relationships between local knowledge, scientific knowledge and policy work.

We understand critical scholarship as a way of questioning the platitudes and myths of society, such as the adoption of the language of alternative development without the accompanying understandings of power. Thinking and researching development critically aims to challenge exploitation and demand just social change (Walzer 2002, Blomley 2007). When ideas become mainstream, they are heard but at the same time tend to be taken for granted, and in many cases co-opted or deployed discursively for political purposes that may be contrary to those of alternative development. The contributors to this book critically examine these taken-for-granted understandings of development and challenge the ways in which ideas and practices of development take place and the role of research in such processes. Lein (2014) and Resurrección (2014), for example, show in different ways in the present volume how there is a need to unpack and interrogate the taken-for-granted knowledges on the relationship between environmental change, climate change and migration. Skotte (2014) questions the ways in which we make categories and structures in order to make sense of our lives. He scrutinizes how we disseminate this knowledge through teaching in universities. He suggests that linking society to teaching, in his case within architectural education, could help unsettle the established knowledge.

Critical scholarship concerns how we do development research. Michael Walzer (2002) discusses the critical researcher's closeness to the field and to the people with whom the researcher produces and coproduces research knowledge. When researching development from an alternative perspective, particularly through ethnographic research on which most of the material in this book is based, we come close to our research subjects by living with them and listening to their stories. The most critical researchers are, in Walzer's view, those who are close – 'only an inch away' – meaning those who are committed to the society whose policies or practices they call into question (see also Lund 2012). However, the practices of engagement with other societies through research may themselves also come under critical scrutiny. Robbins (2006), for example, explores the concept of research as theft – as an unwarranted accumulation of knowledge about the subjects of research for purposes of academic publication and the compilation of reports for international organizations in far off capital cities rather than as commitment to the subjects' political cause.

To trace the history of alternative development as well as explore the innovations in this field, we need to consider its legacy and foundational themes and how these have developed and expanded to include more recent discussions. Here, we consider the field of alternative development research focusing on local communities and participation, poverty, gender and the environment, and extend our discussion to new areas such as globalization, rights-based approaches, conflict, forced migration and climate change. We also address the ways in which feminism, postcolonialism and postdevelopment have shaped the current development discourse. We discuss how understandings of participation and agency have changed over time, as well as the influence of globalization on the alternative development discourse and how these developments have resulted in new discussions of the responsibilities and ethics of development studies and development practice.

Unpacking Development Through Feminism and Postcolonialism

The notion of development, largely created in the Global North, has been challenged within development studies and development practice both in the academy in the South and in the North. From a broader and more radical viewpoint, there has grown a deeply pessimistic view of development, particularly of mainstream development. Thus having anything to do with development at all is seen in some quarters as leading to 'contamination' and doing harm rather than good (e.g. Sachs 1992, Blaut 1993, Escobar 1995). What Booth (1985) defined as an 'impasse in development' came from increasing realization that development in the South had failed. This was accompanied more generally by the critique raised in social sciences, particularly from feminism, postcolonialism, and poststructuralism.

Contributions to this book show that the past thirty years of feminist engagement with development have contributed to the definition of alternative development as a distinct and pluralist field of inquiry and practice (e.g. Attanapola 2014, Eide 2014, Hyndman 2014). Together with postcolonial perspectives, feminism has helped to transform methodologies and theories of alternative development. Emerging in the 1970s, feminist responses to the marginalization of women have largely been formulated within different approaches termed 'women in development' (WID), 'women and development' (WAD) and 'gender in development' (GAD). These frameworks are discussed, elaborated upon and used extensively in the literature (Visvananthan et al. 1997, McIlwaine and Datta 2003), and in this book (e.g. Attanapola 2014). Feminist scholarship has developed increasingly critical stances on the development process and knowledge production. A seminal contribution was Chandra Talpade Mohanty's 'Under Western Eyes: Feminist Scholarship and Colonial Discourses' (1988), where she provides a critique of the hegemony of Western feminism, presented as universalism, and simultaneously points to the uneven power relations between Western women and Third World women. By analysing the ethnocentrism of development theories and particularly the objectification of Third World women, the article stands as a symbol of some of the major development critiques that have emerged under the headings of feminism and postcolonialism. An example of this inspiration is found in Hyndman's chapter (2014).

Scholars from the South had a critical role in shaping contemporary feminism and development studies through critiques based on postcolonialism with a view 'to compel a radical rethinking of knowledge and social identities authored and authorized by colonialism and Western domination' (Prakash 1994: 1475). In this context, postcolonial perspectives may mean both the theorization of the period after formal colonization came to an end and a methodology to interrogate the colonial logics and practices of Euro-American hegemony both historically and contemporarily (Robbins 2006). Postcolonial approaches speak to the violence towards and the marginalization and exclusion of postcolonial subjects and knowledges, and analyse the complexities of development paradigms and geographies (Radcliffe 2005). The postcolonial critique has led us to rethink the way knowledge is produced and the way we practice geography and development studies (e.g. Blunt and Wills 2000, Radcliffe 2005). A postcolonial lens has enabled the destabilization of key assumptions and mainstream thinking within development thinking by 'provincializing' Europe and decolonizing the discourses and practices of development (Chakrabarty 2000, Robinson 2003, Brun and Jazeel 2009). Consequently, the postcolonial lens also enabled a critique of the power relations in development practice, and questioned whose voices and understandings counted and how development scholarship could represent these voices. What has been termed 'the crisis of representation' led to an impasse in development research, and fieldwork was abandoned by some researchers in favour of textual analyses (Nagar 2002). Raju (2002: 174) describes this impasse as an 'apology' by the researchers for their inability to represent their research subjects adequately. An important tension then arises between the need to develop empirically based research and the need to protect vulnerable populations from possible exploitation or harm (Leaning 2001, Robbins 2006).

The radical critique of development thinking and practice resulted in the promotion of 'postdevelopment' with the suggestion that engagements with the Global South needed to take place *outside* the development paradigm altogether. Development was rejected because it did not work, was a westernization project and was understood to bring environmental destruction (Sachs 1992, Blaut 1993, Escobar 1995, Peet and Hartwick 2009, Nederveen Pietersee 2010). Postdevelopment finds much common ground with Western critiques of modernity and technoscientific progress. According to Nederveen Pietersee (2010), it parallels alternative development, but stands outside and distances itself from development because development never releases itself from colonial discourses. The dominance of the giver is always present and as a consequence embodies geopolitics in that its origins are bound up with Western power and strategy for the Third World, enacted and implemented through Third World elites (Sidaway 2008: 16). Mainstream development thinking has been substantially refigured in recent years because of this critique. Notions of agency and locally situated practices, identities and knowledges have been retheorized, as we show in the following.

Interrogating Agency – New Forms of Participation

Key notions in alternative development are social practice, participation and empowerment with the overarching goals of distributional and procedural justice. At the heart of these notions lies the understanding of 'agency'. Contributors to this book analyse and question notions of agency. Agency in this context means people's capacity to make choices and pursue their own goals. It is their ability to act upon forces that restrict their choices. Marginalized subjects tend to be discursively constituted as paradigmatic victims symbolizing inequality, poverty, passivity and helplessness (Chua et al. 2000). They are subjects acted upon rather than acting subjects (Lubkemann 2008). Since Robert Chambers' (1983, 1994) discussion of participatory approaches, the notion of participation has become more nuanced and problematized. The various discussions of agency in this book are inspired by the critique of participation that originates from Cooke and Kothari's (2001a) book showing the ways in which participation has become a taken-for-granted concept and lost its original meaning of 'making "people" central to development by encouraging beneficiary involvement in interventions that affect them and over which they previously had limited control or influence' (Cooke and Kothari 2001b: 5). The critique shows how participation can become a means through which aid projects become more efficient and effective in achieving their aims rather than participation being seen as an end in itself

(Cleaver 2001). Cleaver (2001) shows how, despite theoretical discussions on the relationship between agency and structure in books such as *Battlefields of Knowledge* (Long and Long 1992), participation had travelled into mainstream development without the theoretical ideas that problematized its meaning and use. Here it is important to understand the policy process through which notions of participation, and the practices that were designed to facilitate it, have to pass. There are two examples in the present book. The first is an account by Eide (2014) of the challenges and opportunities in implementing an international consensus on women in development. The second is by Blaikie (2014) on how best to navigate policy away from a normalizing, top-down and state-dominated process towards a more participatory natural resource management by emancipatory research. The critique of participatory approaches has led to a renewed focus on participation as transformative spaces of participation through notions of 'rights' and 'citizenship' (Hickey and Mohan 2004). The current participatory approaches in alternative geographies of development are clearly inspired by these recent debates (Cooke and Kothari 2001a, Cornwall 2002 a, b, Hickey and Mohan 2005, Mohan 2008, Refstie and Brun 2012). We ask how participatory approaches more effectively may produce spaces where citizens can make their voices heard and hold their authorities accountable. Specific and practical measures to facilitate participation in more than name include a wide variety of practical means, including the creation of open access citizens' forums, activist meetings to increase awareness of rights and responsibilities, training (particularly of women) to run meetings (and to resist attempts to recover control by dominant males), promotion of issues of justice through local radio and newspapers, and training in bookkeeping (see Aitken et al. 2014 in this volume).

Contributions to this book make visible the various dimensions and foundations for understanding agency. We discuss the transformative potential of agency (Azmi 2014) and conceptualize agency through coping capacity (Kusakabe 2014). Our taken-for-granted notions of agency and power are deconstructed in the chapter by Vågenes (2014), where indirect and informal powers, bargaining tactics and strategic manoeuvres among Hadendowa women in Sudan are analysed to show the importance of cultural and symbolic capital. From our position in alternative development studies, we argue for the continued importance of participation as a form of agency that is by its nature political and politicized, and whereby agents of the state and citizens interact in new ways (Cornwall 2002b). However, we caution against a new tyranny, whereby agency is essentialized and unquestioned in a 'triumphalist version of agency' (Assad 2000: 27, Shanmugaratnam et al. 2003, Brun and Lund 2014). Resurrección (2014) shows how this essentializing of agency through, for example, focusing on how women are capable of adapting to climate change may also naturalize and reinforce unequal gender divisions of labour when translated into policy and thus add new burdens to women rather than empowerment.

Many of the contexts and processes analysed in this book relate to people living with insecurities and adversities. People struggle to get through their everyday lives, although subject to conflicts, disasters, poverty and marginalization. Their voices are not heard, and they are made invisible by the state and other powerful actors. How can we conceptualize agency in such restricted contexts? Would there still be a possibility of transformative spaces of participation, including the seemingly mundane and practical measures suggested above? The authors in this book reflect on the limits of agency and take a critical approach by not taking agency for granted, but rather discuss the various ways in which agency may be played out on the ground under pressure and in very restrictive environments. Agency is understood in relation to individual experiences and more widely in relation to how people and local communities mobilize to create change. We show that these processes are always multiscalar and involve extralocal actors and processes. We need to understand the limitations of agency and the ways in which people living in adverse circumstances may, for example, choose passivity as their strategy (see also Scott 1985)

A Grounded Approach to Globalization: Multiple Scales of Rights, Flows and Power Relations

While the foundational alternative development writing tended to concentrate on local – and often relatively immobile – communities, the past 30 years of scholarship have seen increased focus on movements, flows and transnational relations between local and higher geographical scales. Similarly, globalization tended to be characterized as a solely top-down process in which states and international organizations play a more important role than individuals and communities, but scholarship has shown that globalization is far from a universal experience and process. Many of the chapters in this book engage with globalization and its multiple connotations from the perspective of individuals, and local and imagined communities who shape and are shaped by globalization. Viewed from a local perspective, the agency of various actors makes their experience of globalization a reflexive one whereby globalizing processes at higher scales shape but are also shaped by local agency. Thus, in the context of alternative development, we consider globalization as a contingent and constructed discourse formed out of the specificities of the people and places involved in shaping globalization processes (Brun 2005). A number of authors have called for a reconceptualization of globalization to pay attention to the role of place and local knowledges in order to overcome the dichotomizing debate of the local versus the global (e.g. Lie and Lund 1995,[1] 2010, Escobar 2001). An example in the present volume illustrating this well is by Aitken et al. (2014) on the negotiation of local and international values in a nature reserve in China. Agyei-Mensah and Wrigley-Asante (2014) provide examples of how global and local forces interact to shape the socioeconomic and

---

1   Also: Lund, R. 'Mobile Livelihoods and Gendered Citizenship: A Study of Indigenous People in India, Laos and China'. Presentation at Gendering Asia Network workshop. Nordic Institute of Asian Studies (NIAS), Copenhagen 8–11 November 2010.

urban landscape. Similarly, Owusu (2014) directs us to the need to understand the global by analysing the local. A helpful perspective that we draw on in this book is the 'grounded feminist approach to globalization' of Nagar et al. (2001) that starts from the lives of a variety of people with diverse relationships to globalization. A grounded feminist approach to globalization should explore the range of social locations (gender, class, ethnicity, race and sexuality) that refract globalization processes and the multiple ways in which globalization is lived, created, accommodated and acted upon in different historical and geographical settings. To ground globalization through analysis of empirical data would mean to trace for each process the socially embedded mechanisms through which it may generate material outcomes in a given context (Brun 2005, Attanapola 2014). Each of the constructs of gender, class and ethnicity signifies specific types of power relations produced and exercised in and through a myriad of economic, political and cultural practices (Brah 1996, Hyndman 2014).

Tracing the links and outcomes of globalization at the local level usually encounters the need for an analysis of the state. Gatekeepers at the state level may be in a position to shape the way in which international capital seizes opportunities for investment and is able to profit from joint ventures or by making large informal payments to gain access to labour, raw materials, building sites for manufacture, port facilities and preferential treatment for investment companies regarding taxation. In the present volume, Kusakabe (2014) compares the Veddhas in Sri Lanka with women fish traders on the Thai–Cambodian border, and stresses the very different relations they have with the state in terms of restrictions and opportunities. Vandsemb (2014) and Azmi (2014) trace the national and local impacts of national policy regarding national identity, gender and displacement in a situation of prolonged civil war in Sri Lanka. Common to these contributions is the way in which the state is shaped and acted upon by actors at both lower and higher scales, such as migrants, the displaced, traders, international development actors and financial institutions.

In their influential book, *Transnationalism from Below*, Smith and Guarnizo (1998) locate transnationalism in the local resistances of the informal economy, ethnic nationalism and grassroots activism. Transnationalism from above, such as transnational capital, global media, supranational political institutions and neoliberal market reforms – also enables and energizes transnationalism from below, such as transnational migration, international labour unions and rights movements. However, as a parallel to our discussion of agency and as the chapters in this book show (e.g. Attanapola and Panda), Smith and Guarnizo (1998) caution against considering transnational spaces as only emancipatory. An important point is that transnationalism is a multifaceted, multilocal process. In this book, we aim to understand how local and global processes are affected by and affect power relations and possibilities for empowerment.

Alternative development has been criticized for being too localized, too little concerned with global process and consequently having little transformative potential in an increasingly global and transnational reality. However, as the chapters show, global flows and transnational connections must be understood in a grounded approach to globalization. Fløysand and Barton (2014) provide an example. Their contribution critically examines to what extent foreign direct investment (FDI) can provide a potential for local economic development and poverty reduction. As a whole, the book helps to unpack the realities of alternative development by placing under critical scrutiny actors in the four overlapping domains identified by Friedmann (1992): the state, the civil society, the corporate economy and the political community.

Development Futures: The Continued Importance of Alternative Development

Alternative development may be seen to have its origin in the early 1970s. The contributions to this book show that despite shifting contexts, changing geopolitics and a very different development agenda, alternative development has continued relevance for securing a continued focus on justice. Alternative development as ways of formulating alternative visions and finding alternative paths is in this book framed within the question of how participatory forms of governance can be integrated in wider projects of redistributive politics and social justice (Hickey and Mohan 2005). We show this by bringing into the discussion the recent influences of feminism, postcolonial scholarship and the increasingly nuanced conceptualizations of agency.

Three sets of challenges and possibilities for alternative development are discussed in the following. First, there is a new multipolarity of actors involved in development. Second, the Millennium Development Goals (MDGs) (UN 2003), which were formulated to be reached by 2015 and have to a large extent shaped the development agenda since 2000, are under scrutiny and new development agendas are being discussed. Third, new spaces for development have become increasingly important.

The first dimension, multipolarity, requires a realization that the Western development agenda and its dominance since the end of the Second World War is now under pressure. Many new actors, such as the emerging economies of China and India, together with private actors, play an increasing role in development assistance and often with less stringent conditionalities and less emphasis on human rights (Tull 2006, Gu et al. 2008). More importantly, we do not know exactly how this shift will affect the support to civil society and community organizations that have been considered keys to alternative development. We find there is a need to maintain a focus on actors that help to voice change and enable participation of marginalized groups. Therefore, we believe alternative development perspectives are still relevant for securing civil society actors a role in just social change.

Second, there are possible shifts following the establishment of the Millennium Development Goals. Some of the MDGs may be reached by 2015, but extreme poverty and marginalization will remain. The next step after 2015 is not clear. How should development goals be renewed? How can we ensure a continued emphasis on distributive justice and empowerment in this process? As the financial crisis has unfolded in the world since 2008, we can identify important discrepancies in the economic status of states and increasing differences and inequalities within countries across the world (Evans 2011). How should new development goals be formulated? Can we find ways in which to institutionalize the politics of redistributive justice?

The third dimension, new spaces of development policies and practice, can be identified in multiple ways. Globalization provides new possibilities for making new alliances, new collaborations and new constellations of researchers, activists, policy-makers and practitioners. It will be increasingly important to maintain alternative notions of development and influence changing development agendas as they evolve in local and global contexts. In the spirit of Ragnhild Lund's work and life, many authors in this book represent long-standing collaborations between actors in the Global North and South that enable a renewed focus on difference and antiuniversalism, and a common stance against ethnocentrism and objectification of marginal subjects. Collaborative research enables relations to develop between researchers and development workers, policy-makers, activists and community groups with a common aim to voice change and achieve distributive justice. While collaborative approaches across boundaries represent a new space for alternative development, we find equally important the new spaces in which alternative development research have taken us in this book. We consider, for example, border areas, everyday politics, transnational spaces, warscapes and spaces of difference – spaces that together provide a more nuanced understanding of the possibilities and continued relevance of alternative development. The chapters exemplify how social change can be conceptualized and how alternative development will continue to be relevant by unravelling marginalization and voicing change.

A Summary of the Chapters

This book reflects on the ways in which alternative development thinking and practice continue to be relevant. The notion of co-constitution of identity as an historically shaped and situated process, constantly being negotiated and renegotiated, and specific in time and space, was one of the key ideas that shaped Ragnhild Lund's discussion of alternative development theory. This would be a place-based, people-centred perspective in a development theory that focuses on gender, social movements and ecology. Such a theory of alternative development would be locally situated, culturally constructed and socially organized. Lund's work deals with local knowledges and practices, poverty, gender, environment and sustainable development, and addresses more specific debates such as forced migration, conflict and climate change. It is clear that this runs counter to much that goes for 'development studies' and mainstream development practice (Corbridge 2008), and this tension is examined in some of the chapters in this book, which we now describe.

After the introduction, Part I of the book is titled 'Knowledge, Policy and Practices of Development' and introduces various ways in which theory may be put into practice. This section examines the questions of how calls for alternative development are heard (or not heard), how ideas and practices move from one scale to another, and how they are turned into practice. In alternative development thinking, we need to ask the question: 'knowledge for whom and knowledge for what?' (Burawoy 2004). In his chapter Blaikie (2014) outlines challenges to policy reform in directions indicated by alternative development and questions the assumed unproblematic link between theory and practice, along the lines of 'truth speaks to power' (Wildavsky 1979), and between new research knowledge and policy implementation. He discusses what it means to conduct useful participatory and engaged research in natural resource management research. In his chapter Skotte (2014) addresses similar issues in the teaching and practice of architecture in the Global South. He argues for the 'humanizing' of architecture, whereby the knowledge of the people who will be the future users of the buildings designed by architects may have a leading role in building design and construction. Skotte illustrates the tensions between the academic and the practical, and the way in which the former may fail to understand local agency and locally useful knowledge production. In the next chapter, Fløysand and Barton (2014) discuss the dynamics between FDI, regional development and poverty in Chile and how marginal actors may press claims for poverty reduction. By considering different scales, it is possible to understand how poverty affects people's access to institutions and services. At the local level, Owusu (2014) discusses access to housing in Ghana by examining the relationship between people in poor residential areas, the state and global governance. In the final chapter of this section, Eide (2014) writes on her experiences from working in the United Nations Development Programme (UNDP) and discusses the 1985 UN Women's Conference and the ways gender was operationalized in development as a result. She exposes the difficulties of engaging a large international bureaucracy in improving the status and life chances of women in development. Here, activism and radical calls for alternative development meet instrumentalism, bureaucratic procedures and engrained gender biases. Together the chapters in Part I suggest two central issues in alternative development. The first is that the study of poverty and deprivation needs to take a multiscalar and contextual approach in order to understand marginalization. The second issue concerns the challenges of turning complex, situated, socially constructed and culturally specific issues into policy at different levels. These are crosscutting issues in the following Parts of the book.

In Part II, titled 'Alternative Geographies of Gender and Development', the contributions focus on local knowledge, and the organization and transformation of societies in terms of gender, globalization and development. Together the chapters show the importance of understanding the local context, local knowledge, identities and power relations as a basis for alternative development but at the same time suggest ways of understanding the multiscalarity of gender, identity and power in development. The contributions give a sense of how alternative geographies of development enable the unpacking of some of the taken-for-granted assumptions we use when studying particular societies and places. In her chapter, Vågenes (2014) critically analyses gender culture in the highly gender-segregated society of the Red Sea Hills in Sudan. She shows how what from the outside, with Western eyes, may be understood as female subordination may in fact represent women's control and influence over their lives. Agyei-Mensah and Wrigley-Asante (2014) pay particular attention to gender, globalization and politics by bringing in different temporal and spatial scales when analysing the history of Accra, Ghana, and its transformation. In this chapter particularly, the reflexive relationship between the global and the local is central. This resonates with earlier work by Lund and her colleague Merethe Lie (Lie and Lund 1995, 2010), who analyse the interplay between actors at global, local and individual levels. Their work has been influential for understanding how a gendered division of labour operates in the global division of labour and how the globalization of industries has created social change locally as well as contributed to the globalization processes. Gender must be understood as one of the core dimensions of alternative development. Attanapola's chapter (2014) on an export processing zone in Sri Lanka analyses how knowledge of rights may increase the demand for accountability by responsible institutions and actors, such as the state and global industry. She considers the role of women's agency and empowerment in responses to rights violations. Also based on women's struggle for social justice in Sri Lanka, Hyndman (2014) argues in her chapter that a feminist analysis of conflict and development must incorporate multiple bases of identity and social relations. She suggests that there is need to move on from gender and development to incorporate other dimensions of identity than gender in order to trace how 'geometries of oppression' (Valentine 2007) affect how identity is created, by whom and when.

    Part III is titled 'Human–Environment Relations, Environmental Discourses and Development'. Together with gender, human–environment relations constitute one of the founding themes of alternative development, placing emphasis on places and the human and environmental contexts where development takes place and the encounters of multiple actors in those places. The four chapters in this section discuss the relations of different actors and stakeholders to their environment. The chapters engage with the key themes of the book by addressing issues such as local practices and knowledge, and the meaning of nature in specific contexts, rights of access to resources and grassroots mobilization, environmental justice and global activism, and questioning the taken-for-granted relationship between environment and social change. An important common denominator in the work of Lund and her colleagues (Lund 1993, Brun and Lund 2008, Owusu and Lund 2008) and the chapters of this book is that place is understood as local articulations of social relations (Massey 1994). People's struggle for rights, access to resources and inclusion in development processes are understood as multiscalar, network-oriented strategies of localization (Escobar 2001). The chapter by Aitken et al. (2014) applies a place- and people-centred perspective related to feminist and poststructural development theories in Fanjingshan Reserve, China. They analyse a participatory mapping project to make visible the resource-use relations between local farmers and an endangered snub-nose monkey species. Panda (2014) shows in her chapter how tensions emerge between local communities and industry. She uses a rights-based approach to reflect more broadly on how rights may be used as a vehicle to mobilize for change on the part of tribal women in Odisha, India, by engaging with mining companies, which, with the connivance of the state and local state, have threatened women's livelihoods and tribal identity. In this chapter, access to and control over local resources and the multiple actors involved are key dimensions for understanding the dynamics between local resources and environment. The chapters by Aitken et al. and Panda help to provide an understanding of ways in which alternative geographies contribute to make visible the complex human–environment relations involved in social change. The chapters of this section together show how alternative development thinking is crucial for critically assessing established and emerging discourses. A prominent theme within and outside development studies in recent years is climate change. Two innovative chapters on climate change and development have been included. Lein (2014) unpacks the notion of 'climate refugees' and analyses the relationship between climate change and migration. Resurrección (2014) addresses the relationship between climate change and migration from a gender perspective. She questions the attitudes of women cast as victims of climate change. Both these chapters critically examine the way categories are formed in development studies and in research on climate change. The chapters critique the politics of knowledge that lead to stereotyping and ignore social complexities.

Alternative development concerns the margins of society and the margins of knowledge. Part IV, 'On the Margins: Conflict, Migration and Development', addresses the notion of agency in situations of conflict and forced migration. The chapters follow the work of Lund and colleagues (e.g. Lund 2003, Lie and Lund 2005, Lund and Agyei-Mensah 2008) in questioning the notion of passive victims, while acknowledging that passivity may also be a strategy in an emergency or a protracted situation of displacement (Shanmugaratnam et al. 2003). The chapters show in various ways how people survive and manoeuvre in landscapes of conflict and not least how boundaries between communities and countries are challenged in such situations. Armed conflict has only recently become part of development studies. Few textbooks in development studies are yet addressing the issue. A key reference to the link between conflict and development is Collier et al.'s book, *Breaking the Conflict Trap* (2003). However, their book is based on statistics at country level and does not include understandings of the local context in which conflict and marginalization takes place. This section therefore fills a gap by introducing local understandings to the dynamics and impacts of conflict – an alternative geography of conflict, forced migration and development that considers the ways in which people on the margins (both socially and geographically) manoeuvre in the oppressive power relations that are formed in conflict settings. Azmi (2014) shows in her chapter the agency of internally displaced women in establishing or reestablishing their livelihoods in Sri

Lanka. Kusakabe (2014) investigates the coping capacity of border fish traders affected by changing restrictions on the Thai–Cambodian border. Vandsemb (2014) considers how spontaneous frontier migration in Sri Lanka challenges the state and state policies. Finally, Brun and Lund (2014) consider how forced migration may be studied as border practices and also how in academia studies of forced migration also cross borders, both disciplinary and in addressing policy discussions and practical dilemmas

Alternative ideas of development can be explored within two differing broad approaches to histories of knowledge. The first examines the social context in which these ideas were produced. The second consists of personal histories of practitioners in the form of biographies and autobiographies and how these personal histories both reflect and affect the development of ideas. The book concludes with an autobiographical discussion of Ragnhild Lund's life and career since the mid-1970s. Penned by Michael Jones (2014), the chapter presents in her own words some of the main themes and developments in her career, which at the same time provides an example of how the field itself has developed over time. Here, from the very outset, voicing change is at the heart of the story. Lund is one of the founding members of Development Alternatives with Women for a New Era (DAWN). Established in 1984, DAWN is a network of feminist scholars, researchers and activists working for economic and gender justice, and sustainable and democratic development. The present book provides examples of how the combination of research, advocacy and development practices are at the core of alternative development.

References

Agyei-Mensah, S. and Wrigley-Asante, C. 2014. Gender, Politics and Development in Accra, Ghana, in *Alternative Development: Unravelling Marginalization, Voicing Change*, edited by C. Brun, P. Blaikie and M. Jones. Farnham: Ashgate, 117–33.

Agyeman, J., Bullard, R.D. and Evans, B. (eds) 2003. *Just Sustainabilities: Development in an Unequal World*. London: Earthscan.

Aitken, S., An, L., Wandersee, S. and Yang, Y. 2014. Renegotiating Local Values: The Case of Fanjingshan Reserve, China, in *Alternative Development: Unravelling Marginalization, Voicing Change*, edited by C. Brun, P. Blaikie and M. Jones. Farnham: Ashgate, 171–90.

Assad, T. 2000. Agency and Pain: An Exploration. *Culture and Religion: An Interdisciplinary Journal*, 1, 29–60.

Attanapola, C. 2014. Ignored Voices of Globalisation: Women's Agency in Coping with Human Rights Violations in an Export Processing Zone in Sri Lanka, in *Alternative Development: Unravelling Marginalization, Voicing Change*, edited by C. Brun, P. Blaikie and M. Jones. Farnham: Ashgate, 135–54.

Azmi, F. 2014. Impacts of Internal Displacement on Women's Agency in Two Resettlement Contexts in Sri Lanka, in *Alternative Development: Unravelling Marginalization, Voicing Change*, edited by C. Brun, P. Blaikie and M. Jones. Farnham: Ashgate, 243–57.

Blaikie, P.M. 2000. Development, Post-, Anti-, and Populist: A Critical Review. *Environment and Planning A*, 32, 1033–50.

Blaikie, P. 2014. Towards an Engaged Political Ecology, in *Alternative Development: Unravelling Marginalization, Voicing Change*, edited by C. Brun, P. Blaikie and M. Jones. Farnham: Ashgate, 25–37.

Blaut, J. 1993. *The Coloniser's View of the World.* London: Grove Press

Blomley, N. 2007. Critical Geography: Anger and Hope. *Progress in Human Geography*, 31, 53–65.

Blunt, A. and Wills, J. 2000. *Dissident Geographies: An Introduction to Radical Ideas and Practice.* Harlow: Prentice Hall.

Booth, D. 1985. Marxism and Development Sociology: Interpreting the Impasse. *World Development*, 13, 761–87.

Brah, A. 1996. *Cartographies of Diaspora: Contesting Identities.* London: Routledge.

Brun, C. 2005. Women in the Local/Global Fields of War and Displacement. *Gender, Development and Technology*, 9, 57–80.

Brun, C. and Jazeel, T. (eds) 2009. *Spatialising Politics: Culture and Geography in Postcolonial Sri Lanka.* New Delhi: Sage.

Brun, C. and Lund, R. 2008. Making a Home During Crisis: Post-Tsunami Recovery in the Context of War, Sri Lanka. S*ingapore Journal of Tropical Geography*, 29, 274–88.

Brun, C. and Lund, R. 2014. Researching Forced Migration at the Interface of Theory, Policy and Practice, in *Alternative Development: Unravelling Marginalization, Voicing Change*, edited by C. Brun, P. Blaikie and M. Jones. Farnham: Ashgate, 287–304.

Burawoy, M. 2004. Public Sociologies: Contradictions, Dilemmas and Possibilities. *Social Forces*, 82, 1603–18.

Chakrabarty, D. 2000. *Provincializing Europe: Postcolonial Thought and Historical Difference*. Princeton, NJ : Princeton University Press.

Chambers, R. 1983. *Rural Development: Putting the Last First*. London: Longman.

Chambers, R. 1994. The Origins and Practice of Participatory Rural Appraisal. *World Development*, 22, 953–69.

Chua, P., Bhavnani, K. and Foran, J. 2000. Women, Culture, Development: A New Paradigm for Development Studies? *Ethnic and Racial Studies*, 23, 820–41.

Cleaver, F. 2001. Institutions, Agency and the Limitations of Participatory Approaches to Development, in *Participation: The New Tyranny?*, edited by B. Cooke and U. Kothari. London: Zed Books, 36–56.

Collier, P., Elliot, V.L., Hegre, H., Hoeffler, A., Reynal-Querol, M. and Sambanis, N. 2003. *Breaking the Conflict Trap: Civil War and Development Policy*. Washington DC: World Bank and Oxford: Oxford University Press.

Cooke, B., and Kothari, U. (eds) 2001a. *Participation: The New Tyranny?*, London: Zed Books.

Cooke, B., and Kothari, U. 2001b. The Case for Participation as Tyranny, in *Participation: The New Tyranny?*, edited by B. Cooke and U. Kothari. London: Zed Books, 1–15.

Corbridge, S. 2008. *Development*. Aldershot: Ashgate.

Cornwall, A. 2002a. *Beneficiary, Consumer, Citizen: Perspectives on Participation for Poverty Reduction*. Sidastudies no. 2 [Online: Eldis]. Available at: http://www.eldis.org/assets/Docs/15280.html [accessed: 24 April 2012].

Cornwall, A. 2002b. *Making Spaces, Changing Places: Situating Participation in Development*. IDS Working Paper 170. Brighton: Institute of Development Studies.

Cornwall, A., Harrison, E. and Whitehead, A. (eds) 2007. *Feminisms in Development: Contradictions, Contestations and Challenges*. London: Zed Books.

Eide, I. 2014. Implementing International Consensus on Women in Development: Context, Policy and Practice, in *Alternative Development: Unravelling Marginalization, Voicing Change*, edited by C. Brun, P. Blaikie and M. Jones. Farnham: Ashgate, 87–97.

Escobar, A. 1995 *Encountering Development: The Making and Unmaking of the Third World*. Princeton, NJ: Princeton University Press.

Escobar, A. 2001. Culture Sits in Places: Reflections on Globalism and Subaltern Strategies of Localization. *Political Geography*, 20, 139–74.

Evans, A. 2011. 2020 *Development Futures* [Online: Actionaid]. Available at: http://www.globaldashboard.org/wp-content/uploads/2020-Development-Futures-GD.pdf [accessed: 23 April 2012].

Fløysand, A. and Barton, J.R. 2014. Foreign Direct Investment, Local Development and Poverty Reduction: The Sustainability of the Salmon Industry in Southern Chile, in *Alternative Development: Unravelling Marginalization, Voicing Change*, edited by C. Brun, P. Blaikie and M. Jones. Farnham: Ashgate, 55–71.

Francis, P. 2001. Participatory Development at the World Bank: The Primacy of Process, in *Participation: The New Tyranny?*, edited by B. Cooke and U. Kothari. London: Zed Books, 72–87.

Friedmann, J. 1992. *Empowerment: The Politics of Alternative Development*. Oxford: Blackwell.

Gu, J., Humphrey, J. and Messner, D. 2008. Global Governance and Developing Countries: The Implications of the Rise of China. *World Development*, 36, 274–92.

Hickey, S. and Mohan, G. (eds) 2004. *Participation: From Tyranny to Transformation? Exploring New Approaches to Participation*. London: Zed Books.

Hickey, S. and Mohan, G. 2005. Relocating Participation within Radical Politics of Development. *Development and Change*, 36, 237–62.

Hyndman, J. 2014. 'No More Tears Sister': Feminist Politics in Sri Lanka, in *Alternative Development: Unravelling Marginalization, Voicing Change*, edited by C. Brun, P. Blaikie and M. Jones. Farnham: Ashgate, 155–67.

IMF 2012. *Poverty Reduction Strategy Papers (PRSP)* [Online: International Monetary Fund]. http://www.imf.org/external/np/prsp/prsp.asp [accessed: 23 April 2012].

Jones, M. 2014. Researching Alternative Development: An Autobiographical Discussion with Ragnhild Lund, in *Alternative Development: Unravelling Marginalization, Voicing Change*, edited by C. Brun, P. Blaikie and M. Jones. Farnham: Ashgate, 307–32.

Kusakabe, K. 2014. Coping Capacity of Small-Scale Border Fish Traders in Cambodia, in *Alternative Development: Unravelling Marginalization, Voicing Change*, edited by C. Brun, P. Blaikie and M. Jones. Farnham: Ashgate, 259–67.

Leach, M., Scoones, I. and Stirling, A. 2010. *Dynamic Sustainabilities: Technology, Environment, Social Justice*. London and Washington, DC: Earthscan.

Leaning, J. 2001. Ethics of Research in Refugee Populations. *The Lancet*, 357, 1432–3.

Lein, H. 2014. The Reemergence of Environmental Causation in Migration Studies and its Relevance for Bangladesh, in *Alternative Development: Unravelling Marginalization, Voicing Change*, edited by C. Brun, P. Blaikie and M. Jones. Farnham: Ashgate, 207–18.

Lie, M. and Lund, R. 1995. *Renegotiating Local Values: Working Women and Foreign Industry in Malaysia*. Richmond: Curzon Press.

Lie, M. and Lund, R. 2005. From NIDL to Globalization: Studying Women Workers in an Increasingly Globalized Economy. *Gender, Technology and Development*, 9, 7–30.

Lie, M. and Lund, R. 2010. Det lokale i det globale: Norsk industri i Asia. *Norsk Sosiologisk Tidsskrift*, 18, 76–94.

Long, N. and Long, A. (eds) 1992. *Battlefields of Knowledge: The Interlocking of Theory and Practice in Social Research and Development*. London: Routledge.

Lubkemann, S.C. 2008. *Culture in Chaos: An Anthropology of the Social Condition in War*. Chicago: The University of Chicago Press.

Lund, R. 1993. *Gender and Place*, Vol. 1: *Towards a Geography Sensitive to Gender, Place and Social Change*; Vol. 2: *Examples from Two Case Studies*. Trondheim: Department of Geography, University of Trondheim.

Lund, R. 2003. Representations of Forced Migration in Conflicting Spaces: Displacement of the Veddas in Sri Lanka, in *In the Maze of Displacement: Conflict, Migration and Change*, edited by N. Shanmugaratnam, R. Lund and K.A. Stølen. Kristiansand: Høgskoleforlaget – Norwegian Academic Press, 76–104.

Lund, R. 2012. Researching Crisis – Recognizing the Unsettling Experience of Emotions. *Emotions, Space and Society*, 5, 94–102.

Lund, R. and Agyei-Mensah, S. 2008. Queens as Mothers: The Role of Traditional Safety Net of Care and Support for HIV/AIDS Orphans and Vulnerable Children in Ghana. *GeoJournal*, 71, 93–106

McIllwaine, C. and Datta, K. 2003. From Feminising to Engendering Development. *Gender, Place and Culture*, 10, 369–82.

Massey, D. 1994. *Space, Place and Gender*. Cambridge: Polity Press.

Mohan, G. 2008. Participatory Development, in *The Companion to Development Studies*. 2nd Edition, edited by V. Desai and R.B. Potter. London: Hodder Education, 45–50.

Mohanty, C.T. 1988. Under Western Eyes: Feminist Scholarship and Colonial Discourses. *Feminist Review*, 30, 61–88.

Nagar, R. 2002. Footloose Researchers, 'Traveling' Theories and the Politics of Transnational Feminist Praxis. *Gender, Place and Culture*, 9, 179–86.

Nagar, R., Lawson, V., McDowell, L., and Hanson, S. 2001. Locating Globalization: Feminist (Re)readings of the Subjects and Spaces of Globalization. *Economic Geography*, 78, 257–84.

Nederveen Pieterse, J. 2010. *Development Theory*. Second Edition. London: Sage.

Owusu, G. 2014. Housing the Urban Poor in Metropolitan Accra, Ghana: What is the Role of the State in the Era of Liberalization and Globalization? in *Alternative Development: Unravelling Marginalization, Voicing Change*, edited by C. Brun, P. Blaikie and M. Jones. Farnham: Ashgate, 73–85.

Owusu, G. and Lund, R. 2008. Slums of Hope and Slums of Despair: Mobility and Livelihoods in Nima, Accra. *Norsk Geografisk Tidsskrift – Norwegian Journal of Geography*, 62, 180–90.

Panda, S.M. 2014. Right to Rights: *Adivasi* (Tribal) Women in the Context of a Not-So-Silent Revolution in Odisha, India, in *Alternative Development: Unravelling Marginalization, Voicing Change*, edited by C. Brun, P. Blaikie and M. Jones. Farnham: Ashgate, 191–206.

Peet, R. and Hartwick, E. 2009. *Theories of Development: Contentions, Arguments, Alternatives*. Second Edition. New York and London: Guilford Press.

Prakash, G. 1994. Subaltern Studies in Post-Colonial Criticism. *American Historical Review*, 99, 1475–90.

Radcliffe, S. 2005. Development and Geography II: Towards a Postcolonial Development Geography? *Progress in Human Geography*, 29, 291–8.

Raju, S. 2002. We are Different, but Can We Talk? *Gender, Place and Culture*, 9, 173–7.

Refstie, H. and Brun, C. 2012. Collaborative Research with 'Urban IDPs' in Uganda: Advocating for Empowerment or Maintaining Marginalisation? *Journal of Refugee Studies*, 25, 239–56.

Resurrección, B.P. 2014. Discourses That Hide: Gender, Migration and Security in Climate Change, in *Alternative Development: Unravelling Marginalization, Voicing Change*, edited by C. Brun, P. Blaikie and M. Jones. Farnham: Ashgate, 219–40.

Robbins, P. 2006. Research is Theft: Environmental Inquiry in a Postcolonial World, in *Approaches to Human Geography*, edited by S. Aitken and G. Valentine. London: Sage, 311–24.

Robinson, J. 2003. Postcolonialising Geography: Tactics and Pitfalls. *Singapore Journal of Tropical Geography*, 24, 273–89.

Sachs, W. (ed.) 1992. *The Development Dictionary: A Guide to Knowledge and Power*. London: Zed Press.

Schofield, P. 2006. *Utility and Democracy: The Political Thought of Jeremy Bentham*. Oxford: Oxford University Press.

Scott, J. 1985. *Weapons of the Weak: Everyday Forms of Peasant Resistance*. New Haven, CT: Yale University Press.

Sen, A.K. 2000. *Development as Freedom*. New York: Anchor.

Sidaway, J.D. 2008. Post-Development, in *The Companion to Development Studies*, edited by V. Desai and R. Potter. London: Hodder Education, 16–20.

Shanmugaratnam, N., Lund, R. and Stølen, K.A. (eds) 2003. *In the Maze of Displacement: Conflict, Migration and Change*. Kristiansand: Høyskoleforlaget – Norwegian Academic Press.

Skotte, H. 2014. Teaching to Learn – Learning to Teach: Learning Experiences from the Reality of an Ever-Changing World, in *Alternative Development: Unravelling Marginalization, Voicing Change*, edited by C. Brun, P. Blaikie and M. Jones. Farnham: Ashgate, 39–53.

Smith, M.P. and Guarnizo, L.E. (eds) 1998. *Transnationalism from Below*. New Brunswick, NJ: Transaction Publishers.

Tull, D. 2006. China's Engagement in Africa: Scope, Significance and Consequences. *Journal of Modern African Studies*, 44, 459–79.

UN 2003. *Indicators for Monitoring the Millennium Development Goals: Definitions, Rationale, Concepts and Sources* [Online: United Nations, New York]. Available at: http://www.undp.or.id/mdg/documents/MDG%20Indicators-UNDG.pdf [accessed: 28 November 2012].

Vågenes, V. 2014. Muted Power – Gender Segregation and Female Power, in *Alternative Development: Unravelling Marginalization, Voicing Change*, edited by C. Brun, P. Blaikie and M. Jones. Farnham: Ashgate, 101–15.

Valentine, G. 2007. Theorizing and Researching Intersectionality: A Challenge for Feminist Geography. *The Professional Geographer*, 59, 10–21.

Vandsemb, B.H. 2014. Spontaneous Frontier Migration in Sri Lanka: Conflict and Cooperation in State–Migrant Relations, in *Alternative Development: Unravelling Marginalization, Voicing Change*, edited by C. Brun, P. Blaikie and M. Jones. Farnham: Ashgate, 269–85.

Visvanathan, N., Duggan, L., Nisonoff, L., and Wiegersma, N. (eds) 1997. *The Women, Gender and Development Reader*. London: Zed Books.

Walzer, M. 2002. *The Company of Critics: Social Criticism and Political Commitment in the Twentieth Century*. 2nd Edition. New York: Basic Books.

Watts, M. 2003. Development and Governmentality. *Singapore Journal of Tropical Geography*, 24, 6–34.

Wildavsky, A. 1979. *Speaking Truth to Power: The Art and Craft of Policy Analysis*. Boston: Little, Brown, and Co.

PART I
Knowledge, Policy and Practice of Development

Chapter 2

Towards an Engaged Political Ecology

Piers Blaikie

Political Ecology and Alternative Development

A good case can be made that alternative development and policy-making do not sit comfortably together. At the risk of caricature, alternative development in this book is characterized by local and participatory decision-making in which the voices of the marginalized have a say. Policies that empower them in ways whereby they can take control of their lives will be 'situated', local and therefore diverse. Set against alternative development is 'mainstream' development, styled at different points in this book as having tendencies of top-down, economistic and 'blueprint' policy-making in the hands of distant, out-of-touch bureaucrats and powerful vested interests in capital cities. It follows therefore that public policy-making in the service of either of these styles of development is an important arena, but it will be a very different one in either case. The former will tend to be diverse, decentralized and less formal, and the latter state-centred, blueprinted and formal. However, these distinctions between alternative and mainstream development hide the importance of interchanges between the two. Alternative development will be reformist in nature; it must state its case and pursue its political ends through a wide variety of potential allies such as social movements, politicians, nongovernmental organizations (NGOs), the press, performing arts, public forums and so on. Yet inevitably alternative development may also be involved in attempts to reform public policy at the state level. Hence calls for alternative development beg some unsettling questions. Who is listening? Who are the 'movers and shakers' in government who need to hear? How should the case be made in terms of political language, argumentation and imagery? What strategy can outsiders devise to play a reforming role in the formal policy process?

This chapter addresses these issues in the arena of environmental politics. The approach suggested here is political ecology (PE). PE provides a powerful tool to provide some answers to the unsettling questions posed above. In the service of alternative development, PE should be engaged in political projects of emancipation and justice across a broad range of actors in ways that are useful and understood by them. There are many reasons not to engage in mainstream 'development' but instead to create spaces for alternative development. This involves the pursuit of procedural justice in the research process itself, involving participation and balanced coproduction in the creation of new knowledge. A linked series of research questions and procedures for an engaged political ecology are suggested.

What is Political Ecology for?

A starting point is to ask the normative question 'what is political ecology for?' It should be borne in mind that PE as a field of study, analysis and critique will mostly be produced in universities and research centres. One set of goals might be to produce emancipating and inspiring outputs to change the way in which people think about environment and society through teaching students, writing newspaper articles and scholarly publications, or engaging in film, theatre and events for whomsoever may chance to witness them. In this sense, PE can inspire students or the broader public. The criteria for evaluating the worth of these outputs are broad and apply to most other academic and artistic work as well. These include notions of elegance and economy of expression, analytical rigour, conceptual innovation, inspiration for students, ethical values, integrity, technical advances that enhance wellbeing, and other more specialized artistic and literary qualities. If there is the active promotion of an explicit political agenda then so be it, but for many academics this is not a central goal of PE.

There has been a long and well-established ethical and political goal in most PE writing: environmental justice. In almost all discussions of the field of PE, environmental justice and injustice are central (a small sample of texts includes: Watts 1983, Blaikie 1985, Blaikie and Brookfield 1987, Peet and Watts 1996, Bryant and Bailey 1997, Stott and Sullivan 2000, Robbins 2004, Neumann 2005, Bryant and Goodman 2008, Forsyth 2008, Schroeder et al. 2008, Rangan and Kull 2009 and many of the case studies in Peet et al. 2011). PE issues are variously identified. The following is a typical and succinct summary:

> situated concerns of environmental degradation, marginalisation, environmental conflict, conservation and control, environmental identity, social movements and differential access to environmental benefits and exposure to environmental harms. (Robbins 2012: 14)

Environmental justice runs through as a central theme in these and many other PE works. A statement of the principles of justice for any engagement adds moral authority and legitimacy, and encourages participants to think through what their involvement contributes to furthering justice. It also can form a rallying point (Agyeman et al. 2003). Environmental justice is defined in many different ways, at different scales and in different sectors but revolves around universal environmental struggles in the Global North and South (Schroeder et al. 2008). The issue of distributive justice (the distribution of environmental goods and bads) is usually a central theme in PE, involving the structuring of access to resources (Blaikie 1985, Sikor and Lund 2009). However, procedural justice is also most important, whereby people have the right to be included in decision-making about the environment in which they live. Here, participatory forums, citizens' associations, freedom of association and of the press, and transparency in decision-making are some of the enabling means by which procedural justice is served (Leach et al.

2010). Procedural justice also encompasses the research process and demands continuous introspection and discussion with research partners. New spaces for an alternative development can be created where the voices of marginalized people (through, for example, combinations of ethnicity, wealth, gender and age) can be articulated and heard (Rocheleau et al. 1996, Lund 1993, 2000, 2008).

While the ethical goal of promoting environmental justice is necessary for PE, it is not sufficient without strategies to reach it. Therefore a focus on the instrumental means of engagement outside universities and research centres becomes essential. Since PE is political, a strong argument can be made that there should be explicit political goals for PE and strategies to reach them (Walker 2006). There has been a long record of arguments for academic work being socially relevant. Jarosz (2004) explores PE as political practice in the classroom and outside it (teaching and learning as political practice and notions of 'public scholarship'). Dreze (2002: 817) comments on the failure to engage with issues outside the academy, saying: 'it is no wonder that "academic" has become a synonym for "irrelevant"'. Burawoy (2005) identifies four categories of academic contribution – professional, critical, policy and public – and enjoins academics to work in all four, not only in the first two (which do not involve stepping outside the privileged position that the university or research institute offers). Walker (2006) poses the question in the title of his article, 'Political Ecology: Where is the Policy?' reviewing arguments for PE being engaged with socially and specifically policy-relevant issues. Corbridge (2008: 503) also urges development studies to be critical, even oppositional, as well as to engage with public policy-making, which 'can be taken as a sign of maturity'. Finally: 'An alternative development imaginary of development can, and in my view should, lead to reflection about what the audience as well as the author can actually do about it' (Blaikie 2000: 1035).

Audiences for, and participants in, PE research are diverse, making generalization difficult. For example, a short term consultancy for a big international nongovernmental organization (BINGO), the setting up of a local urban self-sufficiency group in the Global North growing its own food or the design of a national park anywhere in the world will each require a very different set of skills and approaches. What the notion of justice means across this wide diversity as well the instrumental means to reach it will vary greatly. In all cases, however, PE has to be produced and communicated outside the academy as well as within it. This has implications for the ways in which PE is produced. To take two examples at opposite ends of the spectrum: the procedure can at one end be a commercial consultancy, often relying only on quantitative and secondary data, or at the other end a participatory PE, mutually coproduced by researchers and the subjects of research. Also, there are implications for how it is presented to the different audiences or participants and what they potentially can do about it. So, an engaged and useful PE is seldom a matter of a list of consultancy recommendations (often to gather dust in some distant government office), nor a simple matter of knowledge transfer from people who claim they know to those who supposedly do not.

Navigating the Dangers of Engagement in 'Development'

In the view of many academic critics, a useful and engaged PE can only become contaminated by association with what is widely understood to be the mainstream 'development process'. PE has been produced historically within a broad 'environment and development' framework, principally in the Global South. That is not to deny there has also been a lively repatriation to the North and from predominantly rural settings to urban (e.g. Heynan 2003, Forsyth 2004, Swyngedouw and Heynan 2004). There has also been criticism of this movement (Schroeder 1999, Robbins 2002), but much of the focus of PE still remains on the South. Many academics have a deep distrust and dislike of applied and policy-relevant work (Rocheleau 2008, Batterbury and Horowitz 2010). In the words of one commentator, 'policy has become an uncouth cousin' (Walker 2006: 382). In most cases it is implicitly assumed that engagement is carried out within the conventional, state-dominated mainstream development context. There are a number of reasons for this. First, the saying 'he [*sic*] who pays the piper calls the tune' implies that the research funder dictates where the research focus will be in problem-framing (rule 2, 'cooptation, cooptation, cooptation', in Bill Cooke's formulation (2004: 43)). Furthermore, funders may not only frame and shape the inputs of research through selective funding but also interfere with the outputs of research. There are many cases of funders who act as gatekeepers, publicizing the results while blatantly misusing them, or embargoing or editing the research output where institutional or private interests are threatened by what the researchers had to say (Schroeder 1999, Rocheleau 2008).

Second, the argument goes, funders will expect 'answers to problems' – but only problems they recognize and answers they like. Their assumption is that 'truth speaks to power'. Truthful answers will be provided by the contracted researchers – as long as the questions remain the prerogative of the funder, who may delete, change or selectively quote and misquote those 'truthful answers' it does not like. While this view may seem to stereotype contracted research, there is too much evidence of cooptation, coercion and asymmetrical coproduction to deny that there are dangers of playing a role in another's game, which may not further the cause of environmental justice at all. Paid research often ensures that the paymaster remains the discourse master, who has the power to delete any critical material (for example, revelations of wrongdoing, exposure of private interests that are contrary to stated public ones and generally any unruly deconstructive activities). Research-funding is provided by a wide variety of sources and these are variable in the degrees of freedom they allow for authors to express what they want to. For example, some national research-funding institutions (e.g. the US National Science Foundation or the UK Economic and Social Research Council) may give applicants a greater degree of intellectual freedom than other funding institutions, but there are still guidelines that referees follow, deriving from political criteria of ideological acceptability and notions of 'sound science', and resonating with broad research priorities of government. Therefore, in varying degrees, funding institutions may draw the academic researcher into an act of asymmetrical coproduction, reproducing the answers and 'solutions' to problems set by others. So, the argument runs: 'don't do it!'

Third, another significant change in attitudes to undertaking research in the Global South is that PE (and more broadly development) research can be a postcolonial and expropriating act (Robbins 2006):

> Hundreds of millions of dollars of national and international funds are poured into funding surveys, analyses, and examinations of a great range of environmental 'problems'. University professors, United States Agency for International Development workers, and arguably even reporters, trek across the world's poorest places, interviewing local people and recording their opinions, their resources, and their ideas, later bartering them for salary and prestige. (Robbins 2006: 313)

Robbins posits two possible responses: either stay home or engage the critique to challenge colonial science and understand the local and alternative politics of environmental knowledge.

Fourth and last, there is a conflated, deeply pessimistic view of all mainstream development (with and without inverted commas). It is a profound distrust emanating from the academy and derives from the conviction that having *anything to do with development at all* leads to contamination and doing harm rather than good (see the trenchant attacks on development by such authors as Sachs 1992, Blaut 1993, Escobar 1995, Rahnema and Bawtree 1997, Peet and Hartwick 2009). There is often confusion here between the path of critical research that seeks to promote alternative development and the path of mainstream development. This leads to the naïve assumption that alternative development can keep aloof from engagement with the state and formal processes of policy-making. A chalice is offered to the academic standing on the walls of her ivory tower full of promise to deepen progressive and alternative discourse and practice, but there are mortal dangers in receiving and drinking from it. Cooptation, contamination, asymmetrical coproduction and bureaucratic conspiracy – all with the affix 'co-', denoting 'together with' or 'contact with' – are just some of the poisonous infections that await the engaging academic. Better refuse the poisoned chalice altogether!

These reasons to resist engagement are discussed here *not* to dismiss them, but rather to learn from the dangers of attempts to engage and intervene while still fulfilling all four of Burawoy's (2005) categories of academic contribution (professional, critical, public and policy). For these reasons, among others, navigational hazards on the voyage to useful PE have to be charted and then avoided. The prospect of engaging in policy reform at whatever level will probably involve a wide range of actors, many of whom will be policy-makers, bureaucrats, consultants, politicians and the press. So, an engaged critical political ecology in the service of alternative development will sometimes have to share arenas and channels with mainstream development.

Steps Towards a Useful and Engaged PE

The discussion so far has rehearsed a few of the reasons why some academics resist involvement with a useful PE that requires engagement with others. A critical PE that engages with reform should develop a research and communication strategy mapping out a path from research design and research practice to communication to others, with learning and feedback occurring throughout. Here I suggest how each step may be strategized to give the researcher and other creators of new knowledge (for example activists, womens' groups or charities with experience of environmental injustice) the opportunity of doing useful PE that minimizes the chances of being coopted by others and reproducing environmental injustice (Simon and Narman (1999) discuss many of these issues of engagement in a development context).

There are many ways in which to make PE useful to those who have leverage in decision-making at different levels, such as small informal civil institutions, social movements, federations of resource users, NGOs, national ministries, and sites of negotiation between national governments, BINGOs and international institutions. There may also be research opportunities to work with and for people in a participatory way where they are denied environmental justice, to be able to make their voices heard. Since there is a wide diversity of actors both outside and inside the academy, strategies have to be carefully crafted. There is a common trade-off between improving environmental justice in a demonstrable way for a limited number of people, often in one location, or having a more diffuse and indemonstrable impact on a wider public.

A strategy for furthering environmental justice through research can and should involve developing a cumulative set of documents, papers and records of internal discussions among those undertaking the research. This may involve individuals who will use these documents as an internal memo, a research group, or a larger group in participatory research where the subjects of research become coproducers of the research. This strategy of documentation becomes part of the practice of political ecology itself. The formation of a documentation facility, at whatever level of sophistication and budget appropriate to the size of a PE initiative, is helpful as a cumulative resource for the researcher(s) and partners. It may contain a wide variety of materials such as photocopies of speeches, minutes of internal discussions, tape-recordings of informants and partners in research, newspaper articles, academic journal articles, or photographs that have caught the eye of the research team as well as electronically stored data. This facility becomes a 'dynamic action file' that informs initiatives to make PE useful. This phrase implies that the file is constantly being added to, or items are being removed perhaps to an historical archive as no longer relevant, and developed in terms of actions taken and planned. This idea was first discussed and tried at the International Centre for Integrated Mountain Development in Kathmandu during

2007–2008 for a range of environmental projects and was adopted informally by working groups within the institution.[1]

The compilation of a set of sub-files that address linked but different aspects of a PE research project is suggested here, but, as always in prescriptions of this sort, the form of the documentation must be made to suit specific circumstances. It may be entirely appropriate to use the list of files as a brief checklist only, in which some will be empty. Some files may have very few relevant documents because the research issue is clear, simple and obvious enough not require lengthy reflection. Whether the exercise comprises a brief checklist or a more fully fledged documentation, the issues in designing a useful PE engagement are labelled 'files'.

File 1. The goals of the research programme or project. This file provides the political and moral compass for the engagement. Here the goals of the project in terms of the specific characterization of sustainable environmental justice are initially set. Those who may benefit from improved access to and availability of environmental justice are identified at the outset. This will imply, at an early stage, scale considerations in the research (international such as international environmental accords, national such as land law, and other socially constructed scales such as the gender division of labour or domestic violence and local politics). The research has to engage with the existing goals as stated by the different involved parties with whom the political ecologist is working. This may require evidence-based research, individual testimony or survey work. There may also be written documentation of intent (e.g. policy documents, official guidelines or manuals of practice). If policy documents exist, they are best treated as a 'frozen' frame of a constantly shifting set of contending discourses and actors with particular interests and it is here that more discursive approaches are valuable. If policy goals have already been set and are beyond negotiation, there may still be room for manoeuvre in implementation such as enhanced citizen participation in citizens' panels, broadening of the observer status to a wider public, greater transparency in decision-making, and the public dissemination of counterdiscourses that expose the political content in the creation of environmental policy. National priorities of natural resource use are central starting points for understanding the sources of justice and injustice. Policy documents are therefore often an essential focus. In other cases, where policy-making is more informal, discontinuous or at an earlier stage of codification, there may be more room for negotiation about policy design and objectives. This is particularly the case with smaller organizations that have a limited network of persons with discretionary power to change policy and practice. All PE justice issues need thorough background knowledge of the on-the-ground situation from both a social and natural science perspective.

File 2 contains a *history of the site of struggle for environmental justice.* This history is often an account of the impacts of and responses to government policy

---

1   Blaikie, P.M. and Muldavin, J.S.S. 'Creating a Useful Dossier for Policy Work.' Unpublished paper for Workshop at the International Centre for Integrated Mountain Development. Kathmandu. 21–22 February 2007.

and the on-the-ground realities of environment and social relations. A historical perspective of policy is necessary, with a time line of events both in relation to the formation of policy and to outside events that may have affected policy (e.g. general elections, global economic downturn, or disasters of national or wider importance such as drought, famine, tsunami and floods). Many problem-orientated consultancies invite an ahistorical approach with rationalistic, acultural and ahistorical lists of recommendations – which gather dust unread (at best! – at least they can then do little harm) or can be interpreted by various actors in ways that legitimate their practices. In more specific terms, this file must also record the ways in which the policy environment is changing at present and will do so in the future.

File 3 identifies the *key technical and scientific debates* about the relevant environmental issues. The file should include critical scientific debates embracing a logical positivist epistemology (based on evidence and scientific procedure, e.g. problem-framing, sampling and statistical procedure). Also the politics of this science must be a crucial focus of this file. Issues of the funding and clients of research and the politics of problem-framing may need careful scrutiny. Alternative voices which express alternative knowledges of the environment should be sought, heard and recorded.

File 4 lists the *key actors in decision-making*. Again, a double focus upon local agency (e.g. trade union activity, social movements, federations of local organizations, strikes, deployment of 'weapons of the weak' (Scott 1985)) as well as on official policy and implementation practice is essential. PE research often involves a cast of actors with on the one hand marginalized people who are deprived of environmental justice by reason of class, gender, ethnicity and wealth, and on the other hand administrations at the national level. The question is: who are the key actors at all levels who shape the policy environment? Public opinion must be included, insofar as future policy formulation and implementation is affected by this. Much PE research involves different actors with very unequal power. This fundamental issue of environmental (in)justice is addressed in subsequent files.

File 5. Interfaces between key actors. A significant factor in the operation of programmes and projects by administrations is the degree of discretion exercised by officials and administrators to interpret laws, rules and regulations at different levels and how they do so. What are the effective, operational linkages between actors? Where does real power lie (including the power to do nothing, pass responsibility down the line, as well as to make judgements and act responsibly and intelligently)? A phrase like 'real power' begs many theoretical questions. They will have to be defined and operationalized in terms of the actors in the research process. Are there any illegal and corrupt practices which substantially affect policy goals? This will need discretion and confidentiality in the discussions between researchers and key members of the audience, but must not be swept under the carpet. The notion of a politicized 'organigram' is sometimes useful in order to understand what 'really goes on' in the relevant part of an administration. This can be based on a formal organizational chart, showing levels of an administration and lines of command, on to which is inserted the results of research showing where personal discretion and independence in decision-making lie, where there is evidence of corruption, how certain levels of decision-making get bypassed and so on. It may be necessary to create from scratch or actively to encourage forums to share and discuss multiple perspectives and foster participation of all concerned groups. For example, in research focusing on environmental justice and forests and forest use, the full cast of actors includes local forest users – and care should be taken to include local marginalized people without a voice, both men and women (do not take the village leaders' word for it!) – as well as forest contractors, charcoal manufacturers, plywood and sawmill entrepreneurs, and forestry officials at all levels. To assemble this diverse cast of actors and expect open democratic rights of speaking and being heard without reprisals may be difficult. As Mayers and Bass (1999: v) succinctly put it: 'One of the key elements of a policy process that "stays alive" is the ability to link directly to experiments with new ways of making things work on the ground'.

File 6. Whom do we work with? This concerns choices with whom to work (or as Mayers and Bass (1999: v) term it 'linking the people who change things'). From the beginning of research planning, there may be partners who invited or initiated the research in the first place and who will help to shape the research and the network of people who 'change things'. There is a wide range of potential collaborators (politicians, leaders of social movements, gatekeepers in the media, filmmakers, intellectuals, activists, lobbyists, workers in aid projects, national federations of local groups as well as the more conventional opinion leaders, and policy-makers and senior administrators in capital cities and the offices of supranational institutions). In most cases though, the choice about whom to work with effectively is constrained by feasible access to identified audiences via personal meetings, telephone calls, letters, reports and e-mails (that actually get read). The issue here is whom do we *want* to work with – and whom to avoid? Those key individuals in Files 2 and 3 will have been identified early but informal discussion can usually elicit attitudes (e.g. various stereotypes include aggressive modernizers, the role of expert/scientific/official knowledge, populist/bottom-up approaches, decentralized/centralized etc.). What are the attitudes and approaches of key officials at different levels to their jobs? How amenable are they to reform through talking and working with them? Here individuals matter and the issue of transfers of personnel in the middle of the research can be problematic and can be noted in this file. How can narratives of local people be heard and who will listen and then act differently? Local narratives themselves have to be subjected to the same sceptical examination as those of senior bureaucrats and politicians. They may be told by a patriarchal 'big man' who claims to speak for the local people and does not hear or ignores other views of the poor, the marginalized, ethnic minorities, the distant and women (not all women but especially those who share the multiple identities of suffering injustice). If this is a relevant question, searches in manuals, documentation and critiques of participatory project work are essential in case studies.

File 7. Policy argumentation. How should the research engage with the policy narratives that obstruct or are hostile to those of the project? Are there alternative narratives that may facilitate the goals of the project? Is it possible to use other arguments that may not anger or confuse one's audience but still not compromise its goals? How can the project show its assumptions can be logically or empirically demonstrated? Here evidence-based research is very useful and familiar to the majority of government servants the world over. What is the role of scientific experimentation in persuading key personnel of the benefits of the project? Here there is a burgeoning literature on colonial and neocolonial scientific narratives that have been proven to be plainly wrong. Examples of this literature include nature conservancy (Neumann 2005), land degradation (Blaikie and Brookfield 1987), and forestry and the issues of deforestation, tree cover and canopy density in India and Nepal (Gadgil and Guha 1995, Sivaramakrishnan 1999, Gadgil 2001), West Africa (Leach and Mearns 1996), and Kenya (Tiffen et al. 1995).

This file is also a record of decisions about communicating PE with others. As the research process develops, reflexivity among research partners and learning may guide and transform research goals, methods, outputs and audiences. Is the language appropriate for the audiences of the project? When is English appropriate and when are national or local languages, and for what level of audience (international, capital city, regional or local)? Is the level of sophistication of the argumentation appropriate to the chosen target audiences, or will project outputs (e.g. manuals, scientific papers, films, newspaper articles, broadcasts etc.) have to be rewritten for different audiences? In a recent PE project (Springate-Baginski and Blaikie 2007), local language films were made for the research project by Indian filmmakers featuring local forest issues and shown to a wide rural audience. Monitoring and evaluation of the experiments, discussions, working parties, citizens' forums etc. are essential for a learning research programme. Also, monitoring and evaluation of the impact of the project, sometimes well after the official completion of the research, continue the learning process for audiences and researchers alike and address the questions 'what contribution to environmental justice did we make?' and 'what lessons have been learnt for future engaged PE?'

Conclusion

Existing frameworks, structures and discourses of 'development' and their contradictions pose obstacles (and some opportunities) to an engaged PE. However, most of the misgivings arise from the false assumption that 'development' always implies mainstream development, whereby the centralized state is dominant in shaping policy. Alternative development is made by many organizations and often includes the state. The state passes laws and grants major funds, but also can send in oppressive and violent police to snuff out an alternative development that powerful groups with leverage on the state apparatus do not like. Therefore, officials in government departments, influential policy-makers and politicians have to be addressed by reformers of policy. While cooperation with NGOs, civil society institutions, social movements and charities in both the North and South may arouse less well-earned suspicion of a critical political ecologist, nonetheless it is with formal development institutions that some of the major hazards have to be negotiated. The pursuit of environmental justice as a principal goal of critical and engaged PE research involves not only the creation of symmetrically coproduced new knowledge but its successful communication to those agents who can make a difference.

References

Agyeman, J., Bullard, R.D. and Evans, B. (eds) 2003. *Just Sustainabilities: Development in an Unequal World.* London: Earthscan.

Batterbury, S. and Horowitz, L. (eds) 2010. *Engaged Political Ecologies.* Cambridge: Cambridge Open Books.

Blaikie, P.M. 1985. *The Political Economy of Soil Erosion in Developing Countries.* London: Longman.

Blaikie, P.M. 2000. Development, Post-, Anti-, and Populist: A Critical Review. *Environment and Planning A*, 32, 1033–50.

Blaikie. P.M. and Brookfield, H.C. 1987. *Land Degradation and Society.* London: Methuen.

Blaut, J. 1993. *The Colonizer's View of the World.* London: Grove Press.

Bryant, R. and Bailey, S. 1997. *Third World Political Ecology.* London: Routledge.

Bryant, R. and Goodman, M.K. 2008. A Pioneering Reputation: Assessing Piers Blaikie's Contribution to Political Ecology. *Geoforum*, 39, 708–15.

Burawoy, M. 2005. For Public Sociology. *American Sociological Review*, 70, 4–28.

Cooke, B. 2004. Rules of Thumb for Participatory Change Agents, in *Participation: From Tyranny to Transformation? Exploring New Approaches to Participation in Development*, edited by S. Hickey and G. Mohan. London: Zed Books, 42–55.

Corbridge, S. 2008. The (Im)possibility of Development Studies, in *Development: Critical Essays in Human Geography*, edited by S. Corbridge. London: Ashgate, 481–513.

Dreze, J. 2002. On Research in Action. *Economic and Political Weekly,* 2 March, 817–18.

Escobar, A. 1995. *Encountering Development: The Making and Unmaking of the Third World.* Princeton, NJ: Princeton University Press.

Forsyth, T. 2004. *Critical Political Ecology: The Politics of Environmental Science.* London: Routledge.

Forsyth, T. 2008. Political Ecology and the Epistemology of Social Justice. *Geoforum*, 39, 756–64.

Gadgil, M. 2001. *Ecological Journeys: The Science and Politics of Conservation in India.* Delhi: Permanent Black.

Gadgil, M. and Guha, R. 1992. *This Fissured Land: An Ecological History of India*. New Delhi: Oxford University Press.

Heynan, N. 2003. The Scalar Production of Injustice Within the Urban Forest. *Antipode (Special Issue)*, 35, 980–98.

Jarosz, L. 2004. Political Ecology as Ethical Practice. *Political Geography*, 23, 917–27.

Leach, M. and Mearns, R. (eds) 1996. *The Lie of the Land: Challenging Received Wisdom on the African Environment*. Oxford: James Currey.

Leach, M., Scoones, I. and Stirling, A. 2010. *Dynamic Sustainabilities: Technology, Environment, Social Justice*. London and Washington, DC: Earthscan.

Lund, R. 1993. *Gender and Place*, Volume 1: *Towards a Geography Sensitive to Gender, Place and Social Change*. Trondheim: Department of Geography, University of Trondheim.

Lund, R. 2000. Geographies of Eviction, Expulsion and Marginalization: Stories and Coping Capacities of the *Veddhas*, Sri Lanka. *Norsk Geografisk Tidsskrift–Norwegian Journal of Geography*, 54, 102–9.

Lund, R. 2008. At the Interface of Development Studies and Child Research: Rethinking the Participating Child, in *Global Childhoods: Globalization, Development and Young People*, edited by S. Aitken, R. Lund and T. Kjørholt. London and New York: Routledge, 131–49.

Mayers, J. and Bass, S. 1999. *Policy That Works for Forests and People*. Policy that Works Series, No. 7. London: International Institute of Environment and Development.

Neumann, R.P. 2005. *Making Political Ecology*. New York: Oxford University Press.

Peet, R. and Hartwick, E. 2009. *Theories of Development: Contentions, Arguments, Alternatives*. Second edition. New York: Guilford Press.

Peet, R. and Watts, M. (eds) 1996. *Liberation Ecologies: Environment, Development, Social Movements*. London: Routledge.

Peet, D., Robbins, P. and Watts, M.J. (eds) 2011. *Global Political Ecology*. London and New York: Routledge.

Rahnema, M. and Bawtree, V. (eds) 1997. *The Post-Development Reader*. London: Zed Books.

Rangan, H. and Kull, C.A. 2009. What Makes Political Ecology Political? Rethinking Scale in Political Ecology. *Progress in Human Geography*, 33, 28–45.

Robbins, P. 2002. Obstacles to First World Political Ecology: Looking Near Without Looking Up. *Environment and Planning A*, 34, 1509–13.

Robbins, P. 2004. *Political Ecology: A Critical Introduction*. Blackwell: Oxford.

Robbins, P. 2006. Research is Theft: Rigorous Enquiry in a Post-Colonial World, in *Philosophies, People, Places and Practices*, edited by G. Valentine and S. Aitken. London: Sage Press, 311–24.

Robbins, P. 2012. *Political Ecology*. Second Edition. Chichester: Wiley-Blackwell.

Rocheleau, D. 2008. Political Ecology in the Key of Policy: From Chains of Explanation to Webs of Relation. *Geoforum*, 39, 716–27.

Rocheleau, D, Thomas-Slater, B. and Wangari, E. (eds) (1996) *Feminist Political Ecology: Global Issues and Local Experience*. London: Routledge.

Sachs, W. (ed.) 1992. *The Development Dictionary: A Guide to Knowledge and Power*. London: Zed Press.

Schroeder, R. 1999. Debating the Place of Political Ecology in the First World. *Environment and Planning A*, 37, 1045–8.

Schroeder, R., St. Martin, K., Wilson, B. and Sen, D. (eds) 2008. Third World Environmental Justice. Special Issue. *Society and Natural Resources,* 21, 547–655.

Sikor, T. and Lund, C. (eds) 2009. *The Politics of Possession: Property, Authority and Access to Natural Resources*. Chichester: Wiley-Blackwell.

Simon, D. and Närman, A. (eds) 1999. *Development as Theory and Practice*. Harlow: Longman.

Sivaramakrishnan, K. 1999. *Modern Forests: State-Making and Environmental Change in Colonial East India*. Delhi: Oxford University Press.

Stott P. and Sullivan S. (eds) 2000. *Political Ecology: Science, Myth and Power*. London: Arnold.

Scott, J. 1985. *Weapons of the Weak; Everyday Forms of Peasant Resistance*. New Haven: Yale University Press.

Springate-Baginski, O. and Blaikie, P.M. (eds) 2007. *Forests, People and Power: The Political Ecology of Reform in South Asia*. London and Sterling, VA: Earthscan.

Swyngedouw, E. and Heynan, N. 2004. Urban Political Ecology, Justice and the Politics of Scale. *Antipode*, 34, 898–918.

Tiffen, M., Mortimore, M. and Gichuki, F. 2004. *More People, Less Erosion: Environmental Recovery in Kenya*. Chichester and New York: Wiley.

Walker, P.A. 2006. Political Ecology: Where is the Policy? *Progress in Human Geography*, 30, 382–95.

Watts, M.J. 1983. *Silent Violence: Food, Famine and Peasantry in Northern Nigeria*. Berkeley, CA and Los Angeles: University of California Press.

Chapter 3

Teaching to Learn – Learning to Teach: Learning Experiences from the Reality of an Ever-Changing World

Hans Skotte

Introduction

We all know that the world around us is not really as simple and as rational as we would like it to be. At times it appears confusing and unpredictable. All the same we strive to simplify it, and to create categories and structures in order to make sense of our lives – and thus be able to act strategically. This pursuit of identifying or constructing underlying patterns may be seen as the ultimate mission for academia: to generate knowledge and insight that can provide plausible explanations in answer to our endless line of 'why-questions'. Hence universities around the world are staffed with brigades of the best and the brightest, engaged in generating and disseminating the answers they deem most useful. Yet are they always answers, and are they always useful?

This chapter examines the pedagogical implications of teaching as a function of students' learning, rather than students' learning as the function of teaching. The European Union (EU) has made the former the guiding principle in lifelong learning (EU 2008). This is not a question of mere teaching technique. It calls into question the very 'what' that students learn – and how the 'what-knowledge' is determined by how that knowledge is generated, i.e. its epistemological challenges. So far we have been content to control their learning by grading their ability to generate 'useful answers'. Yet this, at best, tells us only half truths about their capacity to act responsibly in society.

The students' own perspective has been recognized as essential for understanding educational phenomena, not least within the medical professions (Taylor 1994, Säljö 1998, Lyon 2007). It is almost self-evident that by understanding the students' learning perspective we can make better sense of their engagement and their reactions to educational settings, hence making teaching the function of students' experience of learning. In this chapter I primarily refer to experience from the field of architecture and urban planning. For the purpose of simplification and argument, I deal with these as a joint discipline. The chapter is based on my own field experience of working with students 'in the real', driven by attempts to improve the life conditions of marginalized people in the Global South.

Architecture and planning are fundamentally applied fields of study, yet both are solidly planted in mainstream academia. However, it is worth noting that several of its most influential practitioners, such as Le Corbusier and Frank Lloyd Wright, never went to architecture schools or universities. Yet they are the grand heroes of the discipline, corroborating its deep roots in practice. Further, most members of our 'academic brigade' hold their position by merit of what they have done as practitioners, not on account of their academic research or published papers. So why is practice so little recognized as a way for students to gain professional insight and command of the discipline? Why do we not let our students learn the same way as we ourselves have gained our specific insight and skills? '"There's a snobbery at work in architecture",' says Sir Norman Foster, another hero, head of one of the most respected architectural practices of today, in an interview with Jonathan Glancey in *The Guardian*. Foster continues: '"The subject is too often treated as a fine art, delicately wrapped in mumbo-jumbo. In reality, it's an all-embracing discipline taking in science, art, maths, engineering, climate, nature, politics, economics"' (Glancey 2011).

This paradox extends to a large part of the world of research since most research is grounded in practice. We explore the real world around us, and compare the results of our examination with what others have found. We all, irrespective of approach, are examining practice. In pursuit of explanations, we have to engage all our human senses, not only what is commanded by our cognitive powers. This is succinctly shown by Ragnhild Lund when doing research in Sri Lanka immediately after the tsunami disaster of 2004 and during the final years of the civil war. Her field notes 'show how my emotions as a researcher have become the "glue" that gives the descriptions analytical insight' (Lund 2012: 5). By 'building on emotionally sensed knowledge of the research participants', she holds that 'additional insights may be gained and new methods of discovery may be developed' (Lund 2012: 1). In spite of such powerful claims on how knowledge is gained, our teaching remains basically book-bound, linked to what *somebody else* has found out, not on the experiences of those set to learn. This is sharply argued in a visually persuasive case by Sir Ken Robinson (2010). I claim that by holding the real world out-of-bounds, we inhibit the learning potential of our students. This in turn curtails their insight and thus reduces our capacity to instigate and mediate positive societal change. Besides – we take so much fun out of studying.

The Context: Trying to Understand the Dynamics of Urban Change

In no other field is upgrading our understanding and reloading our theoretical canons as urgently needed as in the field of urban development. That the future is urban is now commonly recognized. It is also a common claim that the urban world, particularly the 'urban South', is under-researched. Joint work by leading international planners held research on urban development in the Global South to be one of the five most critical blind spots of the profession (Blanco et al. 2009a,

2009b). The call for relevant, appropriate knowledge and understanding of the dynamics of the urban South was forcefully stated in *A Home in the City*, the report from the Millennium Project Task Force on Improving the Lives of Slum Dwellers (Garau et al. 2005). This task force was established to explore strategies related to the goal of significantly improving the lives of at least 100 million slum dwellers by 2020. This makes up only a small fraction of the 1.4 billion that are projected to be living in slums by 2020. The call was repeated perhaps even more forcibly in *Planning Sustainable Cities* (UN Habitat 2009) and seems more than justified when facing the numbers: whereas the number of people living in rural regions in the world will remain about as it is today, i.e. 3.3 billion, the number of urban dwellers will within a single generation increase by about 1.5 billion, from 3.5 in 2010 to 5 billion in 2030. Almost all of this increase will take place in what the United Nations (UN) labels 'less developed regions' – where we find most of the world's slums. Today almost one third of the urban population in the world lives in slums, well over 900 million people. Slum dwellers make up a majority in almost all cities in Africa, about 30 per cent in Asia, and 25 per cent in Latin America (UN Habitat 2009).

Neither 'the best and the brightest' of formal academic institutions nor the profession of architect–planners fully know how to deal effectively with this problem in a manner that may improve the lives of the ever-increasing number of marginalized urban dwellers. There is growing unease within the profession that the traditional planning practices as applied in rich countries with strong institutions have not only been ineffective but have made things worse (Koenigsberger 1964, Hamdi 1995, 2010, Hamdi and Goethert 1997, Garau et al. 2005, UN Habitat 2009). This may not be due to 'wrong' theories or 'ineffective' methods as such; it may stem from inappropriate conceptualizations of the dynamics of this urban reality, not least pertaining to the decisive role played by 'informality'.

I argue that we should look at the way we arrive at our theories and methods. How do we conceptualize the urban reality of the South, or said more plainly: 'how' do we learn 'what'? 'I see what I see ... – but what am I looking at?' as Nabeel Hamdi (2010: 203) says about this very challenge. So let's start looking!

A Question of Approach

Retaining the analogy of looking, we will approach our urban complexity by gazing through a twin set of lenses, one examining the power over others, e.g. professors over curricula and hence over students, that is embedded in knowledge, and the second the way this knowledge is generated. Obviously these are related. We need not make the often quoted references to Foucault in underpinning the power-of-knowledge issue, nor need we extend the arguments of Illich (1973) or Freire (1972) on the liberating – no, revolutionary – power of knowledge and learning. What is important to reiterate in regard to Illich and Freire, however, is that they show how these dimensions are essentially linked to 'the way we learn' – not by 'being told' but by 'being part of' the learning field. Donald Schön (1983) gives additional strength to that argument through his influential reflection-in-action approach. His case study from an architectural design studio is particularly pertinent in helping us look. Yet Schön may seem more interested in understanding the reiterative yet rigorous process of design rather than how the design process may be influenced by the societal setting of the design object in question. This is at the heart of Boyer and Mittgang's call for architectural education to be more connected to society. They refer to 'the architecture community's long history of failure to connect itself firmly to the larger concerns confronting families, ... communities and societies'. Furthermore: 'architecture students and faculty at many schools seem isolated socially and intellectually, from the mainstream of campus life' (Boyer and Mittgang 1996: xv–xvi). The need for linking the realities of society to architectural education has been consistently propagated by the internationally famous educator Michael Lloyd. His directorships of architecture schools in Africa and Latin America, as well as in the UK and Norway, are based on the principle that students best generate architectural insights and skills by interacting with 'real people'. The same applies in the case of Nabeel Hamdi (2010), from whose work the title of the present article is derived. Teaching as a function of learning seems to guide Hamdi's approach in a way reminiscent of how Richard Sennett (2008: 7) describes the way Hannah Arendt acted as a teacher: 'The good teacher imparts a satisfying explanation; the great teacher – as Arendt was – unsettles, bequeaths disquiet, invites argument'. This is learning as a fought-for discovery – much like research. Angela Brew in her *Research and Teaching: Beyond the Divide* (2006) takes on this very issue by arguing for engaging students in research – as a most efficient and also most appropriate way of learning when learning means 'skills of critical analysis, gathering evidence, making judgments on a rational basis and reflecting on what they are doing and why. These are skills of inquiry' (Brew 2006: 13). However, skills of inquiry will question the very knowledge structure (i.e. power structure) in society – not least that of institutions of higher learning staffed with a brigade endowed with vast amounts of Bourdieu's symbolic capital (Bourdieu 1986). Like any other 'identity group', they defend their entitlements. It is just that those entitlements constitute the very knowledge base we as a society depend on for progress. When Jeevan Vasagar (2010) reports in *The Guardian* that Oxford and Cambridge admit less than 1 per cent of students from 'poor families' (i.e. families with an annual income of less than £25,000), it is but a confirmation of how economic power is linked to knowledge (re)production. Hamdi claims, further to this, that many of the constraints clogging progress stem from lack of willingness to update our knowledge, i.e. our profession's unwillingness to learn (Hamdi 2010). This conforms to the intellectual dispositions of our institutions, their epistemic history and the unconscious aspects of the (academic) field, as Bourdieu would have it, according to Özbilgin and Tatli (2005). Under conditions like these, no wonder looking is difficult, or made difficult.

Looking in on the Philippines, Haiti, Uganda and Thailand

The Department of Architecture and Fine Art at the Norwegian University of Science and Technology (NTNU) has recently extended its practice of conducting courses in the Global South. This is primarily due to students themselves identifying projects they want to develop and implement as part of their architectural and planning studies. They even identify sources of funding themselves, sometimes large amounts of money (which again is entering into a different kind of reality than a university course normally allows for). In addition some official project money has also reached us.

The students thus have designed and built or are in the process of designing and building a school for street children in the Philippines, a bakery for a women's collective in Haiti, a football facility in a village in western Uganda, a library and dwellings for orphans in northwestern Thailand and a community library in Bangkok. In addition, they have initiated community development projects in the slums of Kampala, Uganda. All projects are planned, designed and implemented in cooperation with local stakeholders, be they slum dwellers, local students, cooperatives, family members or local host organizations.

On returning to Norway, all students have submitted written reflections on what they deemed their most important experiences as students of architecture/ planning. What do they feel they have learnt? Unanimously, all come back claiming a wider and more reflected understanding of the societal forces with which they have engaged. 'My attitudes were no longer abstract ideas; they were now relating to real things and people. I had become part of a reality I so far had considered as something outside of myself', one student wrote.

What follows is based on what the students wrote back to me, and thus provides critical signals on how and what to teach, given that it bears relevance to the real-world challenges of the profession of architect-planners. Although their comments stem from significantly different contexts, their common role as architect students on a learning mission invariably carries a significant reference to our teaching approach – in spite of all not giving parallel signals.

The purpose of universities is to enable students to gain knowledge and skills, and to instill values. The European Union has changed 'values' into 'competence' in the European Qualifications Framework for lifelong learning (EU 2008). This accentuates the contentious nature of 'values' inasmuch as values are not taught but discovered, actually much in the same way as knowledge is. This is also what comes back from students working in the real world: a discovery of the values embedded in architecture and architectural practice.

Values

The process of planning and building triggered reflections on the social role of architecture – and of architects. Working in marginalized areas – and with their own hands – provided students with a powerful arena for professional – and personal

– reflection, providing insight that can only emerge from practice, as discussed by Richard Sennett in *The Craftsman* (Sennett 2008). This is also what we feed our students when we call for them to 'use the pencil!' It is not the skill itself we are looking for. We aim at the insight born through the reflections these practical processes generate. 'The workshop format gives us a unique chance to reflect while creating, and in creating by doing. We let the building be the moderator of the dialogue and thus allowing it to respond by telling us the consequences of our actions before we proceed,' wrote one of the Thailand workshop participants.

The multilayered situations emerging when the students work and live within the very place the project is to function also seem to help demystify the 'poor beneficiaries', be they unemployed urban poor or disenfranchised orphans – they become fellow human beings. Even local students experienced places they had previously not explored. One stated that 'it was an eye opener to me not only as a professional but also in my general outlook to life'. She had lived all her life in a city where 60 per cent of the population lives in slums. Significantly, her new-gained understanding, echoed by the other local students, came through attending a practice-based university course, and not from her previous studies of the 'slum problem'.

There is a lot of talk about 'values' in architecture. However, because values require an ethical base, values in architecture easily become self-referential claims concerning 'good architecture', regarding buildings in themselves while ignoring their social relevance or their potential strategic social function. One of the students had previously worked at the office of one of Japan's most celebrated architects, where the emphasis was on 'refining the language of architecture, ... focusing on spiritualizing space and not on the physical and the concrete'. However, 'when the intention behind architecture was embedded in the dialogue with the users and their life world', as was the case in Tacloban in the Philippines, 'architecture took on a much larger societal, contextual and economic relevance'. Of these two complimentary aspects of architecture, the former holds sway in the studio, while the latter defines the reality where the projects are situated. Another student added: 'The abstractions of the studio and the touch of cynicism that steers our efforts on to making our projects look good in a portfolio, holds no value here – and that feels good'. This feeling of 'having been somewhere and ... having experienced having done something ... with highly real people', as one student formulated it, underpins the almost unanimous opinion among the students that they have made a societal contribution – and been able to do so by changing or expanding their understanding of what architecture is about. From the work of John Turner (Turner and Fichter 1972) development practitioners have formulated a general proposition on the role of housing in development. It states that: 'it's not what housing is, it's what housing does' The students' feedback can be summarized as: 'It's not (only) what architecture is, it's what architecture does'. Their understanding of architecture may have changed forever. One of those returning from Bangkok wrote:

Through the nine previous semesters, the focus has always been on designing buildings. Here this was turned on its head. The preparatory phase and the construction were far more important. Not only in relation to time spent, but also in relation to the very nature of architecture. Working closely with the users, then building with them, gives me a whole new understanding of the value of architecture. It's no longer (only) about designing a building, but it's about supporting a small community, both through the project and the process.

Further: 'What we did was real, important and true. Unfortunately, this is a feeling I never have had before at NTNU'. This ultimate claim from a student seems to pull the rug from under our traditional teaching approach, given the university's educational mission of accommodating knowledge, skills and values.

Responsibilities

Learning means 'changing your behaviour'. The fact that the students themselves built what they previously had planned carried immediate and tangible consequences for themselves, for the building, for the people they worked with as well as for those they built for. The feedback from the students is unanimous in pointing out that these responsibilities were what made their activities such an intense learning experience. They were forced to make decisions and had to live with the consequences. However, this has also come to be the most contentious issue in these projects. Some of the students, particularly the students working as architects for a client in Haiti, found this responsibility also to be professionally inhibiting and a strain on the progress of their project. They were simply not qualified to design the structural work for a large building in an earthquake-prone area. They therefore became dependent on others, yet felt they carried full responsibility for the structural safety of the building, as well as for spending other people's money, and the coordination of all the disparate interests of the projects. 'We somehow lost our own role because we tried so desperately to secure the input from the actors "not present" … which meant that we could not fully argue as design architects'. The Haiti group could therefore not afford the 'beautiful mistakes' that had characterized many of the other projects students had taken part in (Skotte 2010). This tells us something about selecting sites and situations that are true 'fields for learning'. Disaster areas may not be the most effective ones. In addition, Sanderson (2010) claims that architects, be they students or experienced practitioners, may not even be the most needed professionals in the immediate wake of a disaster. This warns that disaster areas may not be appropriate learning fields for students.

The fact that the projects were 'in the real' meant that the students had to respond to time, and they had to choose solutions in line with available tools, materials, skills and manpower – and the money available – while coming up with a plan or a building that reflected the architectural or professional ambitions set for the project. 'The fact that the projects I make actually bear real consequences is one of the most important aspects of our studies, and represents a profound learning experience', wrote one. 'We don't have a choice, and we learn to live with it. This is something we will carry with us from now on', wrote another. Many claim that they themselves, and not least the local people they worked with, developed a sense of independence through this burden of responsibilities. 'We saw clearly how their [a women's group's] self-confidence grew in line with increased responsibilities', said one, who commented on how this made the women 'more and more equal to the men working on the site'. Another student reporting from another project on the opposite side of the globe wrote: 'It was rare for people in the community to see women working hand in hand with men .... Here was architecture obliquely contributing to gender equality'. This was from all quarters held to be the result of a reciprocal process where each party had a pronounced need for the others in order to realize the project. 'All [affected families and students] brought on their contributions and experiences. These in turn affected subsequent processes. The families immediately began taking on ownership functions, and eventually they were the ones running the project', reported one student. The self-confidence emerging from these experiences is, as we know from development theory, a prerequisite for positive change. This applies to societal progress as well as to becoming a good architect. 'The learning value of this semester has primarily come from the independence one had to have', said one. Another simply wrote: 'I have become much more self-confident as a student of architecture'. What better learning outcome is there?

Practical Insight

The feedback concerning the practical insight gained was the most obvious. Many of the workshop participants in Thailand, for instance, considered using these weeks as part of their mandatory 'building site practice' since the experience gained here was such that 'we rarely gain as trainees at a Norwegian building site'. Even more critical was the confrontation with materials and with making informed technical choices, aside from the implementation practices. That would entail, for example, learning how to handle a handsaw, nail floorboards, mix mortar, apply screed, or do drainage – not to mention earthquake-safe construction details, and structures withstanding typhoons and monsoon rains. In addition there was the matter of managing a site where the students sometimes were outnumbered one to five, or one to 10 on some of the building sites. In several cases they also had to deal with official bureaucracy, often a slippery and time-consuming endeavour. 'In (this) reality, problems I never had to think about earlier may have become the most important ones', said one with reference to what he had learnt and not learnt in his years of university training. Rightly so; the task is impossible to handle in an auditorium or a studio.

    This practice-based approach revealed new aspects of the creation of architecture and realizing plans. Some could finally come to grips with the claim Professor Ole Jørgen Bryn made in a course at NTNU: 'architecture is created in stride', in an almost dialectic progression. Others quoted Professor Sami Rintala 'who seems to be of the opinion that a building is partly created as we go along and that this dynamic process must be respected in order to let it influence the building, and thus make it better'. This was not, as we have seen, in line with the Haitian experience, thus showing how contextual some of our general architectural statements are. Yet from all corners the most glaring fact stemming from this practice-based approach was perhaps the total absence of the usual 'architectural chatter', or as one student said, 'I find it a good thing to solve problems through consideration of specific and actual matters, without too much philosophic prattle', much as Norman Foster formulated. The architecture was made concrete. Throughout there has been a liberating and 'clean' approach to architecture gyrating towards 'an architecture with people' (referring to the fact that when the national and international architectural press presents buildings, they invariably are empty – without people). 'Here it is the people in the house that "fill" the architecture. Without people, architecture is not beautiful'. This mature statement comes from a fourth-year student claiming his insight stemmed from his encounter with a challenging reality.

It is impossible not to refer to Sami Rintala's comments to the students in Thailand when they were designing the in-between space between the internal floor and the external ground. The students proposed different solutions using different materials, to which Sami retorted: 'The hardwood boards are difficult to get hold of. Bamboo we have not really worked with. But here's so damn much stone …!' A few days later a couple of large stones made up a beautiful in-between space. This realization of what architecture is all about was echoed by a student: 'Every now and then I get the feeling that architecture is becoming increasingly difficult, but then again I am similarly struck by how easy it really is. Be self-confident. You don't need to design yourself difficulties'.

Cooperation and Communication

'Architecture is a team sport' comes out of most of the reflections. The students referred to the interdependence between all actors involved. The following two statements, from two different parts of the world, highlight this discovery:

> [The workshop] has contributed to my looking at architecture as something more than what is created on a designer's table. It is something created in cooperation where many more are involved besides the designer(s).
> To experience architecture as something greater than merely a physical object is something I have never experienced before.

What is important and all the more revealing about these statements is that they come from students who have spent several years at a university without discovering this fundamental quality of architecture. When Renzo Piano, one of our time's most respected architects, explains why he named his firm 'Renzo

Piano Building Workshop', he sounds like my students: 'The term "building workshop" deliberately expresses the sense of collaboration and teamwork that permeate our design process' (Peltason and Ong-Yan 2010: 159). It may not be a surprise then that practitioners time and again complain about the lack of 'appropriate' competence of fresh architects graduating from our architecture schools (repeatedly voiced inter alia by members of the Governing Board of the Faculty of Architecture, NTNU, representing the practitioners).

The 'art of communication' was touched upon by all the students. One after another claimed that the most palpable result from the project period was recognizing the critical role of communication. One student simply said that it was 'within the realm of communication and role definition I learnt the most'. Another student in another part of the world claimed, 'I learnt a lot about my capacity as communicator, and believe I will make a much better job the next time I'm in a situation like this. I feel that communication and making decisions in the right way were the greatest practical challenges of the project'. Communication is a practice-born activity, as are tactical skills, both crucial qualities for achieving any societal objective, yet none have a noticeable academic standing in traditional schools of architecture. Things may be changing. Strelka, the new private postgraduate academy of architecture in Moscow, has on behest of Rem Koolhaas, one of the 'giant stars' of contemporary architecture, made a significant part of their two-year programme focus on modes of communication.

Traditional face-to-face communication between students and teachers was minimal – as planned (except the workshop in Ban Tha Song Yang, Thailand, in January 2009, where two teachers participated). In more difficult cases, such as Haiti, the students both missed it and needed it, whereas others were satisfied with being left on their own. 'One must not let the more experienced [teachers] take over the course', said one. This supports a claim made by Patricia Lyon (2007) that the capacity to learn from such challenges is significantly linked to the students' ability to manage them.

Architecture's Strategic Dimension; 'What Architecture Does'

We have so far seen how students reorient their conceptual approach to 'what architecture is', as well as how they try to conceptualize 'what architecture does'. This stems from their experiencing the strategic dimension of forging a project, called by one group 'the architecture of the process' ('It ain't what you do, it's the way that you do it', to cite the title of a jazz tune from 1939 written by Cy Oliver and Trummy Young). Several examples highlight the process of working with local actors, but students also come back pointing to a wider understanding of the timeliness of their process-driven interventions: 'understanding that a room obviously will not be used for the purpose you set, however much you want it to. Probably [the building] will be extended and changed innumerable times in its lifetime. The need for "resilience" has become more apparent'. I interpret this as seeing the consequences of the local momentum they have created through the process, what I call 'a transfer of agency' (Skotte 2004) to the local stakeholders. 'In our case the very nature and meaning of architecture came to be embedded in the relationship that emerged through the process of [planning and building] the study centre'. Again we see 'architecture with people', as we do when another said that 'the building is a result of a dialogue and is subservient to human relationships'. They have literarily set the 'ball rolling'. Traditional studio work does not normally make this into a significant issue. Time as a dimension in architecture is in fact a highly contentious issue (Habraken 1972, 1998, Butenschøn 2009).

What seems to be the most rewarding experiences stem from the outcomes of the momentum their 'rolling balls' have created, as in this example from the Philippines: 'The most interesting point for me was when it dawned on us what impact our building would have on its [physical and social] surroundings'; or the statements coming from the groups working with planning interventions in Kampala: 'The repercussions of our actions were not hypothetical; they were taking place in front of us. But the best moments were when they started taking shape irrespective of us. Getting redundant never felt so good!' A similar reaction was reported from the upgrading processes the Bangkok group initiated: 'Already during construction the closest neighbor started upgrading all on his own', and after the students had completed their project, others from the community 'made several personal adjustments to what we've done'. Ultimately it is such spin-offs that justify our students' wading into poor people's lives.

So What's in it for the 'Bright Brigades'?

What the students brought back is the very essence that could inform the teaching of architecture and planning, all the more so when we see how our teaching modes and approaches might change when they are seen as a 'function of learning'. An immediate challenge would be to engage teachers who are willing to 'get their hands dirty' or are willing to 'leave the high ground and wade into the swamps', as the great Indian architect Charles Correa once told me paraphrasing Donald Schön (1983). In practical terms, professors and the like would be called to advise on construction matters and plain skills of building. Since most of the teaching staff is recruited from practice – and not academia – most of them should be prepared to shift their teaching focus. The other requirement emerging would be to advise on processes, group dynamics, negotiating skills, etc. Coming from practice, many of the staff have Norwegian experiences to draw from. In addition we would have to provide teachers or advisors that hold more site-specific experiences. It is obvious that negotiations are conducted differently in Norway and India, or Brazil and Kenya. Further, the teaching staff would have to be able to advise on where and under what circumstances a 'learning arena' would be appropriate for students' engagement. This could open up for closer cooperation with architecture schools or universities of the Global South. First, they would be better prepared to advise on local issues, and second, it could open up for participation by local students, thereby expanding the learning outcome for students from the North. We have in the Kampala feedback seen how this also changed the local students' perceptions of their own environment.

However, do we see 'learning by doing' as a relevant and thus appropriate approach to grasping the momentous challenges of the urban South? It is relevant, but not sufficient. However, the students experienced initiating processes that seem to have energized the principal stakeholders in 'claiming their rights', as one would say in the development industry. This is at the core of the rights-based approach (RBA) to development. Most probably without knowing it, the students have brought architecture and planning into this wider realm of societal change. For something more to happen, however, and this is RBA's Achilles heel, these interventions have to be linked to the general urban planning processes. Linking them is imperative for any effective improvement to the functioning of the cities as well as to the condition of their marginalized populations. It is a matter of 'bottom–top planning', the convergence of the 'bottom-up' and 'top-down' approaches, or what Nabeel Hamdi calls the 'emergence and design' approach (Hamdi 2004). Through the emergent dimension of their intervention, the students have contributed to the bottom.

It is truly beyond our students' capacity to link the two domains. Any bottom–top linkage depends on local energies. Our contribution lies in our cooperation with local institutions whose responsibilities are to provide appropriate knowledge and skills aimed at closing the gap.

I have deliberately not problematized the design issue, most certainly a crucial dimension in the study and teaching of architecture. This is simply because the students hardly mentioned it. Nor did they touch upon the wide range of other skills and competences architects must master in order to function as professionals worth their salary.

There is a strange nerve that seems to join these two aspects. It seems so in our cases and it certainly seems so from looking at what other design–build programmes have achieved elsewhere in the world, and there are many.[1] Several have achieved an extraordinary high standing within the profession. Numerous architectural prizes have been bestowed upon them, and their student works, particularly Rural Studios from Auburn University in Alabama, have been regularly featured in the international architectural press. So has Tyin Tegnestue, the first group of NTNU students that worked in Thailand in 2008–2009. They have probably brought more laurels to the Faculty of Architecture and Fine Art

---

1   E.g. 'Reality Studio', Gothenburg; 'Global Studio', Sydney/New York; 'Yale Building Project' and the 'Howard S. Wright Neighborhood Design/Build Studio' in Seattle; 'Studio 804' in Kansas; and 'Rural Studio' in Auburn, Alabama. When it comes to planning intervention programmes that initiate on-site activity, I know of none besides ours. Most such programmes merely present planning proposals.

than any other person or achievement.[2] Some of their reflections are quoted in the text. The design dimension in this approach therefore seems to hold its own. In spite of virtually having had no design tutoring, the students came up with innovative designs that also landed them top grades once back in the system.

Obviously there are aspects not covered in such courses, just as there are aspects the students do not find at the home university in spite of being challenged by them in the field. Nevertheless, our students bring back a different understanding of what architecture is and does, which in turn should have an impact on how other home-grown programmes are conducted. First, it takes architectural education closer to practice – not, as the current educational winds blow, towards the academic publish-or-perish tradition. Second, it could justify joint design-and-build programmes with students from other disciplines. Crossdisciplinary learning is officially recognized at our university through our 'Experts in Team' programme. Not only does this interdependency allow all students to learn more, but they learn to tackle challenges that architecture–planning students cannot on their own, as indicated from the Haiti feedback. Third, and perhaps most importantly, it highlights adaptability: our need constantly to find new ways, improvise, innovate, acknowledge serendipity, solve the problems at hand with the means at our disposal, strategize, and be flexible. Admittedly these are difficult challenges for the academic brigades. Yet, recognizing the overriding global challenges, do we as architect–planners have a choice? The students think not. I am not worried as long as they come back, as they do, with statements like the following:

> As in Kampala, Ibanda was illustrative of the great resource universities have at their disposal in the form of people with drive, energy, curiosity, a willingness to learn and a willingness to act. There is tremendous potential for this resource to engage directly with real issues and contribute to real solutions. As we are expected constantly to adapt, learn and disseminate knowledge to keep up with change, the old distinctions between the learned and the learners seem more obsolete by the day.

It is from students like these that teachers learn.

---

2   Their Thailand work won the *Best of TIDA, Eco and Conservation Award 2010*, presented by the National Thai Society of Architects, they won the *Museum & Library of the Year Award* from ARCH Daily, the world's largest architectural web site, and have had their work published in some of the world's most prestigious architectural magazines, e.g. *Architectural Review* and *Casabella* as well as having been featured in numerous books.

References

Blanco, H., Alberti, M., Forsyth, A., Krizek, K.J., Rodrígues, D.A., Talen, E. and Ellis, C. 2009a. Hot, Congested, Crowded and Diverse: Emerging Research Agendas in Planning. *Progress in Planning,* 71, 153–205.
Blanco, H., Alberti, M., Olshansky, R., Chang, S., Wheeler, S.M., Randolph, J., London, J.B., Hollander, J.B., Pallagst, K.M., Schwarz, T., Popper, F.J., Parnell, S., Pieterse, E. and Watson, V. 2009b. Shaken, Shrinking, Hot, Impoverished and Informal: Emerging Research Agendas in Planning. *Progress in Planning,* 72, 195–250.
Bourdieu, P. 1986. The Forms of Capital, in *Handbook of Theory and Research for the Sociology of Education,* edited by J. Richardson. New York: Greenwood, 241–58.
Boyer, E. and Mittgang, L. 1996 *Building Community: A New Future for Architecture Education and Practice.* Princeton: Carnegie Foundation for the Advancement of Teaching.
Brew, A. 2006. *Research and Teaching; Beyond the Divide.* Basingstoke and New York: Palgrave Macmillan.
Butenschøn, N. (ed.) 2009. *Tid i arkitektur: Om å bygge i fire dimensjoner.* Oslo: Akademisk Publisering.
EU 2008. Recommendation of the European Parliament and of the Council of 23 April 2008 on the Establishment of the European Qualifications Framework for Lifelong Learning. *Official Journal,* C 111, 6.5.2008 [Online: Europa, Summaries of EU legislation]. Available at: http://europa.eu/legislation_summaries/education_training_youth/vocational_training/c11104_en.htm [accessed: 25 May 2011].
Freire, P. 1972. *Pedagogy of the Oppressed.* New York: Herder and Herder.
Garau, P., Sclar, D.E. and Carolini, G.Y. (eds) 2005. *A Home in the City: UN Millennium Project Task Force on Improving the Lives of Slum Dwellers.* London: Earthscan.
Glancey, J. 2010. Norman Foster at 75: Norman's Conquests. *The Guardian,* 12 June.
Habraken, N.J. 1972. *Supports: An Alternative to Mass Housing.* London: Architectural Press.
Habraken, N.J. 1998. *The Structure of the Ordinary: Form and Control in the Built Environment,* edited by Jonathan Teicher. Cambridge, MA: MIT Press.
Hamdi, N. 1995. *Housing without Houses.* London: IT Publications.
Hamdi, N. 2004. *Small Change: About the Art of Practice and the Limits of Planning in Cities.* London: Earthscan.
Hamdi, N. 2010. *The Placemaker's Guide to Building Community.* London: Earthscan.
Hamdi, N. and Goethert, R. 1997. *Action Planning for Cities: A Guide to Community Practice.* Chichester: John Wiley & Sons.
Illich, I.D. 1973. *Deschooling Society.* London: Penguin Books.

Koenigsberger, O. 1964. Action Planning. *Architectural Association Quarterly*, 79(882), 306–12.

Lund, R. 2012. Researching Crisis – Recognizing the Unsettling Experience of Emotions. *Emotion, Space and Society*, 5, 94–102.

Lyon, P.M. 2007. Students' Experiences of Learning in the Operating Theatre, in *Transforming a University: The Scholarship of Teaching and Learning in Practice*, edited by A. Brew and J. Sachs. Sydney: Sydney University Press, 95–104.

Özbilgin, M. and Tatli, A. 2005. Understanding Bourdieu's Contribution to Organization and Management Studies. Book Review Essay. *Academy of Management Review*, 30(4), 855–77.

Peltason, R. and Ong-Yan, G. (eds) 2010. *Architect: The Pritzker Prize Laureates in their Own Words*. London: Thames & Hudson.

Robinson, K. 2010. *RSA Animate – Changing Education Paradigms* [Online: YouTube]. Available at: http://www.youtube.com/watch?v=zDZFcDGpL4U [accessed: 12 January 2011].

Säljö, R. 1998. Thinking With and Through Artifacts: The Role of Psychological Tools and Physical Artifacts in Human Learning and Cognition, in *Learning Relationships in the Classroom*, edited by D. Faulkner, K. Littleton and M. Woodhead. London: Routledge, 54–66.

Sanderson, D. 2010. Architects are Often the Last People Needed in Disaster Reconstruction. *The Guardian*, 3 March.

Schön, D.A. 1983. *The Reflective Practitioner: How Professionals Think in Action*. New York: Basic Books.

Sennett, R. 2008. *The Craftsman*. New Haven and London: Yale University Press.

Skotte, H. 2004. *Tents in Concrete: What Internationally Funded Housing Does to Support Recovery in Areas Affected by War; The Case of Bosnia-Herzegovina*. Dr. ing. thesis. Doctoral theses at NTNU, 2004:61. Department of Urban Design and Planning. Trondheim: Norwegian University of Science and Technology (NTNU).

Skotte, H. 2010. Å lære med hendene og tenke med hodet. *Arkitektur-N*, 71(4), 62–6.

Taylor, P. 1994. Learning About Learning: Teachers' and Students' Conceptions, in *Achieving Quality Learning in Higher Education*, edited by P. Nightingale, and M. O'Neil. London: Kogan Page, 62–76.

Turner, J. and Fichter, R. (eds) 1972. *Freedom to Build: Dweller Control of the Housing Process*. New York: Macmillan.

UN Habitat 2009. *Global Report on Human Settlements 2009: Planning Sustainable Cities*. London: Earthscan.

Vasagar, J. 2010. Percentage of Poor Pupils Admitted to Oxbridge? 1%. *The Guardian* [Online, 22 December]. Available at: http://www.guardian.co.uk/education/2010/dec/22/percentage-poor-pupils-oxbridge-one-percent [accessed: 14 January 2011].

Chapter 4

Foreign Direct Investment, Local Development and Poverty Reduction: The Sustainability of the Salmon Industry in Southern Chile

Arnt Fløysand and Jonathan R. Barton

Introduction

The rate of Foreign Direct Investment (FDI) has risen dramatically during the past three decades; developing countries' inward stock of FDI amounted to about 30 per cent of their GDP in 2009, compared to just 12 per cent in 1980 (UNCTAD 2011). This has led to a great deal of optimism that FDI can provide a potential for economic development and poverty reduction. However, this potential depends on how FDI interacts with the environment in which the investments take place (Lall and Narula 2004, Moran et al. 2005). To discuss these types of interaction, we propose an analytical framework approaching FDI as consisting of capital, actors and knowledge, or what we call the *capital–actor–knowledge complex.*

When the literature on FDI and its performance for local development is reviewed, both *FDI as progress* and *FDI as dependency* can be observed (Jakobsen et al. 2005). Our point of departure is that research on foreign direct investment and local development should account for such positions, but also go further. There are reasons to question how mainstream studies of FDI relate to alternative geographies of development and discuss the relations between FDI and poverty reduction and how FDI sustains biological diversity and the ecological systems it capitilizes on.

In the following, we analyse the complexity of Norwegian FDI in the salmon industry in southern Chile. In recent years, Chile has been the second largest producer of farmed salmon for the international market, behind Norway. The 'boom' of the Chilean salmon aquaculture industry has come about through economic liberalization and inflows of foreign direct investment since the 1980s. The present chapter uses the booming Chilean salmon aquaculture industry to shed light on capital–actor–knowledge dynamics in relation to FDI as progress or dependency, poverty issues and sustainable development. The industry for a long time seemed to exemplify FDI as progress, demonstrating capacity for poverty reduction, but this was reversed by a virus that hit production in 2007, twisting the case towards FDI as dependency. Since this seems to coincide with more political space for previously marginalized stakeholders such as labour unions, environmental organizations and local authorities and for stricter environmental regulation of the industry, we conclude that this particular case gives hope for a change in the practices of the sector towards sustainability. However, it remains to be seen if a larger share than now of the profits from the FDI will be returned to the local community and lead to poverty reduction on a long-term basis.

Repeated fieldwork has been undertaken in Chile since 2006, using participatory observation, document analysis and around 70 semi-structured interviews. Data collection has in particular focused on managers of Norwegian firms in the salmon industry of Chile, workers, other civil society stakeholders and local municipal authorities. Reports and newspaper articles have provided additional information.[1]

The Complex Dynamics of FDI for Local Development

The Capital–Actor–Knowledge Complex

The recent explosion of FDI reflects significant transformations in economic and political processes. Yet the complex dynamics of FDI and the outcome for poverty and sustainable development are poorly understood. In financial analysis FDI is defined as a cross-border investment where an investor intends to establish a lasting financial interest and exert an effective influence on the activities of the investment object. In our view, this definition of FDI fails to capture how FDI engages with socioeconomic development. Regarding FDI as a *capital–actor–knowledge complex* (Fig. 4.1) corresponds to an elemental model in geography that sees space in terms of nature (capital representing materiality in our approach), social relations (actors and their positions in networks) and meaning (knowledge and discourses). For each of the elements we propose an accompanying analytical approach: for knowledge, discourse analysis; for actors, social field analysis; and for capital, development analysis.

Knowledge refers to systems of meaning and the way these flow in networks and paths limited by networks. When this is investigated through discourse theory, it is assumed that power relations can be investigated through what Neumann (2001: 38) calls the 'existing conditions for action' or what Foucault (2003: 42) calls the 'rules of formation' of statements. Studying FDI as discourse means investigating the 'rules of formation' involved in FDI, in other words, the knowledge environment and exchange of meaning among institutions and organizations that produce FDI. Within the capital–actor–knowledge complex, the study of FDI as discourse focuses on knowledge–capital relations. Applying discourse theory to

---

1   The analysis draws on two research projects based at the Department of Geography, University of Bergen, and funded by the Research Council of Norway.

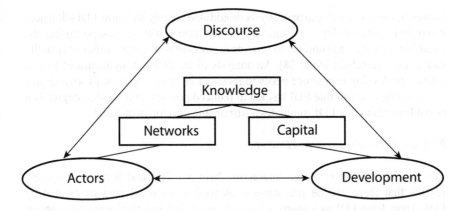

Figure 4.1   The capital–actor–knowledge complex

a study of FDI enables a focus on how power relations are articulated through narratives on FDI and how such narratives produce and reproduce power relations.

Actors refer to individual persons or groups of persons involved in the capital transaction, both directly and indirectly, by being involved in networks tied to the transaction and its effects. Actors and their position in networks are investigated through theory of practice. Recent theories of local development assume that economic actors are simultaneously embedded within different systems of relations operating at different geographical scales (Amin and Cohendet 1999, Fløysand and Jakobsen 2011). An FDI can be seen to be situated in a system of networks at different geographical scales. Some FDIs are embedded in local networks constituted by local specialized suppliers and dependent on global networks of customers and nonspecialized suppliers. Some FDIs may be embedded in systems of global networks that leave little room for establishing relations to local firms. In this case, the importance of networking within the FDI will be modest and local raw materials and the specialization of the local labour market and services may be the only economic benefits of the region. Other FDIs may be strongly embedded in networks that are local, promoting local innovation and local development. It is necessary to map and analyse the following: the networks that the actors involved in an FDI take part in; the untraded assets and rules of conduct in these networks; how the interplay in and between the networks can explain the capabilities of innovation in the region; and the relations between composition of networks, capability of innovation and local development.

Capital refers to the material world and real economic values transferred in relation to the FDI and is investigated through capital–development analysis drawing on concepts from economic theory. A focus may be on reasons for governments to regulate and attract FDI. One such reason is that FDI can be seen through development effects (e.g. direct and indirect employment, or knowledge transfers) and dependency effects (e.g. increased reliance on foreign capital, or resource drain).

However, capital–development analysis should also comply with how FDI influences ecological sustainability, i.e. 'meeting human needs without compromising the capability of nature to continue to provide ecosystem services in sufficient quantity and quality' (Gladwin 1998: 38). An analysis of the complex dynamics of FDI in relation to development hence needs to focus on questions such as: to what degree is it possible to claim that FDI has led to reduced poverty; and to what degree is it possible to claim that FDI stimulates a sustainable development path?

FDI, Local Development and Poverty

The available analysis of the interaction between FDI and local development reveals that there can be real dangers as well as clear advantages created by FDI. Apart from FDI as a source of economic capital and employment of labour, an important way in which the economy of a developing country may benefit from FDI is through spillover effects. This refers to the effects of FDI in raising productivity in local firms, increasing flows of knowledge in local networks and strengthening the innovation capabilities of regions and their institutions.

Local effects of FDI have traditionally been measured as job creation and local purchases of goods and services (Dunning 1993). However, Andersson and Forsgren (1996) argue for a shift in the understanding of FDI in relation to multinational corporations (MNCs). FDI should not only be understood as the source of a branch-plant economy that has mainly backward supply linkages to the local economy (Watts 1981, Phelps 1992). It is also necessary to emphasize the flow of knowledge and competence caused by FDI in a region (Ivarsson 1999). Very often, FDI provides MNCs with local linkages to the competence and technology of local firms (Dunning 1993), while the spillover effects of FDI may be the diffusion of competence and skill from the foreign firm to economic agents in the region (Ivarsson 1999). It is claimed that FDI benefits the local economy since MNCs can fully exploit economies of scale and scope. This means that a region with FDI can adjust more quickly to changes in technology and demand (Dunning 1993).

Other contributions discredit a presupposed direct link between FDI and local benefit, and assume that there can be both positive and negative local effects of FDI. Bellandi (2001) argues that close ties between the subsidiaries of MNCs and local firms are most likely to develop when the local production culture (referring to local codified standards and knowledge in clustering industries) is neither 'too weak' nor 'too strong'. A 'too strong' culture can block for network-sharing and exchange of technology and ideas. Borensztein et al. (1998) note that FDI contributes to economic growth but only when the local economy is sufficiently able to absorb the advanced technologies. The main conclusion from these studies is that a country is more likely to benefit from multinational investment if it is integrated into the country's development and technological plans (Milberg 1999). However, while there are examples of countries, such as China and Malaysia, that have been successful in regulating the impact of FDI, there are also studies that show spillover effects to be difficult to regulate even when this is attempted. Agosin and Mayer (2000) investigate an important aspect of spillover effects by asking whether FDI in host countries 'crowds in' further investment by local firms, or 'crowds out' existing investments of these firms as a consequence of increased competition and hence lower profits. The research suggests that between 1970 and 1996 there was strong 'crowding in' in Asia, 'crowding out' in Latin America and more or less neutral effects in Africa. The study concludes that the positive impacts of FDI on domestic investment are not assured. Te Velde (2003: 4) asserts that 'while FDI may have been good for development (e.g. we find positive correlations between FDI and GDP, or productivity, or wages) this masks the fact that different countries with different policies and economic factors tend to derive different benefits and costs of FDI'. Machinea and Vera (2006: 38) also note that the effects of FDI on GDP growth are highly variable, with negative impacts in the primary sector, positive impacts in manufacturing, and ambiguous impacts in services:

> In certain enclave-based primary activities, no such effects [positive effects on GDP] are generated and, in some cases, the only impact is the depletion of natural resources and massive capital outflows in the form of royalties and dividends.

They argue for more selective 'use' of FDI by national governments in order to maximize the positive domestic impacts that may exist (Machinea and Vera 2006: 43):

> In any event it seems clear that – as in the case of trade liberalization – the benefits of FDI have been exaggerated ... As part of their creation of an FDI-friendly environment, the countries of the region should make an effort to attract the kinds of FDI that will have a greater impact on terms of linkages and R&D resources.

Returning to Jakobsen et al. (2005), FDI as progress, is characterized by extensive local effects such as vertical linkages, knowledge spillover, spin-offs, innovation networks and technology transfer. FDI as dependency is characterized by situations in which the local economy is dominated by the FDI, but involves negative effects such as social and employment dislocation, environmental disruption, corruption and 'out-crowding'. Generally, profits are returned to the investing country.

The concept of poverty has moved beyond simple economic indicators and towards acceptance of multidimensional perspectives on its conceptualization, formation and reduction.[2] The Millennium Development Goals, officially established following the Millennium Summit of the United Nations in 2000,

---

2   Spicker, P. 2003. 'Eleven Definitions of Poverty.' Unpublished chapter in 'Approaching Poverty Manual: Poverty Reduction for Practitioners', CROP/Sida Pilot Training Course.

focus on tangible dimensions including the absolute poverty line (extreme poverty is measured with the $1 a day yardstick, arguing that per capita consumption of $1 a day represents a minimum standard of living), together with human development indicators such as literacy levels, levels of access to health services and access to basic services such as water and sanitation (UN 2003). These kinds of indicators were popularized in the 1990s through the UNDP's Human Development Index (HDI) (UNDP 2011). The recognition of qualitative as well as quantitative poverty indicators has made it possible to take account of and integrate a multiplicity of social interests and demands. The best illustration is the huge recent increase in participatory poverty assessments (PPAs) carried out in developing countries to improve the effectiveness of public policy aimed at poverty reduction. The largest global participatory poverty assessment was the World Bank funded 'Voices of the Poor' study (Narayan et al. 1999). A stated objective of the exercise was to capture the wider dimensions of poverty so that poverty reduction policy and programmes would be based on the 'experiences, reflections, aspirations and priorities of poor people themselves' (Narayan et al. 1999: 3).

Towards an Analytical Framework

The theoretical discussion has presented some concepts representing recent academic and practical progress in redefining foreign direct investments, the local effects of FDI, poverty and sustainable development. The challenge is to combine the concepts in practical research to discover the dynamics and links between them. The challenge for a geographical study of FDI is to determine how the capital–actor–knowledge complex is embedded in spatial scales and how this influences the outcome of FDI for power, local development and poverty, in other words how events and processes at local, regional, national and global scales combine to create complex dynamics (Fløysand et al. 2005).

The following study of the salmon industry in southern Chile aims to discover whether the FDI complex there leads to FDI as progress or FDI as dependency; also how this outcome relates to the poverty situation in terms of poverty line figures, HDI scores and signs of increased collective action of poor people; and how ecologically sustainable the industry is and how this feeds back into the FDI complex (Figure 4.2).

The Case of the Salmon Industry in Southern Chile

A Booming Industry

Salmon aquaculture in Chile was promoted from the late 1970s in order to diversify the Chilean economy away from its traditional dependence on copper exports (Barton 2006). Until 2007, the growth rates in the sector, in terms of investment, site expansion and export volumes and values, all mirrored a booming industry.

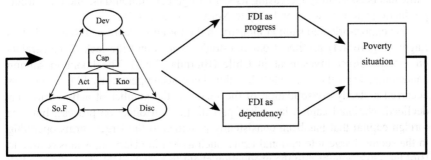

Figure 4.2   The dynamics between an FDI complex, progress or dependency, poverty and sustainable development

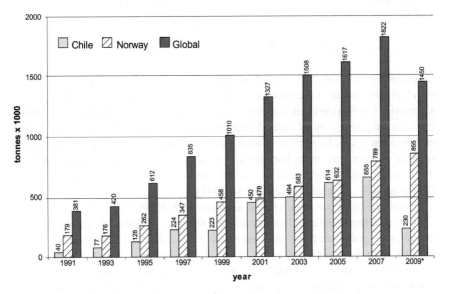

Figure 4.3   Chilean, Norwegian and global salmonid production

Sources: Salmon Chile and for 2009 Norwegian Seafood Export Council

Chile has been steadily increasing its share of global salmonid (salmon and trout) production and is now second only to Norway (Figure 4.3).

The export-oriented salmon industry in the formerly depressed Región de Los Lagos (Figure 4.4) profited from free-trade treaties and the lack of constraints on foreign direct investment in Chile (Bjørndal 2001). The boom in salmon aquaculture from the late 1980s has thus been accompanied by FDI. There has occurred a steady consolidation of the sector as the number of active firms has declined. National capital has taken part in this consolidation process, but it is foreign capital that has been consistently provided to the largest firms operating in the sector. Large international firms, such as Marine Harvest, Mainstream and BioMar, have bought into the production chain from fish-meal production to fish production (Phyne and Mansilla 2003). FDI rose swiftly in the 1990s and the first years of the new millennium. By 2004, six foreign firms accounted for 35 per cent of total exports (by volume); the size of these firms relative to their Chilean counterparts can be seen in the fact that the remaining 65 per cent of exports were accounted for by 26 firms (Revista Aqua 2007). A significant share of Chilean fisheries and aquaculture FDI took place in region X, Región de Los Lagos, where 39.8 per cent of all fisheries and aquaculture FDI were concentrated between 1974 and 2005 (FIC 2006).

Foreign Direct Investments as Progress or Dependency?

Discussion of development in the Región de los Lagos has centred on the economic development related to the salmon cluster, a term that in itself immediately brings associations with a dynamic business environment. The large literature describing the case as a situation of FDI as progress stresses that production, export and employment levels have risen substantially over time, and that spillover effects, such as successively more homegrown technology and competence, as well as the formation of supply services for the salmon industry, have been generated (Bjørndal 2001, Vergara 2003, Montero 2004).

However, there are more critical accounts as well, and while these agree that Chilean neoliberal policies have performed strongly in terms of macroeconomic factors, they point out that little change has been registered in redistribution and social resource allocation (Barrett et al. 2002, Barton 2002, Olavarría 2003). According to Barrett et al. (2002: 1952), 'there is substantial evidence that surplus labor, low wage levels, and poorly enforced or nonexistent health and safety standards are conditioning factors in the growth of the salmon industry.' González (2008) holds that the aquaculture industry in Chile faces weak trade union systems and inadequate participatory mechanisms.

Our research on the issue of FDI and development confirms this double-sided picture. The impact in terms of linkages and R&D resources seems to be modest in the case of Norwegian FDI in Chile. Discussing circumstances which restrain innovations and processes of learning amongst Norwegian controlled

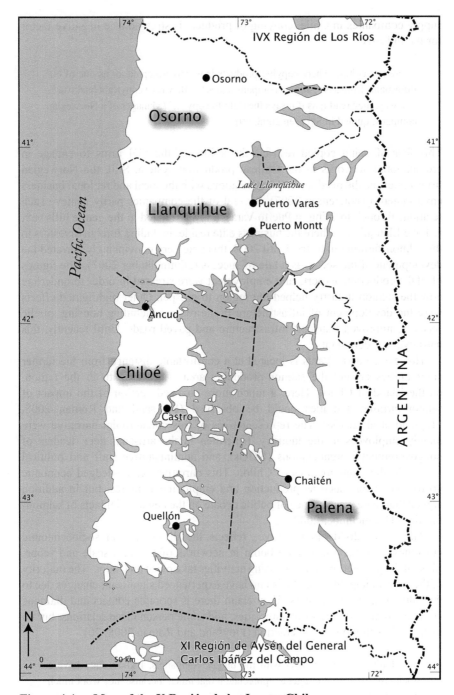

Figure 4.4   Map of the X Región de los Lagos, Chile

supply companies in Chile, copying of products is stressed as a negative factor for the industry:

> There is a culture where copying products is widely accepted, or as one of our informants from a Norwegian company states '... they read copyright backwards in a way, they read it as if it says the right to copy ...' (Manager of a Norwegian company, quoted in Svensen 2008: 69)

This fear of being copied restricts the ability of the FDI firms to engage in crossnational networks in the Chilean production system. Still, the Norwegian firms underline the need for building relations with the local and regional business environment (Thorstensen 2007). Several of the companies partly achieve such relations through locating in Puerto Varas, a town situated in the gentle hills next to Lake Llanquihue, where the business elite reside, including foreign investors in the salmon-farming industry. Until 2008, this elite very convincingly narrated the development of the sector as a tale of success (Himmelhuber 2007). The figures of FDI, production, exports and employment were stressed in order to underline how the salmon industry helped the region to prosper. Other highlighted effects were the development of infrastructure in the region, including housing, public services, improved transport infrastructure and paved roads. Until recently, this narrative was dominant.

However, Puerto Varas is located at a comfortable distance from the farmer communities around the salmon-producing areas further south in the region, on the island of Chiloé. Here, a much more nuanced version of the impact of salmon-farming activities could be observed (Fløysand and Román 2008, Fløysand et al. 2010a). The representatives of this counterpoint narrative were mainly employees in the industry, fishermen in the artisanal fleet, leaders of nongovernmental organizations (NGOs), and administrative staff and political leaders in the municipalities of Chiloé. This narrative acknowledged economic growth indicators such as production and employment figures, but in addition argued that there was absence of public debate on the negative impacts of salmon-farming (Himmelhuber 2007).

The above divide in narratives reflects the differences in socioeconomic outcomes from the aquaculture boom in networks of different scale and scope. Most of the production and processing activities take place on Chiloé. The majority of the ten municipalities on the island have experienced significant changes due to the salmon-farming activities. They claim there is sporadic contact and dialogue between companies and local authorities, and some personalized relations, but no reciprocal institutions for interaction (Fløysand and Román 2008: 70).

> We don't have any contact with the bigger companies. I know they have a social role, but we don't get to appreciate it that much, because they work much like islands. Their headquarters are much like the mountains on those islands. (Interview, municipal officer)

The officials in these municipalities were less than optimistic about the possibilities for local spillovers in the wake of FDI and industrial growth, mainly because of problems existing within three areas of action. First, they were critical of the passivity of the state in generating investment in local capacities for absorbing spillovers, citing a need for a type of labour education that could prepare local workers for better jobs. Second, they signaled the importance of establishing arenas and meeting places for the public and private sectors, where dialogue and collaboration could be institutionalized. Third, they complained about their own lack of capacity to foster associations between authorities in different municipalities where the aquaculture industry operated, whereby common visions and territorial solutions could be worked out, and which could form a platform for engaging with the industry on a more equal footing (Fløysand et al. 2010a). Still, it is claimed that the industry led to reduced poverty in the places where production, processing and multiplier activities are located (Montero 2004). As evidence, the salmon producers' association, SalmonChile, refers to government statistics (SalmonChile 2011):

> Based on the results of the Socioeconomic Survey of the Ministry of Planning for the years 2000 and 2003, it appears that salmon-farming communes in the X region reduced their rates of poverty and homelessness by 13% and 42% respectively, well above the national average reduction of 6% and 10%. (Authors' translation)

Sustainable Reduction of Poverty?

Using the concentration of production as a proxy for the local intensity of production and its impacts, it is possible to compare localities according to the Human Development Index for 1993 and 2004 in order to observe whether the industry has brought significant benefits to the municipalities where it operates, compared with other municipalities in the region where salmon-farming is not present, or its presence is marginal.

At the provincial level, it is evident that the provinces that cluster the salmon industry perform considerably better than those that do not. In terms of the differences in HDI between 1994 and 2003, the 11 municipalities that have the most concessions (over 25 each) have all increased their HDI by over 20 points. However, there is little notable difference in terms of health and education within provinces or between provinces. When the scale is raised to the regional level, the generally low levels of development in the region can be observed clearly, despite the existence of one of the most dynamic sectors of the national economy and a Chilean flagship in the global economy. In 2006, the region registered one of the highest rates of national poverty, with 17 per cent of the population registered as poor (MIDEPLAN 2007).

Labour in the salmon industry has become increasingly organized, indicating increased participation and collective action by low-income groups and the organizations working on their behalf to exert an influence on political and economic processes (Webster and Engberg-Pedersen 2002). A strike in August 2006 by workers in Mainstream Chile (the largest, salmon-producing firm, which is Norwegian-owned) pointed to the poor wages and labour conditions, and high fatalities of divers in the salmon aquaculture sector generally. The Mainstream workers staged a sit-in at the company headquarters in Puerto Montt. Workers at the firms AquaChile and Toralla (Spanish-owned) also went on strike. Following the Mainstream strike, the NGO Foundation Terram, with support from Oxfam principally, announced the creation of an observatory of labour and environmental conditions (OLACH) on the island of Chiloé. The NGOs have been pressuring the aquaculture sector intensely since 2000.

Their reasons for pressuring the firms can be seen in recent regulatory sanctions. In the inspection of 500 firms by the regional Labour Directorate, over 60 per cent of firms were not in compliance with norms for labour conditions (Fundación Terram 2006). The NGOs were also highly critical of the environmental management of fish farms (Buschmann 2002, Gutiérrez 2005). The regional environmental agency imposed sanctions on 13 fish farms for exceeding permitted production volumes; they exceeded the authorized levels on average by 136 per cent (Fundacion Terram 2006). Concerns for how overproduction and other practices threatened the ecological sustainability of the production regime of salmon in Chile were also expressed by OECD/CEPAL (Barton and Fløysand 2010). Their recommendations were (OECD/CEPAL 2005: 29):

> ... to improve environmental and sanitary protection in aquaculture (in relation to eutrophication, salmon escapes, lake ecology equilibrium, antibiotic use, epidemiological vigilance, eradication of infectious disease, among others), particularly the strengthening of capacity to meet norms and regulations; to apply the 'polluter pays' principle in the aquaculture industry in the context of the Environment Law; to generate a precise plan of coastal zoning of aquaculture; to adopt integrated environmental management in coastal areas. (Authors' translation)

This questioning of the ecological sustainability of the FDI-driven salmon production regime and its socioeconomic outcomes received its legitimation with the emergence of the ISA (infectious salmon anemia) virus in 2007. Despite warnings from both scholars and NGOs that lack of regulation, overcrowding and chemical use were endangering the industry, Chile had failed to learn lessons from similar crises due to rapid growth in the Norwegian, Canadian and Scottish salmon aquaculture industries (Barton 1997, Barrett et al. 2002, González 2008). The spread of the virus instigated a crisis in the salmon-farming industry and resulted in production-site quarantine, fish slaughter and divestment by firms. In a now familiar pattern, the crisis has hit the worker communities hardest, as some 10,000 workers had lost their jobs by April 2009 according to industry figures (Fløysand et al. 2010b).

Capital–Actor–Knowledge Dynamism Revisited

An inevitable consequence of the crisis is the dismantling of the business. Output and employment will probably never again reach 2007 levels. A much more positive outcome of the crisis is that it has sparked demands for better environmental management. The painful experience of the industry has made the firms realize they have a financial interest in taking ecological considerations. A shift towards better governance of the sector can also be noticed. For example, the involvement of major international NGOs, such as Oxfam, has linked unions to roundtable conferences that were formerly dominated by the industry. This has entailed a gradual upscaling in organizational structure of the former firm-based labour movement by regional-based federations and a national confederation. Through this, the unions have managed to expand the networks in which claims are being pressed from local to regional and national governmental levels (Oseland Ellevseth 2010; Oseland Ellevseth et al. 2012), making it easier for poverty reduction policies and programmes to be based on the 'experiences, reflections, aspirations and priorities of poor people themselves' (Narayan et al. 1999: 3). Other actors representing civil society have also started to organize in order to gain a stronger say. At national level, new legislation has been adopted, providing stricter regulations, limitations on licence period and giving workers the right to organize.

Conclusion

In this chapter, we have addressed FDI, local development and poverty reduction. We have defined FDI as a complex consisting of actors, capital and knowledge; discussed how FDI has been linked to questions of local development and poverty reduction; and suggested an analytical framework that also includes sustainable development. Drawing on previous studies, we have used this to understand the dynamics of the salmon industry in Chile. Since the late 1990s, there have been huge changes in the economy of the Región de Los Lagos, on the island of Chiloé in particular, caused by a booming salmon-farming industry. Partly these developments are linked with FDI, in particular by Norwegian MNCs. The region in the boom period exemplified a modest case of FDI as progress. FDI also seems to have had a positive influence on the poverty situation in the areas where the production took place. Recently, these positive changes have been reversed because the industry was not operating with ecological sustainable productions standards. One explanation behind this is the double-edged nature of neoliberal policy: it attracts FDI, but does not promote governmental regulation. It is this tension between attracting investment and regulating it to encourage sustainable practice that is the dilemma of the government. The private sector has other imperatives and objectives, despite claims of sustainability.

The case supports our theoretical assumptions that deeper insight into the contextual dynamics between FDI, development and poverty is impossible without paying attention to the complex processes of capital, actors and knowledge and how such processes influence marginalized actors' ability to press effectively claims for poverty reduction and sustainable development. The case also demonstrates that through increased engagements from civil society (e.g. unions and NGOs), through the establishment of a regulation system sensitive to contextual conditions, and through more strategic uses of FDI by national governments, problems resulting from FDI can be met. To conclude, our case indicates that such harmonization is needed in order to ensure better relations between FDI and national welfare improvements, leading to poverty reduction and guiding the particular capital–actor–knowledge complex operating in the Chilean salmon industry towards a more sustainable development path.

References

Andersson, U. and Forsgren, M. 1996. Subsidiary Embeddedness and Control in the Multinational Corporation. *International Business Review*, 5, 487–508.

Amin A. and Cohendet, P. 1999. Learning and Adaptation in Decentralised Business Networks. *Environment and Planning D: Society and Space*, 17, 87–104.

Agosin, M.R. and Mayer, R. 2000. *Foreign Direct Investment in Developing Countries: Does it Crowd-In Domestic Investment?* UNCTAD Discussion Paper 146 [Online]. Available at: http://archive.unctad.org/en/docs/dp_146.en.pdf [accessed: 6 November 2012].

Barrett, G., Caniggia, M.I. and Read, L. 2002. 'There Are More Vets Than Doctors in Chiloé': Social and Community Impact of the Globalization of Aquaculture in Chile. *World Development,* 30, 1951–65.

Barton, J.R. 1997. Environment, Sustainability and Regulation in Commercial Aquaculture: The Case of Chilean Salmonoid Production. *Geoforum*, 28, 313–28.

Barton, J.R. 2002. State Continuismo and Pinochetismo: The Keys to the Chilean Transition. *Bulletin of Latin American Research*, 21, 358–74.

Barton, J.R. 2006. Sustainable Fisheries Management in the Resource Periphery: The Cases of Chile and New Zealand. *Asia Pacific Viewpoint*, 47, 366–80.

Barton, J.R. and Fløysand, A. 2010. The Political Ecology of Chilean Salmon Aquaculture, 1982–2010: A Trajectory from Economic Development to Global Sustainability. *Global Environmental Change*, 20, 739–52.

Bellandi, M. 2001. Local Development and Embedded Large Firms. *Entrepreneurship & Regional Development*, 13, 189–210.

Bjørndal, T. 2001. *The Competitiveness of the Chilean Salmon Aquaculture Industry*. SNF Working Paper no. 7/2001 – Centre for Fisheries Economics Discussion Paper no. 37/01. Bergen: Foundation for Research in Economics and Business Administration.

Borensztein, E., De Gregorio, J. and Lee, J-W. 1998. How Does Foreign Direct Investment Affect Economic Growth? *Journal of International Economics*, 45, 115–35.

Buschmann, A. 2002. *Impacto ambiental de la salmonicultura en Chile: La situación en la décima región de Los Lagos*. Análisis de Políticas Públicas 16. Santiago: Fundación Terram.

Dunning, J.H. 1993. *Multinational Enterprises and the Global Economy*. Wokingham: Addison-Wesley.

FIC 2006. *FDI Statistics* [Online: Foreign Investment Committee]. Available at: www.foreigninvestment.cl/index/fdi_statistics.asp?id_seccion=2 [accessed: 3 November 2007].

Fløysand, A., Haarstad, H., Jakobsen, S-E. and Tønnesen, A. 2005. *Foreign Direct Investment, Regional Change and Poverty: Identifying Norwegian Controlled FDI in Developing Countries*. SNF Report no. 4/05. Bergen: Institute for Research in Economics and Business Administration.

Fløysand, A. and Jakobsen, S-E. 2011. The Complexity of Innovations: A Relational Turn. *Progress in Human Geography*, 35, 328–44.

Fløysand, A. and Román, A. 2008. *Industria salmonera, sistemas de innovación y desarrollo local: El punto de vista de las municipalidades de Chiloé*. Bergen: Departamento de Geografía, Universidad de Bergen.

Fløysand, A., Barton, J.R. and Román, A. 2010a. La doble jerarquía del desarrollo económico y gobierno local en Chile: El caso de la salmonicultura y los municipios chilotes. *EURE*, 36, 123–48.

Fløysand, A., Haarstad, H. and Barton, J.R. 2010b. Global-Economic Imperatives, Crisis Generation and Spaces of Engagement in the Chilean Aquaculture Industry. *Norsk Geografisk Tidsskrift–Norwegian Journal of Geography*, 64, 199–210.

Foucault, M. 2003. *Archaeology of Knowledge*. London: Routledge.

Fundación Terram 2006. *Condiciones laborales de la industria salmonera*. Santiago: Terram.

González, E. 2008. Chile's National Aquaculture Policy: Missing Elements for the Sustainable Development of Aquaculture. *International Journal of Environment and Pollution*, 33, 457–68.

Gladwin, T.N. 1998. Economic Globalization and Ecological Sustainability: Searching for Truth and Reconciliation, in *Sustainability Strategies for Industry: The Future of Corporate Practice*, edited by J.N. Roome. Washington DC: Island Press, 27–54.

Gutiérrez, C. 2005. *Moratoria a la Salmonicultura*. Santiago: Oceana.

Himmelhuber, C. 2007. *On the Road to Sustainable Development? Rural Development and the Discourse on the Impact of Salmon Farming Activities in Quellón on Chiloé*. Master's thesis in geography. Bergen: Department of Geography, University of Bergen.

Ivarsson, I. 1999. Competitive Industry Clusters and Inward TNC Investment: The Case of Sweden. *Regional Studies*, 33, 37–50.

Jakobsen, S-E., Rusten, G. and Fløysand, A. 2005. How Green is the Valley? Foreign Direct Investment in Two Norwegian Industrial Towns. *Canadian Geographer*, 49, 244–59.

Lall, S. and Narula, R. 2004. Foreign Direct Investment and its Role in Economic Development: Do We Need a New Agenda? *European Journal of Development Research*, 16, 447–64.

Machinea, J. L. and Vera, C. 2006. *Trade, Direct Investment and Production Policies*. Santiago: CEPAL.

Milberg, W. 1999. Foreign Direct Investment and Development: Balancing Costs and Benefits, in *International Monetary and Financial Issues for the 1990s*. United Nations Conference on Trade and Development, Vol. XI. New York: United Nations, 99–115.

MIDEPLAN 2007. Info Pais [Online]. Available at: http://sir.mideplan.cl [accessed: 5 June 2007].

Montero, C. 2004. *Formación y desarrollo de un cluster globalizado: El caso de la industria del salmón en Chile*. Santiago: CEPAL.

Moran, T.H., Graham, E.M. and Blomström, M. 2005. *Does Foreign Direct Investment Promote Development?* Washington DC: Institute for International Economics.

Narayan, D., Patel, R., Schafft, K., Rademacher, A. and Koch-Schulte, S. 1999. *Voices of the Poor*, Volume 1: *Can Anyone Hear Us? Voices from 47 Countries* [Online: World Bank]. Available at: http://publications.worldbank.org/index.php?main_page=product_info&products_id=21661 [accessed: 15 November 2012].

Neumann, I. 2001. *Mening, materialitet, makt: En innføring i diskursanalyse*. Bergen: Fagbokforlaget.

OECD/CEPAL 2005. *OECD/CEPAL, Evaluación de Desempeño Ambiental 2005*. Santiago: CEPAL.

Olavarría, M. 2003. Protected Neoliberalism: Perverse Institutionalization and the Crisis of Representation in Postdictatorship Chile. *Latin American Perspectives*, 30, 10–38.

Oseland Ellevseth, S. 2010. *The Political Space of Labour Movements in Times of Crisis: A Study of the Labour Movement on Chiloé, Chile*. Master's thesis in geography. Bergen: Department of Geography, University of Bergen.

Oseland Ellevseth, S, Haarstad, H. and Fløysand, A. 2012. Labour Agency and the Importance of the National Scale: Emergent Aquaculture Unionism in Chile. *Political Geography*, 31, 94–103.

Phelps, N.A. 1992. Branch Plants and the Evolving Spatial Division of Labour: A Study of Material Linkages Change in the Northern Region of England. *Regional Studies*, 27, 87–101.

Phyne, J. and Mansilla, J. 2003. Forging Linkages in the Commodity Chain: The Case of the Chilean Salmon Farming Industry 1987–2001. *Sociologia Ruralis*, 43, 108–27.

Revista Aqua 2007. *Exportaciones de salmónidos, enero a febrero 2006-07* [Online]. Available at: http://www.aqua.cl/estadisticas/ESTADISTICAS114. pdf [accessed: 1 June 2007].

SalmonChile 2011. *Preguntas frecuentes: Laborales y Comunidad* [Online: Asociación de la Industria del Salmón A.G.]. Available at: http://www. salmonchile.cl/frontend/seccion.asp?contid=366&secid=130&secoldid=130 &subsecid=360&pag=1#1 [accessed: 28 November 2012].

Svensen, T. 2008. *Innovasjon og læreprosesser: Et case studie av tre norske leverandørselskap etablert i chilensk fiskeoppdrettsindustr*i. Master's thesis in geography. Bergen: Department of Geography, University of Bergen.

Te Velde, D.W. 2003. *Foreign Direct Investment and Income Inequality in Latin America: Experiences and Policy Implications*. London: Overseas Development Institute.

Thorstensen, M. 2007. *FDI, nettverk og kulturelle barrierer – en studie av norske oppdrettsaktører i Puerto Montt, Chile*. Master's thesis in geography. Bergen: Department of Geography, University of Bergen.

UN 2003. *Indicators for Monitoring the Millennium Development Goals: Definitions, Rationale, Concepts and Sources* [Online: United Nations, New York]. Available at: http://www.undp.or.id/mdg/documents/MDG%20 Indicators-UNDG.pdf [accessed: 28 November 2012].

UNCTAD 2011. *Inward and Outward Foreign Direct Investment Stock, Annual, 1980–2010* [Online]. Available at: http://unctadstat.unctad.org/ReportFolders/ reportFolders.aspx [accessed: 10 November 2011].

UNDP 2011. *Human Development Index (HDI)* [Online: Human Development Reports]. Available at: http://hdr.undp.org/en/statistics/hdi/ [accessed: 10 November 2011].

Vergara, M. 2003. *Acuicultura en Chile*. Santiago: Editorial Techno Press.

Watts, H.D. 1981. *The Branch Plant Economy: A Study of External Control*. London: Longman.

Webster, N. and Engberg-Pedersen, L. 2002. *In the Name of the Poor: Contesting Political Space for Poverty Reduction*. London and New York: Zed Books.

Chapter 5

Housing the Urban Poor in Metropolitan Accra, Ghana: What is the Role of the State in the Era of Liberalization and Globalization?

George Owusu

Introduction

Rapid urbanization and urban growth in much of the developing world are leading to ever-expanding and sprawling cities and towns with limited services and infrastructure. Nowhere are these challenges more severe than in sub-Saharan Africa, where urbanization has failed to keep pace with economic growth, resulting in the inability of local and national governments to provide services and infrastructure needed by the urban population (Tipple 1994, BBC 2007, K'Akumu 2007).[1] While the pace of urbanization in sub-Saharan Africa can be compared with that of China and India (described as the industrial giants of the future), economic growth rates of African countries do not come near that of these two countries (BBC 2007).

A key challenge confronting urbanization in sub-Saharan Africa is inadequate housing, especially among low-income urban dwellers. This challenge becomes clearer when one looks at the broad definition of housing as including physical shelter with related services and infrastructure as well as the inputs such as land and finance required to produce and maintain it. In other words, housing covers the solutions geared towards improving the shelter and the environment in which it exists (GoG/MWRWH 2009). This broad definition indicates that when housing is reduced to shelter or living space only, dwellings tend to be built without regard to the environment and services needed to support the inhabitants (GSS 2005). In many African countries, not only are the environment and institutions poor, but also the quantity and quality of housing available is woefully inadequate. Many analysts have described the housing challenge confronting many cities in Africa as a crisis (Tipple 1994, Songsore 2003). It is argued that the housing crisis, or

---

1   This point was also raised by Erguden, S. 'Low-Cost Housing Policies and Constraints in Developing Countries.' Unpublished paper presented at the International Conference on Spatial Information for Sustainable Development, Nairobi, Kenya, 2–5 October 2001.

housing poverty, is the result of decades of neglect of the sector as a critical area for development. There has been a series of fragmented policies rather than a holistic, comprehensive vision to deal with the complexities of the housing challenge (NDPC 2005, GoG/MWRWH 2009).

Ghana, like many African countries, is experiencing rapid urbanization. The proportion of the total population living in urban areas, which was about 8 per cent in 1921, rose to 23 per cent by 1960, 32 per cent in 1984, about 44 per cent in 2000, and was estimated to be about 51 per cent in 2009 (GSS 2005, ISSER 2007, UNFPA 2007). It is projected that this rapid growth of the urban population is unlikely to slow down until the year 2025, when the urbanized population would have reached almost 63 per cent. However, a feature of this sharp increase in the level of urbanization is limited infrastructure, including housing. Nowhere is the housing challenge so severe than in the largest metropolis and national capital, Accra. The existing housing condition in Accra is a result of rapid urban growth fuelled by increased population growth (both natural growth and rural–urban migration), exacerbated by economic liberalization and globalization. Increasingly, the effects of liberalization and globalization are reconfiguring housing supply-and-demand dynamics, resulting in increasing land and property values and rent. This is pushing some middle-income Ghanaians to slums and other poor neighbourhoods of Accra. Yet the effects of globalization on housing have been given little attention in the urban housing literature on developing countries (Sajor 2003, Owusu 2008).

The effects of rapid urbanization and urban growth resulting from national policies and economic globalization, and the consequent impact of poor housing access and marginalization on weaker members of society, call for alternative geographies of development. This perspective, strongly emphasized by Ragnhild Lund, must be pro-urban poor and entail multiple voices, strategies and solutions grounded within the local context (Lund 1994, 2002, Owusu et al. 2008). The goal is to achieve an equitable and socially viable development agenda that enables the poor to participate in and enjoy the fruits of development (Lund 1994).

This chapter takes its inspiration from Lund's work on pro-poor development and the need to understanding globalization by analysing the local. It examines housing among the poor in Accra during the last three decades of economic liberalization and the resultant consequences of economic globalization. The chapter provides a brief overview of Ghana's housing policy, followed by a discussion of the scale of the urban housing need. It then examines the new global face of the housing challenge in Accra and its implications for housing the urban poor. The chapter ends with a call for an alternative development approach towards providing housing for the urban poor.

Ghana's Housing Policy Since Independence: A Brief Overview

Ghana's housing policy has been characterized as fragmented and piecemeal (NDPC 2005, GoG/MWRWH 2009). Housing development has never been taken seriously as a critical sector for national development. It is possible to delineate three time periods within the postindependence era, each with distinct housing policies and strategies: the immediate postindependence era of public housing provision (late 1950s–early 1980s); the structural adjustment and economic liberalization era (mid-1980s–early 1990s); and the poststructural adjustment and economic recovery era (mid-1990s–present). These periods are characterized by distinct political and socioeconomic developments that produced different patterns of urbanization with different implications for housing supply and demand in Ghana.

Immediate Postindependence Era

The immediate postindependence era, from the late 1950s to the early 1980s, marked the period in which there was active and direct involvement of the state in the provision of mass housing. Two state institutions, the State Housing Corporation (SHC) and the Tema Development Corporation (TDC), were established for the purpose of providing mass housing. While the TDC was established for the special purpose of developing houses for the industrial city of Tema as part of a major industrialization drive, the SHC was created to develop residential units in all regions of Ghana. In addition, state-owned financial establishments such as the Bank for Housing and Construction (BHC) and the First Ghana Building Society were established to provide financial support for public housing. This active involvement of the state in housing provision continued through the 1970s under the various military regimes of the time. Special mention can be made of the construction of low-cost houses in district and regional administrative capitals under the Supreme Military Council (SMC) regime of General I.K. Acheampong.

Though the private informal sector provided the bulk of the housing (estimated at about 80 per cent) during the late 1950s to early 1980s (Songsore 2003), the impact of state provision was significant, largely due to the period's low level of urbanization. However, the worsening economic conditions of the country due to economic mismanagement and political instability, especially in the late 1970s, led to resources allocated to public housing agencies running dry (GoG/MWRWH 2009, Nsiah-Gyabaah 2009). These state housing agencies therefore became a drain on public resources and were unable to pursue their core mandate of public housing provision (Songsore 2003, Nsiah-Gyabaah 2009).

Much of the intervention by government in the housing sector in the immediate postindependence era targeted workers in the formal sector, leaving low-income earners and the very poor in the large informal sector. Moreover, a number of policy measures such as a cap on rent and poor macro-economic performance affected private individuals' capacity to provide housing. The cap on rent was designed to make housing affordable but it had the unintended consequence of dissuading private developers from creating rental housing units. This situation made it more difficult for urban low-income groups to find rental accommodation (Tipple 1987, Songsore 2003, *The Statesman* 2007).

Structural Adjustment and Economic Liberalization Era

Following the economic crisis that began in the late 1960s, lasting through the 1970s and peaking in the early 1980s, Ghana embarked in 1983 on an economic recovery programme (ERP) and structural adjustment programme (SAP) supported by the World Bank and the International Monetary Fund (IMF) to restore macro-economic stability and growth (Nugent 2004). The ERP/SAP entailed among other things economic liberalization and privatization, and withdrawal of the state from various sectors of the economy. In line with the economic policy of the time, government housing policy took a dramatic departure from that of the proceeding decades, becoming mainly geared towards facilitating and creating an enabling environment for private sector participation in housing delivery. A number of policy documents on housing since the mid-1980s have strongly emphasized the role of the private sector in housing delivery. This period marked the emergence on the Ghanaian housing market of private real estate companies under the umbrella of Ghana Real Estate Developers Association (GREDA).

Backed by economic stability and growth, improved macroeconomic conditions and incentives for foreign investors, Ghana's door was opened to global capital and investments. However, the spatial impact of economic liberalization and influx of global capital and investments has not been uniform, favouring large cities such as Accra. Quarterly updates from the Ghana Investment Promotion Centre on foreign direct investment (FDI) in Ghana reveal that on average over 80 per cent of all non-mining FDI is concentrated in the Greater Accra region (GIPC 2009–2010). This situation has made the metropolis attractive to rural–urban migrants as well as to foreign actors such as migrants from elsewhere in West Africa, expatriates working with international nongovernmental organizations (NGOs) and multinational corporations (MNCs), and members of the Ghanaian diaspora living in Europe and elsewhere. This has led to rapid urban growth and outward expansion of the city, and an intensification of the housing crisis.

Poststructural Adjustment and Economic Recovery Era

The post-ERP/SAP era (early 1990s to date) has witnessed the intensification of the forces of liberalization and globalization, and the continuous expansion of large Ghanaian cities, especially Accra and the second city, Kumasi (Konadu-Agyemang 2001, Yeboah 2001, 2003, Grant and Nijman 2002, Grant and Yankson 2003, Owusu 2008, 2010, Owusu et al. 2008, Grant 2009). Housing was affected by a sharp rise in land prices and property values in the core built-up areas and the fringes of the city of Accra. As land and property prices rose, private real estate developers have consistently targeted the middle and upper classes of resident Ghanaians as well as non-resident Ghanaians living in the diaspora, especially Europe and North America. The increases in rent have resulted in a sharp rise in density levels, leading to overcrowding in poor neighbourhoods of the city as increasing numbers of low-income Ghanaian residents are priced out of the land and housing market.

Meanwhile, government housing policy has largely remained unchanged from the ERP/SAP era. The Draft Housing Policy of 2009 emphasizes the need to create an enabling environment to strengthen the private sector participation in housing delivery for low-income groups (GoG/MWRWH 2009). However, this is unlikely to materialize without appropriate incentives to the private sector. This is because the private sector has never played any meaningful role in housing delivery for the urban poor (ISSER 2008). Faced with the high cost of land and building materials as well as town planning regulations setting housing standards that preclude the use of local technology and raw materials, private real estate companies have delivered housing units to the segment of the urban population that can afford them. These are the urban rich and to a limited extent the middle-class as well as Ghanaians living in North America, Europe and other parts of the developed world who want to acquire property back home (Owusu 2008). The pricing out of low-income groups from the housing market has further marginalized the poor with serious consequences for development.

The Scale of Urban Housing Need

Data on urban housing in Ghana indicate a large deficit in the housing stock. However, the existing data vary and in many instances are difficult to verify. Based on the reported total of 2.8 million residential units in the 2000 Population and Housing Census, the Ministry of Water Resources, Works and Housing (MWRWH) estimated that Ghana needs 70,000 housing units annually, of which only 35 per cent are currently supplied (NDPC 2005). On the other hand, the Draft Housing Policy indicates a national housing deficit in excess of 500,000 units, with an annual requirement of 120,000 units, of which only about 33 per cent are actually supplied (GoG/MWRWH 2009). These figures indicate that the annual supply of housing units in Ghana falls far short of demand, with between 65 and 70 per cent of the national requirement remaining unsatisfied.

Other estimates of the housing deficit present a much grimmer picture. Using the average household size and the number of households per house, Mahama and Adarkwah (2006) estimated the housing deficit in Ghana from a little over 36,000 units in 1960 to over 1.2 million in 2002. They argued that, with a population of about 20 million in 2000 and an ideal of 5.0 persons per household per housing unit, 3.7 million housing units were required rather than the official figure of about 2.9 million, a deficit of over 1.5 million. Mahama and Adarkwah (2006) further noted that at present the minimum two-bedroom housing unit per household is certainly a luxury beyond many households.

Table 5.1   Housing stock in Accra Metropolitan Area, 1960–2000

Year	Housing stock	Estimated housing stock deficit
1960	22,663	52,536
1970	40,802	100,579
1984	64,441	150,935
2000	131,355	237,297

Source: Derived from Population and Housing Census reports and Accra Metropolitan Assembly 2006 (ISSER 2008).

The Institute of Statistical, Social and Economic Research (ISSER) estimated in 2008 the housing deficit in Accra from census data. Table 5.1 shows that even though the housing stock in Accra has grown from 22,663 in 1960 to 131,355 in 2000, the deficit has shown a consistent increase over the period, reaching over 237,000 units in 2000. Logically, the inability of housing supply to meet effective demand leads to upward movement of rents and property prices.

At the heart of the housing crisis is the high increase in land prices and property values, a situation which the conventional literature on urban housing in developing countries has blamed on high rural–urban migration and urban growth (Sajor 2003, Owusu 2008). However, cities in the Global South such as Accra are linked to global capital. Endogenous institutional and policy factors at the national and city levels 'intersect with globalizing forces in constituting particular characteristics of the property market and in re-ordering existing spatial patterns in a way that creates redistributive shifts affecting groups of residents' (Sajor 2003: 714). High land prices in Accra, combined with formalization of land transactions as well as the changing of tenancy agreements into monetary terms, are resulting in a situation where usufruct and inheritance rights to land under customary tenure regimes are no longer guaranteed, as many indigenous people belonging to landowning families are left to compete for land with in-migrants; in the struggle for land, the losing parties are the poor indigenes and migrants, simply because they cannot afford the price of land (Owusu 2008).

In the absence of local or central government provision of public land for housing, and with the lack of regulation of land use to secure adequate space for housing development for the urban poor, increasing numbers of residents of Accra are moving to low-class high density areas of the city with depressing conditions and overstretched infrastructure and services. The Accra Metropolitan Assembly (AMA) estimated in 2006 that about 60 per cent of the city population were living in poor low-class areas. It summed up the conditions of housing in low-income areas of the city as follows:

> Almost all low-income areas are built up with little room for expansion. This is particularly so in the indigenous areas of the inner city. Conditions are generally depressed with poor supporting social and engineering infrastructure. Buildings in low-income depressed localities have poor quality material such as mud, untreated timber and zinc roofing sheets for walling. The housing environment is characterized by haphazard development, inadequate housing infrastructure, poor drainage, erosion and high population concentrations. (AMA 2006: 111)

Global Face of Housing in Ghana: A New Reality?

The literature on urban housing in developing countries has largely viewed the availability and affordability of urban housing land as a process resulting from the interplay of endogenous factors relating to rapid rural–urban migration and urban growth. The influence of factors at the international or global level is rarely considered.

As indicated in Figure 5.1, city and national factors such as weak urban governance institutions, outdated city planning and buildings codes, inadequate infrastructure, national policy of concentration, compulsory land acquisition and economic liberalization have directly and indirectly constrained land and housing supply in Accra. However, these factors have been amplified by conditions at the global level. It is estimated that between 2001 and 2008 Greater Accra received about 84 per cent of total foreign direct investment inflows to Ghana (ISSER 2009). The consequence of this global capital flow is that undue pressure is exerted on land and housing in Accra, leading to higher prices with increasing social and spatial differentiations in the quality of housing for the poor.

Several studies have noted that, although the process of incorporating major Ghanaian cities and towns into the global economy has a long history, it has increased in intensity due to the neoliberal economic and political policies carried out since the mid-1980s (Konadu-Agyemang 2001, Grant and Nijman 2002, Owusu 2008, Grant 2009). For Accra, the national policy of concentration backed by private and public infrastructure investments have made it the most attractive destination for FDI. Corporate organization headquarters, including MNCs, as well as NGOs have established themselves there. As a result, the city has attracted not only resident Ghanaians but also the Ghanaian diaspora population, foreign expatriates living in Ghana and an ever increasing number of West Africans, especially Nigerians, Liberians and Sierra Leoneans (Owusu 2008).

The recent discovery and production of offshore oil from Ghana is likely to attract more multinational corporations into Accra and lead to oil-induced urbanization (Obeng-Odoom 2009). International oil organizations such as Tullow Oil and Komos Energy have established themselves in Accra. The presence of these

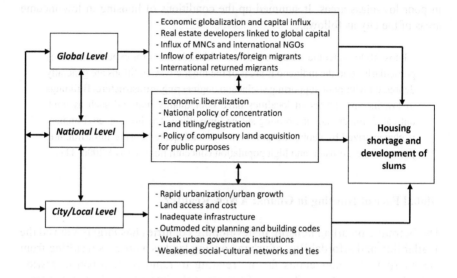

Figure 5.1   Factors shaping housing supply and demand in Accra

Source: Adapted from Owusu (2008: 181).

MNCs will further increase the demand for land and housing, and the consequences of high rent and property values are likely to intensify in the years ahead. Unless pragmatic measures are put in place, the market-driven shifts in land prices will seriously reduce access to land for housing low-income groups in Accra.

As private real estate developers have concentrated on groups of Ghanaians and non-Ghanaians linked directly and indirectly to global capital, there has been an intensification of the trend toward imported building materials as opposed to local production (Songsore 2003, *The Statesman* 2007, Wellington 2009). There is a strong preference for foreign building designs and architecture. Increasingly, new residential, office and commercial buildings are being built with glass and other materials largely imported from Europe, China, North America or elsewhere. This has resulted in new housing types in the Ghanaian urban landscape. To a large extent, the 'self-contained housing type' has come to represent the order of the day. This low-rise housing type with designed glazed glass, normally containing between two and five rooms, is ideal for a nuclear family. On the other hand, developers are increasingly shying away from the compound house, described by Afram (2009) as Ghana's traditional house. The compound house usually comprises dwellings consisting of many small rooms (housing several households) with an open courtyard and shared facilities such as toilets, bathrooms and kitchens. This type of dwelling unit is very popular with low-income groups because it is affordable and allows the sharing of facilities with known groups and individuals (Afram 2009). However, available data indicate that the percentage of compound house types in the total housing stock of Ghana declined from 62 per cent in 1990 to 42 per cent in 2000 (Grant 2009). At the same time, the self-contained housing type (detached and semi-detached houses, flats or apartments and related kinds) has increased significantly. The changing trend towards self-contained housing reflects in part the weakening of the extended family and a preference for nonsharing in-house facilities such as toilets and kitchens.

Towards an Alternative Housing Development Approach

Ghana faces a daunting task as it seeks to compete for global capital, leading to the ever-increasing integration of its largest metropolis, Accra, into the world economy. Consequently, an ever-shifting topography of land values quickly and efficiently excludes all potential users who are unable to meet the price of a given parcel of land (Sajor 2003). At the same time, there is the need to address the high demand for land for housing resulting from the rising rural–urban migrant population in Accra. The difficult balancing act is how to ensure the inflow of global capital and their actors, with resultant rising land values, while guaranteeing decent housing for low-income groups in the city.

So far the market alone has not guaranteed equity and redistribution of land towards low-income groups in Accra. The weak capacity of the Accra Metropolitan Assembly (AMA), especially in terms of financial resources and managerial expertise, limits its potential to promote a pro-poor and socially viable urban development agenda that enables low-income groups to access affordable land and housing. For these reasons, the role of the state as a mediator in determining housing development outcomes is critical. Actions and policies of the state at the national and city levels are very important in determining the extent to which poor and low-income groups will feel alienated from the city. As Thomi (2000) noted, the state in general can be regarded as a regulatory framework for a given territory and its population. National sector policies, whether positive or negative or whether implicit or explicit, can have a great impact on the development of urban centres, but in reality the state may have very limited capacity to implement national sector policies at all.

A new type of urban development involving the active engagement of the state and the poor is required. The state and local authorities have largely conceived the growth and development of Ghanaian towns and cities from the development experience of urban centres in the west. Such an approach has resulted in the development of building and planning regulations largely alien to the Ghanaian situation. For example, the Town and Country Planning Act of 1945 was transplanted from the pages of the British Town and Country Planning Act of 1932 (Konadu-Agyemang 2001). It has been argued that the planning schemes, layouts, minimum lot sizes etc. outlined in the 1945 Act are simply unrealistic in the present Ghanaian context (Konadu-Agyemang 2001, Grant and Yankson 2003, Owusu 2008).

An alternative housing development approach must be responsive to the criticisms and failures of the present market-driven approach to land and housing delivery. A new approach must give the urban poor and low-income groups a voice and choices. It should also give meaning to development as an issue of welfare distribution, equality and democratic rights, facilitating the raising of the productivity of the poor and hence leading to poverty reduction (Lund 1994). More importantly, it should reflect the true meaning of housing and thus provide a strong justification for adopting the basic needs approach. Housing involves not only physical shelter but also access to basic public services such as drinking water, sanitation, health, and education. These are key features of the basic needs approach as articulated under the broader alternative development paradigm. Many of the difficulties faced by the urban poor are linked to a greater or lesser extent to the quality, location and security of housing:

> A roof and an address in a habitable neighbourhood is a vital starting point for poor urban people, from which they can tap into what the city can offer by way of jobs, income, infrastructure, services and amenities. Decent shelter provides people a home; security for their belongings; safety for their families; a place to strengthen their social relations and networks; a place for local trading and service provision; and a means to access basic services. It is the first step to a better life. For women, property and shelter are particularly significant in terms of poverty, HIV/AIDS, migration and violence. (UNFPA 2007: 38)

The housing issue is at the core of urban poverty. Housing improvement can be a critical tool for national development. Adequate housing provides important social benefits to both the occupants and wider society. These benefits include better health, fewer behavioural problems, especially among children, greater educational attainment and increased labour force participation (Newman 2008). It is therefore important that the state in Ghana takes measures to facilitate the delivery of land and housing to low-income groups. The goal here will be to encourage the inflow of global capital while attempting through taxes and other instruments to channel some of the benefits of globalization toward providing the needs of the poor.

Within the context of a globalized and liberalized environment, the state is urged to engage directly and indirectly in the housing sector through measures such as securing land banks for public housing, providing incentives through legal frameworks that allow private real estate developers to devote part of their resources for social housing, and developing public social housing. Other measures should include promoting self-help housing construction, upgrading existing informal and slums settlements, and promoting research and development (R&D) in local building materials and encouraging their use.

In an alternative urban housing development approach, the proposed measures to address the urban housing crisis in Ghana must be pursued within the framework of improving urban governance and promoting a pro-poor urban development agenda as strongly articulated in the Draft National Urban Policy of

2011. This draft policy, yet to be passed by Parliament, provides a comprehensive framework for the development of all towns and cities in Ghana, including the Accra metropolitan area (GoG/MLG&RD 2011). Taking into account that the needs of individuals and groups for housing vary, a 'one-fits-all' approach should be avoided. An urban housing development agenda should accommodate multiple voices and solutions, leading to the creation of an inclusive urban society where all categories of society live in harmony. More importantly, such an alternative development approach would allow the voices of the poor to be heard above powerful market forces – a development perspective that runs throughout the work of Ragnhild Lund.

References

Afram, S.O. 2009. The Traditional Ashanti Compound House: A Forgotten Resource for Homeownership for the Urban Poor, in *Proceedings of the 2009 National Housing Conference*, edited by Council for Scientific and Industrial Research (CSIR)/Ghana Institute of Architects (GIA). Kumasi: CSIR–BRRI, 74–84.

AMA 2006. *Medium-Term Development Plan 2006–2009*. Accra: Accra Metropolitan Assembly

BBC 2007. Dazzled by the City. *BBC Focus on Africa*, July–September 2007, 16–19.

GIPC 2009–2010. *Quarterly Operational Performance Reports* [Online: Ghana Investment Promotion Centre]. Available at: http://gipcghana.com/library.php?pageNum_pub_list=1&totalRows_pub_list=24&id=16 [accessed: 4 January 2013].

GoG/MWRWH 2009. *Draft Housing Policy 2009*. Accra: Government of Ghana, Ministry of Water Resources, Works and Housing.

GoG/MLG&RD 2011. *Draft National Urban Policy 2011*. Accra: Government of Ghana, Ministry of Local Government and Rural Development.

Grant, R. 2009. *Globalizing City: The Urban and Economic Transformation of Accra, Ghana*, New York: Syracuse University Press.

Grant, R. and Nijman, J. 2002. Globalization and the Corporate Geography of Cities in the Less Developed World. *Annals of the Association of American Geographers,* 92, 320–40.

Grant, R. and Yankson, P. 2003: Accra. *Cities*, 20, 65–74.

GSS 2005. *Ghana Population Data Analysis Report: Socio-Economic and Demographic Trends,* Vol. One. Accra: Ghana Statistical Service.

ISSER 2007: *The State of the Ghanaian Economy in 2006*, Accra: Institute of Statistical, Social and Economic Research.

ISSER 2008. *Situational Analysis of Selected Slum Communities in Accra and Sekondi–Takoradi*. A Report submitted to CHF-Ghana. Accra: Institute of Statistical, Social and Economic Research.

ISSER 2009. *The State of the Ghanaian Economy in 2008.* Accra: Institute of Statistical, Social and Economic Research.

K'Akumu, O.A. 2007. Strategizing the Decennial Census of Housing for Poverty Reduction in Kenya. *International Journal of Urban and Regional Research,* 31, 657–74.

Konadu-Agyemang, K. 2001. Structural Adjustment Programs and Housing Affordability in Accra, Ghana. *The Canadian Geographer,* 45, 528–44.

Lund, R. 1994. *Development Concept in the Light of Recent Regional Political Changes and Environmental Challenges.* Doctoral Lecture, April 27, 1994. Papers from the Department of Geography, University of Trondheim No. 139. Trondheim: University of Trondheim.

Lund, R. 2002. Ethics and Fieldwork as Intervention. *Acta Geographica Trondheim,* Series A(2), 214–19.

Mahama, C. and Adarkwah, A. 2006. *Land and Property Markets in Ghana.* London: RICS.

NDPC 2005. *Growth and Poverty Reduction Strategy (2006–2009),* Vol. I: *Policy Framework.* Accra: National Development Planning Commission.

Newman, S. J. 2008. Does Housing Matter for Poor Families? A Critical Summary of Research and Issues Still to be Resolved. *Journal of Policy Analysis and Management,* 27(4), 895–925.

Nsiah-Gyabaah, K. 2009. The Urban Housing Challenge and Prospects for Meeting the Housing Needs of the Urban Poor in Ghana, in *Proceedings of the 2009 National Housing Conference,* edited by Council for Scientific and Industrial Research (CSIR)/Ghana Institute of Architects (GIA). Kumasi: CSIR-BRRI, 15–25.

Nugent, P. 2004. *Africa since Independence.* New York: Palgrave Macmillan.

Obeng-Odoom, F. 2009. Oil and Urban Development in Ghana. *African Review of Economics & Finance,* 1, 17–39.

Owusu, G. 2008. Indigenes' and Migrants' Access to Land in Peri-Urban Areas of Ghana's Largest City of Accra. *International Development Planning Review (IDPR),* 30, 177–98.

Owusu, G. 2010. Social Effects of Poor Sanitation and Waste Management on Poor Urban Communities: A Neighbourhood-Specific Study of Sabon Zongo, Accra. *Journal of Urbanism: International Research on Placemaking and Urban Sustainability* 3, 145–60.

Owusu, G., Agyei-Mensah, S. and Lund, R. 2008. Slums of Hope and Slums of Despair: Mobility and Livelihoods in Nima, Accra. *Norsk Geografisk Tidsskrift–Norwegian Journal of Geography,* 62, 180–90.

Sajor, E.E. 2003. Globalization and the Urban Property Boom in Metro Cebu, Philippines. *Development and Change,* 34, 713–41.

Songsore, J. 2003. The Urban Housing Crisis in Ghana: Capital, the State Versus the People. *Ghana Social Science Journal* (New Series), 2, 1–31.

The Statesman 2007. A Brief History of Housing in Ghana [Online, 27 January]. Available at: http://www.ghanaweb.com/GhanaHomePage/NewsArchive/artikel. php?ID=117756 [accessed: 20 January 2010].

Thomi, W. 2000. The Local Impact of the District Assemblies, in *A Decade of Decentralisation in Ghana: Retrospect and Prospects,* edited by W. Thomi, P.W.K. Yankson and Zanu, S.Y. Accra: EPAD Research Project/MLG&RD, 229–67.

Tipple, A.G. 1987. *The Development of Housing Policy in Kumasi, Ghana.* Newcastle upon Tyne: University of Newcastle.

Tipple, A.G. 1994. The Need for New Urban Housing in Sub-Saharan Africa: Problem or Opportunity? *African Affairs,* 93, 587–608.

UNFPA 2007. *UNFPA State of World Population 2007: Unleashing the Potential of Urban Growth.* New York: UNFPA.

Wellington, H.N.A. 2009. Gated Cages, Glazed Boxes and Dashed Housing Hopes – In Remembrance of the Dicey Future of Ghanaian Housing, in *Proceedings of the 2009 National Housing Conference,* edited by Council for Scientific and Industrial Research (CSIR)/Ghana Institute of Architects (GIA)Kumasi: CSIR-BRRI, 69–73.

Yeboah, I. 2001. Structural Adjustment and Emerging Urban Form in Accra, Ghana. *Africa Today,* 7, 61–89.

Yeboah, I. 2003. Demographic and Housing Aspects of Structural Adjustment and Emerging Urban Form in Accra, Ghana. *Africa Today,* 10, 106–9.

Chapter 6

Implementing International Consensus on Women in Development: Context, Policy and Practice

Ingrid Eide

Introduction

The World Conference to Review and Appraise the Achievements of the United Nations Decade for Women: Equality, Development and Peace, held in Nairobi, Kenya, in 1985 is now considered an historical event. A quarter of a century has passed since this significant event in the political history of the advancement of women, significant also in the history of the United Nations (UN). The Nairobi conference followed two other world conferences on women, one in Mexico in 1975 and the other in Copenhagen in 1980. In 1995 the Fourth World Conference on Women was held in Beijing. The number, locations and time span of these conferences are characteristic of how the UN operates as a world organization. Once a theme has been defined as important, problematical and of global concern, a process evolves whereby member states, relevant units of the UN system, nongovernmental organizations (NGOs) and other stakeholders may mobilize.

This chapter takes the 1985 Nairobi Forward Looking Strategies for the Advancement of Women (FLS) (UN 1986) as a point of departure for a review of the efforts of the United Nations Development Programme (UNDP) to develop and implement its own comprehensive corporate strategy. One element in this strategy was a project review form that operationalized concerns and concepts central to women and development studies at the time. This chapter concludes by discussing the use and utility of this review form in development cooperation.

The Emergence of a Woman's Focus in the United Nations

In 1972, the UN General Assembly proclaimed 1975 as the International Women's Year, 'to be devoted to intensified action to promote equality between men and women, to ensure the full integration of women in the total development effort and to increase women's contribution to the strengthening of world peace' (UN 1986: 5).

The activities before and during the International Women's Year resulted in a demand for a full decade similarly devoted to the advancement of women. Member states established preparatory committees, and a series of conferences was planned, one conference in each continent. National and international documentation was collected and circulated. A new knowledge base on the situation of women became available for participants in political debates. Women researchers and activists began regarding the UN as their arena.

Even if the formal conferences were the domain of diplomats and politicians, they included a significantly higher number than normal of women officially appointed and instructed by their governments as members of national delegations. The less formal NGO meetings, on the other hand, included thousands of women in parallel events. At a time when the UN was marred by the Cold War, it developed a political space that was open to a new constituency of women.

Three themes – equality, development and peace – reflected priority concerns of a divided world. Equality issues were emphasized by the Western group, peace was a more common reference point of the Eastern group, while development was the obvious priority of the group of developing nations. Including all three was itself a compromise. Over the years, with the end of the Cold War, it has become more acceptable and more fruitful to see how these themes are interrelated, as reflected in the Security Council Resolution on Women, Peace and Security (UN 2000). In 1985, however, the Cold War, the arms race and the threat of nuclear war dominated international relations, while developing nations experienced economic crises, indebtedness and 'structural adjustment programmes'. The Bretton Woods institutions, the World Bank and the International Monetary Fund (IMF), appeared more influential than the UN system. The Development Decade of 1980 was described as the 'lost decade for development'. UN debates were compared to marathons – exhaustive but unproductive. In a bipolar international structure, the UN appeared to be marginalized. Perhaps in an effort to circumvent great power politics, the UN emphasized its thematic thrust and its corresponding documentary and informative functions. Significant global challenges were brought to the fore, such as disarmament, environmental degradation, desertification, ill health, illiteracy, habitation and population, with the advancement of women as a crosscutting issue. For the UN and its majority of loyal, hopeful member states, it became particularly important to agree on thematic policies and their implementation by all relevant actors.

The 1985 Conference discussed and adopted 'The Nairobi Forward-Looking Strategies for the Advancement of Women (FLS)' (UN 1986). The 372 paragraphs first give the historical background, and then there is one chapter for each theme: 'Equality', 'Development' and 'Peace'. Obstacles, basic strategies and measures for implementation of the basic strategies at the national level are proposed. The reference to the national level is necessary, as the purpose is to make an impact on the daily lives of women. It is also necessary as the UN is an intergovernmental organization and its members are sovereign states, whose policies, practices and indeed ideology on women's roles differ. There are a few paragraphs where the

USA or a group of developed countries – principally the Organization of Economic Cooperation and Development (OECD) – have formulated reservations, asked for a vote or abstained from voting. The paragraphs provide extremely ambitious visions of social change, including the need for a 'New International Economic Order', as well as general statements about imperialism and neocolonialism. The Holy See consistently reserved its position on women's reproductive rights. Against this background the adoption of the FLS by consensus in Nairobi, and without a vote by the UN General Assembly the same year, proved that the decade-long deliberations had been fruitful (UN 1985).

During this process, member states and their delegates at the UN reached a new level of understanding of the roles and rights of women. This was expressed as expectations concerning follow-up by the UN system. The FLS had a long chapter on 'International and Regional Co-operation'. UN member states, particularly developing countries, emphasize the multiple functions of the UN system in development cooperation. The 'like-minded' donor nations, including the Nordic states, were active and consistent in their attention to FLS follow-up. A system-wide commitment to implementation was agreed upon.

Implementing the Strategy

The present author had the privilege of being Director of UNDP's Division for Women in Development from April 1987 to July 1989. The team was international: in addition to myself – a Norwegian sociologist – UNDP recruited a demographer from India, an economist from Haiti and a junior professional economist from Germany. The secretaries were from Jamaica and Guyana. UNDP, with its headquarters in New York, is defined as a coordinator of UN development cooperation and is largely field-based with a resident representative and a field office in the capitals of most developing countries. In 1987 the number of field offices was 112, covering cooperation with 152 countries. The budget of USD 1 billion could not possibly match the ambitions of the UN system or the needs of the countries concerned. Nevertheless, the resident representatives were considered important multilateral actors, locally and internationally, particularly if designated as coordinator of UN activities more generally.

Among the 112 resident representatives in 1987, only eight were women, while 25 per cent of the professionals were women. The total number of staff was 1,000 internationals and 4,000 nationals. In addition UNDP had 8,500 internationally recruited experts and consultants from 130 countries and nearly 5,000 national experts. Fellowships for training reached 9,500 and 1,200 UN volunteers were at work in more than 100 countries.[1] This was a complex organization, bureaucratic and hierarchical, vast and dispersed.

---

1   Figures based on UNDP's Annual Report 1987.

In 1986 there occurred a change of agency head; William Draper, a US citizen, became Administrator of UNDP. He signalled that he would focus on four themes: the private sector in the development process, the involvement of NGOs, attention to the environment, and *women in development, (WID)*. By this move, WID became a manifest concern of the top management. To follow up the new focus on WID, UNDP decided to reorganize its work: a director at the level of senior management was recruited and a division of three international staff members established. The division was mandated to assist in ensuring and monitoring throughout UNDP's programmes and projects a substantially larger role for women, both as active participants at all levels and as beneficiaries of such projects. Close liaison and working relationships internally and externally were emphasized in the mandate. Organization-wide networks of focal points were to be established and training programmes for staff and others were to be arranged. Guidelines, monitoring and evaluation were to contribute to achieving the objective of a substantially larger role for women in development. These points and functions are in accordance with FLS, although there is no explicit reference to FLS in this one-page mandate.

The International Development Fund for Women (UNIFEM), established in 1975, was defined as in 'autonomous association' with UNDP. Member states, always keen to criticize possible overlap and duplication in the UN system, were worried. Perhaps UNIFEM could have absorbed these functions? However, UNIFEM was a separate fund with its own purse and portfolio of small, women-specific projects. The WID Division was defined as internal to UNDP. It was not intended to compete in the market of scarce donor funding for 'women's projects'. A UNDP evaluation had found that in a sample of projects only 4 per cent of the funding involved women. Asked what amount of funding I would aim for, I had to answer: 'All – and nothing'. The Division was there to ensure that the entire UNDP honoured the proclaimed focus on WID and, as a consequence, integrated women as 'participants and beneficiaries' in all activities.

A comprehensive programme of staff training had to be organized to reach this goal. We considered factors of *time* (urgency), *space* (field operations), *numbers* (thousands of staff members), *resources* (limited), *professional integrity and status* (highly educated, experienced, career-oriented staff operating in a hierarchical structure), and concluded that staff training would have to be both formal and informal, and clearly relevant for the theme as well as the tasks at hand. We should aim to make it intellectually challenging and sufficiently intriguing to make an impact on the thinking of the persons involved as individuals or teams.

Senior management of UNDP and other UN funds, such as the Joint Consultative Group on Policy (JCGP), comprising UNDP, the United Nations Population Fund (UNFPA), the United Nations Children's Fund (UNICEF) and the International Fund for Agricultural Development (IFAD), had already attended a series of lectures on the roles of women in various sectors, and a report was produced and widely disseminated. The event legitimized that attending training on WID was appropriate and the report functioned as reference documentation.

Our plan, which we followed, was to work our way in the organizational hierarchy, and also to arrange training seminars in the field, first by covering all regions and some sub-regions, then, whenever feasible, individual field offices. Training was mostly organized as regular staff training by the Division of Personnel, an arrangement that regularized and legitimized this WID initiative. As far as possible, we cooperated with other UN organizations and involved participants from national governments and NGOs.

'Conditionality' was one of the buzzwords of the development debate at that time and part of the disputed World Bank and IMF adjustment programmes. Instead we said UNDP's aim was to qualify trainees for 'informed advocacy' on WID. WID as a theme might still be seen by some as too political, even personal and emotional, and hence 'nonprofessional'. By making the WID training inclusive and transparent, we hoped to avoid a feeling that WID was a new conditionality, or a top-down initiative. On the contrary, it was an effort to involve the various partners in seeing more clearly and fully the realities of people, women and men, in programmes and projects, thereby creating a common ground for corrective or new action. UNDP staff had a strong identity as development professionals. To make them feel comfortable in their role as WID advocates and practitioners, a shared terminology was introduced. It consisted of ten basic concepts from current development studies, supposedly relevant for UNDP.

Between Theory and Policy

The concepts, all presented with a brief discussion, were the following: *gender* and gender relations; *statistical categories* and gender gaps; *women's activities* and their combination of productive and reproductive work; their participation within *systems of activities*, such as production chains in agriculture or fisheries; *time budgets*, which often proved women to be both overburdened and underutilized; *resources* and women's access to and control over resources; *household economies*; *private and public arenas*; *participation*; and, finally, *change*, with a brief discussion of trends that were functional or dysfunctional for the advancement of women. A more detailed presentation of how these concepts were defined as relevant for development practitioners is found in Eide (1991).

According to our mandate, women in their roles as participants in and beneficiaries of development projects and programmes were our point of departure. We clarified the different implications of women being a 'participant' and/or 'beneficiary'. If women participate without benefiting, they are exploited; if they are neither participating nor benefiting, they are bypassed and possibly marginalized; if they only benefit, their involvement will probably not be sustainable. In other words, UNDP should see and involve women in both capacities – as participants and as beneficiaries. Similarly we analysed the male–female division of labour in reproduction and production systems, and in present production systems compared to what was foreseen in a project proposal. The purpose of the terminology and these simple analytical exercises was increased awareness of the actual and potential social and economic roles and activities of women and men. In addition, we wrote 'guidelines by sector' to exemplify dimensions relevant for WID in project design and drafting of country programmes.

The diversity of the organization in both staff and functions, and the diversity of the countries involved, required attention to what was considered 'culture-specific'. The terminology and guidelines should provide rudimentary frameworks for seeing and analysing culture-specific factors in a given setting where a development project intervened. Did the intervention confirm and upgrade women in their traditional functions, or were they bypassed or given additional tasks that just added working hours? If nontraditional, gender-neutral opportunities were introduced, did women have access to them? Realistically, a project or even a programme launched in a given country could only exemplify an approach to 'the advancement of women'. If relevant and sustainable in the local context, it might have a multiplier effect by being replicated. A buzzword among development practitioners at the time was that projects should be 'catalytic'.

The FLS had already insisted that women were *contributors* to development, probably in opposition to the common view that, in general, women represented *costs* as daughters, sisters, wives or mothers. Dowries, bride prices, selective abortions and infanticide were frequent and difficult issues at the time. However, FLS had referred to women in their roles as planners, producers, traders, consumers, operators, owners, transporters and managers. Women were now seen as *agents of development* and not only as 'vulnerable groups – women, children and the handicapped', which had been a common phrase. FLS had a separate chapter on 'Areas of Special Concern', where 14 categories of women were listed. Victimized women were not overlooked, but they were no longer to dominate the image of women. Moreover, these women were often both *victims* as well as *agents of change.*

We made a special effort to visualize this. All too often pictures of women had been used for ornamental purposes in UN publications. We proposed a photo exhibition and a poster series depicting a wide range of active, productive women, traditional and modern, in fields, factories, offices and laboratories around the world. We showed women at work alone or with male partners. The material was available in UNDP's archives. Gender-sensitive photographers and journalists in the field had provided it over the years, but a lack of general awareness or demand for attention to gender issues had left it largely overlooked.

Knowledge – Policy – Practice

While the focus of training was clearly on women and the FLS commitment to WID, the approach was *gender analysis.* Trainees were to be sensitized to the roles and relations of men and women over the life cycle, including their interdependence in reproduction and production, and their unequal access to and control of resources. This meant that male roles, and men, were included and not alienated. On the other hand, 'gender' was an abstract concept and a word sometimes difficult to translate. 'Sex' is ambiguous and denotes categories, while 'women' and 'men' are real, concrete and obviously diverse depending on stage in life cycle, class, culture, time and space. The participants' experiences from their countries of origin could more easily be mobilized when we spoke about women and men rather than 'gender'.

In addition to attending lectures and group work on 'WID theory', participants brought their own desk material to the sessions for discussion: what were the tasks they struggled with, what was a women's perspective or women's interests in a project or other activity, and what were the entry points for WID relevant action? Was gender analysis fruitful? This was an effort to bridge theory and practice. We invited a 'reality check' regarding the relevance of what was taught and defined as policy at conferences distant in time and place. The FLS as well as national level initiatives during the Decade for the Advancement of Women (1975–1985) had already made good use of WID studies from development research institutes around the world. It was a challenge for UNDP in the late 1980s to familiarize its staff and counterparts with this knowledge base and to ensure that the individual and collective commitment to FLS was enacted in a professional manner. It was a 'triple approach': theory, policy and practice.

At Headquarters, UNDP had institutionalized a related exercise in its Action Committee. Here projects were presented for approval to senior management by the staff member responsible and commented upon by committee members, the heads of regional bureaus, the Administrator and Assistant Administrator, and the directors of the NGO and WID Divisions, respectively. This was where the WID Director on a weekly basis could voice critical remarks, point out alternative action, or just enquire if and how women were integrated as participants and beneficiaries. Little by little, such comments were expected and 'preempted' by the presenters, mostly proving that WID was taken seriously and understood. Projects were very rarely rejected.

The discussions of the Action Committee were by definition a superficial kind of review, as 20 to 30 project document briefs could be received late Friday and meetings held the following Wednesday. Yet they provided regular opportunities for discussion with senior management and other staff on why and how women should be integrated in specific development activities that in sum covered the entire range of UNDP's involvement around the world. The Action Committee was probably the most important arena for informal WID training and for immediate feedback to the WID Division on the viability of its approach.

Another arena involved resident representatives on mission at headquarters. The Administrator had decided and put in their schedule that they should all spend at least half an hour with the WID Director in her office. This was an opportunity to discuss more directly their plans for a follow-up of the organization's emphasis on WID. In preparation for these meetings I had looked at available statistics for the country in question as well as its country programme. A well-designed poster with statistical tables, prepared for the 1985 Nairobi Conference by the UN, was particularly useful. It had information about 172 countries concerning 13 indicators relevant for understanding gender issues in development. Three indicators covered population composition and distribution, two covered education, training and literacy, three showed economic activity, and three gave information on marital status, fertility and maternal death rates. The thirteenth indicator gave the numbers of male and female representatives in national legislative bodies. All indicators gave information about *'gender gaps'* – the relative percentages of males and females, or the percentage of females in the respective totals.

This statistical poster proved extremely helpful at the first meetings with resident representatives. They were often unfamiliar with 'gendered' statistics and the range of indicators relevant for understanding gender issues in their country. It probably also helped overcome latent or manifest opposition to the need for the advancement of women. How to interpret and deal with hard facts in the form of statistics was a better entry point for cooperation at this level. UNDP produced in 1985 its own book of *Social, Monetary, and Resource Tables* (the SMART book). At first the statistics were 'gender-blind', but later editions had, wherever possible, data with gender breakdowns. In 1990, UNDP launched its series of Human Development Reports (UNDP 1990–2011), with a new index and a better understanding of development processes, initiatives and opportunities. It documented that populations must be seen as both male and female because gender gaps and gender relations have a lot to do with general development trends.

Based on formal and informal interaction with staff in decision-making and training, it was possible to develop a policy paper that reflected many of the FLS points. UNDP accepted the paper as its own *corporate strategy*. To confirm our respect for the diversity of UNDP and facilitate the use of the policy paper, we had it translated into Arabic, French and Spanish. The policy paper maintained WID as a priority theme, the Division for WID as UNDP's focal point and the Director as part of top management. The Division was also to collaborate at Headquarters with a system of focal points across the organization and liaise with other organizations dealing with WID. At the field level, resident representatives were responsible for concrete action to integrate women in projects and programmes, assisted by local focal points, both international and national. The focal points were intended to provide resource persons on WID, able to collect and provide relevant documentation. We expected that over the years UNDP's field offices would build a knowledge base including a small library that could be consulted by other agencies and organizations committed to implement WID policies. Our hope was that sharing information locally would prevent duplication and waste, and also encourage cooperation with local resource persons. Considering the information and communication technologies of a quarter of a century ago, it is easy to understand that it was a formidable task to keep this extensive network alert and motivated for concerted action. In an effort to promote an exchange of ideas between field offices, we produced a small newsletter, *WID-Link*. This initiative was appreciated, but very few contributions were written by field office staff.

UNDP always insisted that it was field-based and action-oriented. The basic policy goals were to encourage concrete action through the inclusion of women's concerns in appropriate stages of UNDP's programme and project formulation. How could we ensure that UNDP implemented system-wide its policy and achieved its goals? As part of the policy paper, we developed a *project review form* and requested its use for all projects during identification/formulation, at time of approval and for annual reviews and evaluations. A separate form was to be used for each project stage and copies were sent to the WID Division.

It was a one-page form only, intended as a reminder or checklist. Attention to WID was intended to be unavoidable. The form could also, potentially and eventually, be used in assessing staff compliance and effectiveness, and provide data for future reports and evaluations. It was an instrument to make UNDP accountable.

Three types of questions were included. First: 'Do women typically work in the (sub) sector of this project in the geographic area where it will be located?' If yes, we asked for the estimated proportion of work by men and women. If no, we asked if it would be against prevailing sociocultural norms of the country to involve women in this project, and would they, please, explain. The second question was: 'In your view have issues of relevance to women been adequately reflected in the project objectives, outputs and activities?' If yes, we asked for descriptions of how; if no, explanations of why. We also asked for recommendations to reorient a project that did not involve women. Similarly, we asked if the project, as planned, would involve women as direct recipients and/or beneficiaries. The third question asked about the respective number of women and men involved in the project as experts, consultants, evaluation team members, fellowship holders, trainees and others.

This initiative was not a welcome exercise, as it meant extra work and rethinking, and often required information not readily available. The fact that it was expected to increase the relative number of women 'substantially' could also provoke conflict over scarce opportunities. We tried to encourage colleagues to do as best they could, and assured them that educated guesses were better than mere anecdotes, or negligence. Many found the exercise useful. After a training session where the review form had been introduced, a resident representative said that she would visit a cattle project and enquire if they had included small cattle (goats and sheep) that typically were the property of women. Another had reviewed a project proposal on fisheries. 'Where are the women?' she had asked, and was told that yes, they had added a health component for women and children. 'Important enough', she answered, 'but how were women involved in the fisheries?' It then appeared that women were in fact fishing in that particular area, and of course trading fish products, but were bypassed by the planned project.

The Division was expected to ensure and monitor compliance with UNDP's WID policy. We had reported annually to UNDP's Governing Council on the WID mandate, policy and progress. The project review forms gave us some of the information needed to document accountability. When we reported to UNDP's Governing Council in 1989, we had received 1,258 forms, representing about one-quarter of all UNDP projects at that time. The sample was probably not representative, but large enough to give valuable information. In 1,090 forms there was contained information on the proportion of women and men working in the project sector. Twelve per cent of the projects were in sectors where women dominated, 20 per cent in sectors where the participation of men and women was about equal, while 68 per cent of the projects were in sectors where men constituted at least 60 per cent of the work force. Further analysis showed that the higher the female participation in a given sector, the more significant was the reflection of WID considerations in project design and the integration of women as participants and beneficiaries in the project. However, women were also registered as participants and beneficiaries in 20 per cent of the projects where they constituted a minority of the work force. This indicated that women, to some extent, had entered new areas by way of a project, and perhaps contributed to diversification of roles and activities of women in the country concerned. At the other extreme, we found projects that failed to take women into account, despite the fact that women contributed significantly to the respective sectors. We concluded that bypassing female producers, while improving access to new resources for male producers, would result in a relative setback for women and probably in a loss of potential total productivity. We did not have baseline data, but gained the impression that increasing proportions of women were being included in as experts (20 per cent), consultants (16 per cent), members of evaluation missions (32 per cent), recipients of fellowships (20 per cent) and trainees (48 per cent). However biased this reporting may have been, due to the sample of projects, these results indicated to us that development cooperation was no longer a male domain, and that women were increasingly involved as beneficiaries, participants and managers of projects (UNDP 1989).

Our intention in the WID Division was to continue this approach to monitoring and to modify the review form. My successor as director of the division did not share this opinion. Many respondents to a questionnaire to field offices supported her view that it should be discontinued, even though it had been an integral part of adopted WID policy. A longer time series would have given more detailed information on how far UNDP was accountable over time in its focus on Women in Development, WID. How FLS and UNDP's corresponding policy were implemented has changed with the changing composition of staff, 'political climates' and priorities inside and outside of UNDP. This is reflected in UNDP's reports to its governing bodies.

Conclusion – From Policy to Practice

The purpose of this account has been to share my experience of how a common concern in research and politics, notably the need to review and advance the position of women in development processes, was absorbed by the UN system, and how it was translated into action within and by way of a complex organization,

UNDP. UNDP could only relate to one part of the FLS, namely development. This was politically convenient, but also in accordance with the division of labour of the UN system and UNDP's mandate. My task was to prove that women, if adequately involved, were not an added burden, but a positive factor for UNDP's success. At times this invited 'instrumentalism' rather than 'feminism'. On the other hand, we arrested excesses in instrumentalism. I remember distinctly an early dispute with a respected colleague who had written 'our task is to harness women for development'. A gender-sensitive, professional terminology helped change such language and perceptions. The focus on Women in Development created a new visibility for women wherever UNDP was involved, and for some women also new opportunities.

References

Eide, I. 1991. Women in Development (WID): A 10-point Framework for Analysis. *Forum for Development Studies*, 18, 21–31.

UN 1985. *General Assembly: A/RES/40/108 13 December 1985, 116th Plenary Meeting: Implementation of the Nairobi Forward-looking Strategies for the Advancement of Women* [Online: United Nations General Assembly]. Available at: http://www.un.org/documents/ga/res/40/a40r108.htm [accessed: 8 January 2012].

UN 1986. *Report of the World Conference to Review and Appraise the Achievements of the United Nations Decade for Women: Equality, Development and Peace, Nairobi, 15–26 July 1985*. New York: United Nations.

UN 2000. *Security Council Resolution 1325 on Women, Peace and Security*. New York: UN Security Council.

UNDP 1985. *Social, Monetary, and Resource Tables*. New York: United Nations Development Programme.

UNDP 1989. *Implementation of Decisions Adopted by the Governing Council at Previous Sessions: Women in Development*. Report of the Administrator, DP/1989/24. New York: United Nations Development Programme.

UNDP 1990–2011. *Human Development Reports* [Online: UNDP]. Available at: http://hdr.undp.org/xmlsearch/reportSearch?y=*&c=*&t=*&lang=en&k=&or derby=year [accessed: 8 January 2013].

PART II
Alternative Geographies of Gender and Development

PART II

Chapter 7

Muted Power – Gender Segregation and Female Power

Vibeke Vågenes

Introduction

The Hadendowa Beja form a distinct ethnic group in eastern Sudan. Traditionally the Hadendowa are nomadic pastoralists, but repeated drought throughout the latter part of the twentieth century deprived numerous families of large parts of their herds and many families settled in villages and towns. As a group, the Hadendowa have become marginalized economically, politically and perhaps also culturally in the context of the nation of Sudan. The Beja tribes are known from medieval historical sources to be a distinct group of fierce and independent nomads (Vantini 1975, Holt and Daly 1979). Their territories stretch across the eastern parts of Sudan from the sloping hills running towards the Nile to the Red Sea coastline (Figure 7.1). The Beja can be described as a tribal confederation and the Hadendowa are the largest group among them. The Beja are Muslims.

When we look more closely into the daily life of the Hadendowa in the Red Sea Hills, it may appear a male-dominated culture. Women are absent from the market places and central shopping streets. Figures 7.2 and 7.3. show an example where both men and women are present at a religious festival but completely segregated. Female seclusion organizes Hadendowa living and people confirm that this is a male dominated, patriarchal culture, in which men have the decision-making power in families and clans.

This chapter focuses on relations between men and women, especially relations between husband and wife. The marital relationship is analysed in order to examine the power game between men and women, with special focus on female power. The term 'power game' is intentionally applied here, as this study approaches power from a constructivist perspective, whereby it is assumed that people engage in social relations within which power is continuously constructed and negotiated. Power negotiations involve people in social settings (Bourdieu 1977, Bourdieu and Thompson 1991, Macleod 1991). The purpose of this chapter is to examine how female power is evident and at the same time subaltern and hardly discernible to the observer. Female power is muted, but evident if we look for it. Certainly, when I came to the areas of the Hadendowa, it took me a while to understand that women are equipped with a set of powerful means of manipulating situations in their interest. It is crucial to understand feminine power in the light of gender

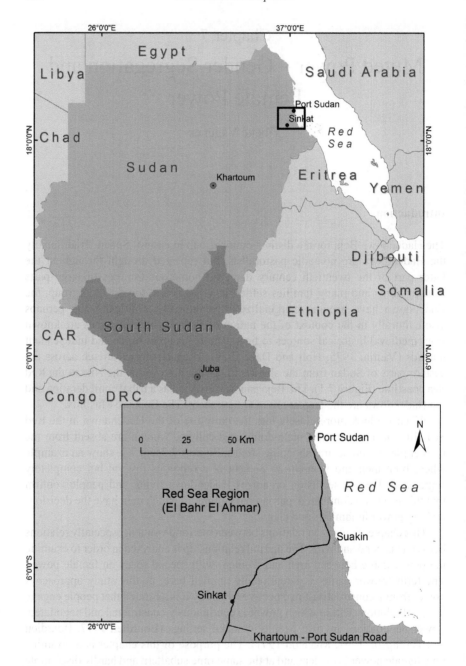

Figure 7.1   Location of the study area in Sudan

segregation and the ideals of honour and shame according to which Hadendowa social practice is valued.

Female empowerment is a major concern in development research and strategies (Lund 2002, Attanapola 2008) and critical voices are currently raised against the presumed universality of approaches towards female empowerment (Syed 2010). For a more sensitive approach to understanding female power, it is necessary to regard gender relations and power inequities in their specific social and cultural context. A superficial impression of Hadendowa society from the outside might suggest that the women's situation is that of the subordinate and submissive gender; in many ways the women *are* dominated by men. On the other hand, it may also be the case that the Hadendowa women do not feel that way. The following text elucidates aspects of how women exercise power among the Hadendowa.

My study is a result of several years of engagement in the Red Sea Hills among the Hadendowa initiated by a research programme following a devastating drought and famine period. I arrived in a community that was destabilized and marginalized. The Hadendowa had previously only been the subject of research to a limited extent, concerning the gender system and women in particular. Local ideology prevents male strangers entering the realms of women and previous studies had not been carried out by female researchers. This chapter is based on my PhD dissertation (Vågenes 1998) but only a fraction of the ethnographic details and analytical arguments presented there can be included in this text, with the danger that some of the arguments may appear partial.

Gender and Power

Power is imbued in a social relation where one of the parties has more ability to act or to influence events than other parties. A conventional understanding of power is, simply explained, that A has power over B to the degree that A can force B to do something B otherwise would not do (Bjørge 2009). In this understanding, a person holds power more or less by virtue of position, age, gender, title and so on.

However, theorists such as Foucault and Gramsci have changed our conceptions of power. Foucault (1972) pointed out that knowledge is critical in the constitution of power. Power is dynamic and always at stake in negotiations between people in social relations (Katz 2009). Discourses are critical in the construction of power. The Foucauldian perspective sees even the subordinate party as actively participating in reproducing power. Gramsci's perception of hegemony similarly takes a dynamic, constructivist perspective, and he emphasizes that power is a dialectical interaction between the subordinate and dominant parties (Gramsci et al. 1971, Macleod 1991). Power exists because of inequality between social actors in a relation and to understand the inequality in power in a discursive perspective it is necessary to take into consideration the continuous creation of inequalities that takes place in social practice and relations between individuals.

**Figures 7.2 and 7.3   A religious festival in Sinkat town with segregated
                           attendance of men (above) and women (below)**
                           Photos: Vibeke Vågenes, 1989

Power takes many forms and can be analysed from various perspectives. Pierre Bourdieu directs our attention to capital and the volume of capital an individual holds in the social field in question (Bourdieu 1977, Bourdieu and Thompson 1991). Capital may take the form of economic, social, cultural and symbolic capital, of which the latter is of great relevance to the case in question here. Women and men engage in symbolic capital in different, yet interdependent arenas. To a large extent this game centres on honour and shame, both individual and collective. From the female side it is to a large extent a matter of managing family honour as well as negotiating the reputation of the husband. However, women also negotiate by addressing the boundaries between honour and shame according to the norms and ideals of an honourable Hadendowa woman. Hence, their individual honour, or symbolic capital, is also at stake, but in this game they may harm a man's symbolic capital.

The following account shows how men are rendered vulnerable because of gender segregation. Because a woman may affect the reputation of her husband in social arenas where he is without direct access, he may be compelled to follow his wife's requirements if he is to secure his symbolic capital.

Research on women in Africa and the Middle East has revealed examples of their indirect and informal powers, bargaining tactics and strategic manoeuvres (Kenyon 1991, Macleod 1991, Mazawi 2006 [2002]). In societies typically publicly dominated by male power, it seems that women exercise considerable influence in matters of family affairs. Women in many cultures may, however, be almost invisible in domains of political and economic arenas of public life. This is without doubt the case of the Hadendowa. Even though women are not represented in public arenas, they may hold power in indirect ways – female influence may be disguised in publicly-agreed ideals of the submissive and honourable woman.

Somewhat in conflict with the apparent lack of feminine power, some studies point out that women themselves are frequently opposed to reforms and modernity, which could modify their subordinate position and eventually open up more public participation for women. Women may fiercely defend the traditional values of men as leaders of family and community and women as dependents. To the outsider, especially the Western observer, this may appear as a paradox as participation of women in public life and decision-making is among the central elements of female empowerment strategies. In discourses concerning modernity it may be tricky to acknowledge the seeming paradox that women are disempowered yet still empowered.

Material Capital

Hadendowa territory is ecologically marginal: mainly semi-desert with great variation in precipitation both in space and time. Water is scarce, and most of their region receives less than 200 mm rainfall annually (Manger 1996). Generally, Hadendowa families live in absolute poverty and during the last two decades of the twentieth century drought and famine have led to massive loss of herds. Migration to urban centres followed in the wake of calamities, by both labour migrants and families who settled where it was possible to find alternative sources of income.

On marriage, the groom gives his wife the material for the tent where they will live; this is part of the dowry. The tent is erected close to her mother's tent. At a time her parents find suitable, or when they think she is ready to consummate the marriage, the husband will move in together with the girl but the tent remains her property. Only the woman may dismantle and erect her tent, giving her some power in the marriage. The groom is also obliged to present other gifts to his coming wife, clothes and jewellery being the most expensive. In addition a woman receives animals (mostly sheep and goats) from her parents: This represents her security. Normally she will entrust her animals to her husband, but in the case of conflict between the spouses she can demand her animals (or an equal number of animals) back. Sometimes a wife asks another relative to manage her herd.

Livestock is owned by both men and women. Under normal circumstances, women are not economically active and men totally dominate trade and regular employment. Many women engage in handicraft production but for selling their products they normally have to depend on children or other relatives while other petty trading in the neighbourhood occurs on a small scale. However, a woman basically relies on her husband for sustenance and a man is obliged to provide for his female dependents. This is essential for a man's honour, an honour code that is clearly expressed in the community.

In times of hardship, or if disease or accidents affect a man, he may be rendered unable to provide for his wife (or wives), children and other dependents. If the family of a man is hungry and poorly dressed, this may be detrimental to his honour. Honour, or the reputation of a man, is crucial to his symbolic capital. Lack of economic capital can produce a dramatic reduction in a man's symbolic capital, and he is highly vulnerable when his ability to provide sustenance for his family is weakened.

Social Capital – The Power in Relationships

To the Western eye, the Hadendowa may seem extraordinarily close to their parents. A man seeks advice from his mother throughout his life, and she remains an emotional support for him whereby he can express his problems in secrecy. In many cases, wives feel that their husbands care more for their mothers than for marital partners. This is a common source of jealousy and marital conflict. Relations between siblings may also be very close and affectionate, sometimes leading to conflicts over scarce resources such as when a man gives money, food or other support to his sister when she needs it rather than to his wife. These are examples of relationships that are described as 'close'; being close is significant. These bonds are emotional and intimate, representing space where personal problems may be discussed – one can always rely on support from people to whom one is close and in a culture where much is kept away from public space, these bonds are important communication channels. Also, in times of hardship, when rituals such as funerals or circumcision are to be celebrated, or when a man wants to get married, economic solidarity is crucial. Support is most likely to come from the close family, and managing social relations to the family is crucial for every individual, which is an argument used for the clear preference for close cousin marriage. The network of close relations is also needed to gain support if a man or woman is in some kind of conflict or in other ways needs to mobilize people around them. The marriage bond is not a close relationship, although the spouses may grow closer as they become elderly.

Many issues are subject to negotiation. Gender issues may pertain to marriage, place of living and general sustenance, and in order to conduct successful negotiations it is crucial to mobilize support among friends and relatives. For a woman it is vital to be able to use social relations to put pressure behind her demands on her husband and his family.

Women rely on support from their networks when they experience conflicts in their marriage. Divorce is relatively common, although if the wife is the one who is opting for divorce she depends on support from her father, brothers and other male relatives. However, female relatives also influence decisions by using their networks.

'Women create men', it is said, with reference to their socializing role. Women are seen as central in managing and maintaining kinship relations; if a woman fails to show up at a wedding, mourning or circumcision celebration, the clan relationships are questioned, and the solidarity that the Hadendowa regard as important for the survival of the group may appear to be weakened. However, responsibility for more distant family or clan relations tends to be entrusted to the men. Thus, women are essential in maintaining the internal affairs of the family groups. They are important in the functioning and reputation of the group.

Symbolic Capital and Gender Ideals

The Hadendowa gendered individual is created throughout life by the application of several signs in ceremonies and daily life. The baby's gender is demonstrated from birth. If the baby is a boy, the placenta is carried out to a large, remote, tree where it is hung from a branch, while participating women sing about the fierce warrior he will become. In the case of a girl, the placenta is silently buried in the ground of the tent owned by the matrilineal grandmother. These ritualized acts, all performed by groups of women, signify the spatial segregation of the boy and the girl to an outside and an inside sphere.

From the time when a boy is circumcised around the age of seven or eight, he will not be allowed to sleep inside the tent with his parents (or mother), sisters and minor siblings. From then on he will increasingly be given duties of looking after livestock. Most children attend school for some years, at least in urban centres; however, there is resistance to the education of girls, although attitudes towards female education seem to be changing. Girls help with household chores but their spatial freedom of movement is not severely restricted until the age of puberty, from which age a girl dresses in a *fouta* (the dress of the woman) and her behaviour becomes more like an adult's. This may be the age when some girls are married and start living with their husband.

The new bride is the ultimate symbol of ideal feminine shyness and concealed beauty. During the days of the marriage celebration, she is hidden and concealed all the time. When the couple move in together, she will still spend most of her time in her mother's tent and with her close female friends. At their first meetings she is supposed to resist her husband's physical approaches; her mouth is closed, and her eyes turned down.

A woman avoids public spaces and spends her time with female friends and relatives in work and leisure; only her husband, brothers and sons may enter her tent, while neighbouring women may come and go all day. Thus, there is a strong segregation between male and female space. Women manipulate these ideals in the sense that following the norms and traditions allows them to increase their symbolic capital through their reputation as a *dayit* woman (one who is honourable, following the traditional ideals). Their behaviour and reputation also reflect upon the family and clan, hence the pressure is strong to behave according to these ideals.

A man is ideally a strong, independent and fierce warrior and often carries his sword in everyday, public settings. Hadendowa men are supposed to defend the land and honour of their clan and family, as well as their dependents. An honourable man does not depend on others; however, he is expected to be generous and show self-restraint. According to Manger (1996), management of male honour centres on being a 'responsible man' (taking care of his family, controlling his women etc.), and becoming the father of many children. Producing numerous offspring may lead him to become the founder of a new *diwab* (clan lineage), and hence his name will live on.

Male time is spent in company with male companions. However, a man's relationships to his parents and siblings remain very strong throughout his life. It seems that the tie to the mother is especially emotional and close, representing a space for warmth and support, although he will never speak about his mother or sisters to others.

With this layout of gendered social spaces, it seems clear that to break the segregation is a powerful statement. Women who are dissatisfied with their spouses may not be able to express their dismay except to their mother, sisters and close friends. They may, however, use spatial movements to express their views; a woman may visit her mother or sister for long periods. This is typically found during conflicts over the place to live after some time of married life, when the husband demands that they move to his relatives. By being away from her household chores, the wife tells the community that her husband is not able to control her and that she is not happy with him. His reputation is harmed; either he is a poor husband or he is just unable to control his wife.

Hadendowa Gender Segregation and the Marital Relationship

More than many other ethnic groups, the Hadendowa strive to uphold gender segregation in everyday life. From the time a baby is born, gender differences are marked and as the child grows older, segregation is gradually introduced and clothing, hairstyles, jewellery, range of movements and social relations change according to age and social status.

Segregation is strong by the time the child reaches puberty. Many girls are already married at this stage, although they continue to live with their parents until puberty. Marital relations are arranged by male members of the families, and in most cases close agnatic ties are the preferred bond. Cousin marriage is predominant in the Middle East and Muslim parts of Africa (Holy 1989) – solidarity and lifelong security of the parents are believed to be among the reasons behind this choice.

Asking a wife about her husband is difficult, as it is not considered *dayit* (honourable, estimable) to speak about the husband. A newly-wed woman will simply draw her *fouta* (dress) across her lower face, blush and become silent when asked about her husband. In this way she literally closes her body. If she talks about her husband, people might believe she is in love or likes him too much – this is shameful (*aib*) – but as the couple grow older, the woman will talk more freely about her husband. Likewise a man will never mention his wife to anyone, except maybe to his mother or other close relatives (parents, siblings). He will never utter her name in public.

During the first phase of cohabitation, the spouses ideally only spend the night together. Daytime is spent apart; she with her female relatives and neighbours; he, with male neighbours and relatives. Later, when children are born, the husband spends most of his time outside the camp. After drinking coffee around sunrise, a man leaves the camp and spends his day at work, either herding livestock, going to markets or to a workplace elsewhere.

Gender segregation is obvious in social activity and spatial arrangement. The camp is female space during the day; women carry out household chores in the company of female relatives and neighbours, no men are present, and women and children visit each other's houses throughout the day. On the other hand, the market, truck stops and central town streets are male spaces. Even village or town houses are arranged so that women occupy the back of the house and yard, while the male rooms and yard face the main street. In rural camps there is normally a communal house for men, where they may spend the day in company with other men. This is also the place for coffee and meals for many men.

The men may come home for a meal or two during the day and if this happens, the husband will announce his approach to allow visiting women to leave his wife's tent before he enters. When the husband enters their home, his wife becomes silent and in most cases covers her head and mouth – she will even avoid looking at him. This is all in accordance with the highly held ideal of female shyness; the closed, virtuous wife obediently follows her husband's directions, takes care of the children and looks after the household. The contrast is striking between the talkative, joking woman in female company, and the closed, silent wife in company with her husband.

The 'closed woman' is symbolically significant in this context. A woman is closed by genital mutilation; her body is closed in by the *fouta*, a big piece of cloth wrapped around body and head. Women dress in bright colours of red, yellow, orange or pink while the male colour is white, with only a dark or blue vest on top of the trousers and *jallabyia* (long shirt worn by men). There is a complex and striking set of symbols signifying gender (Vågenes 1998) and in this connection body posture can also be mentioned. The male body is erect, strong and upwards in orientation; the female body, on the other hand, is ideally closed (with symbolic resemblance to a container) and soft. Except when in solely female company, a woman will guard her body movements from too much gesturing. She will control and restrict her body, directing her looks downwards, and will be silent when in company with a man who is not her brother or perhaps her father. Very few women will eat or drink in front of their husband – a woman serves her husband and older sons first, and later eats in company with the younger children. Men also consume meals outside their home in the company of male friends and relatives. Here is a quote from an elderly woman:

When I was *kwaddat* [a girl of pre-menstruation age] my mother gave my father the food and I wanted to join my father. My mother prevented me from going with him. She said: 'Don't eat with men' and she held my hand tight. I cried and my father said: 'Leave her, she can go with me, what will the men do to her?' Thus, I was allowed to eat with men. This is the ignorance of the child.

The strongly gendered time–space of life in the camps and villages regulates how women posit their bodies, how they speak and what they will speak about. In female company women share news and gossip, stories and information but most of what women speak about is guarded from male hearing and the strictly gendered division is evident in communication and themes for conversations. Among women, news about other women is discussed, involving issues concerning how women are treated by their husbands, brothers or sons. On the other hand, male spaces, like the streets and the market, are dominated by other kinds of news and information. Information does, however, cross the gendered borderlines by means of close relationships, e.g. those between brothers and sisters, or mothers and sons.

The *Zar* Patient

When asked, a woman will say that the husband is the one who takes decisions in the family. Nevertheless, women have their way in many cases, even though it is sometimes clearly in contradiction with male interests or entails great difficulties. One such case will be referred to here. It is a relatively common case, where a woman finds herself in a marriage where she is not content and in this case it was a young, beautiful woman married to a much older man. He was said to be disabled, and unable to generate a proper income and give his wife a good life.

This young wife was diagnosed with a *zar* problem (*zar* is a widespread belief in spirits in Sudan). The spirits hold specific identities, and they may inhabit the bodies of women from where they cause several problems – this young wife's problem was hiccups. She could not stop hiccupping and as this was diagnosed as a *zar* problem, the spirit had to be satisfied according to the prescriptions from the *zar sheik*. In this case the *zar* demanded a party and a gold necklace for the wife – in most cases the spirit demands that a party be held, with food for the guests, drumming, dancing and rented paraphernalia. A *zar* party is something most husbands dread because it involves considerable expense and the husband is obliged to raise the funds needed. If the costs supersede his available resources, he will have to ask his relatives for support. In this case, women told me that the husband had been forced to ask her relatives for help to cover the expenses which was considered a severe loss of honour for the husband.

The women around the *zar* patient vividly discussed her case. Some believed there were no other causes than the *zar* taking possession of her body, while others doubted the woman. They claimed that the spirit might be real, but that the whole problem was caused by the wife wanting a divorce from her husband. In this case, the power was shifted from the husband and the family who had arranged the marriage to the woman herself, who was 'possessed'. The husband had lost his power in the sense that he appeared a weak man in everybody's eyes and so neighbours and participants in the event agreed that she would have her divorce.

Muted Female Power?

This case is not unique. It is one of the ways women manipulate their situation by means of the segregation between genders. Men fear what takes place in the female spheres. In the female spaces, women are more outspoken than men, who will try to live up to ideals of the proud, silent and independent man, rendering men vulnerable to gossip and rumours.

Several researchers have remarked that women are keepers of the group's honour and are heavily involved in preservation of group unity; this seems to be so in the case of the Hadendowa, where crucial among female responsibilities is their role in maintaining bonds between relatives. Considering the preference for close marriages within clans, individuals may be related in several ways. Kinship is the basis for support when someone is in need due to some calamity. It is usual to hear women claiming they will always support close relatives because 'we drink from the same breast', pointing to the strongest of all bonds – that between mothers and children.

When there is a family occasion, women travel to be with their close female relatives. Or they may travel to stay with sisters, mothers and cousins if they feel the need for extra comfort. There is a line between maintaining the needs of the wife and the family, and disobeying the will of the husband concerning his wife's absence from their home. Women use absence to impose their will. For example, there are numerous conflicts over where to live – with her or his parents. According to the norm and tradition, the couple should move to his family after approximately one to two years of married life. Nevertheless, it is more common to find matrilineal relatives as neighbours (Vågenes 1998). Likewise there are normally conflicts involved when a man wants to take a second wife. Polygamy, although not frequent, is still found among the Hadendowa. Some people even feel sorry for the polygamist, because he can expect trouble from his first wife. In many cases, she will put pressure on the husband and if she manages to do so without disrupting her personal symbolic capital, her power increases. One female strategy may be to leave the husband alone for long periods; she may move to her mother or sister. It is unlikely that a husband will complain directly to his in-laws in such a situation, as this is considered not honourable.

Hence, there is much power vested in the gender segregation of space, time, labour and networks. The power of female spaces depends on the exclusion of men. If the borderlines between gendered discourses and social spaces are moved or broken, there is reason to believe that the power of women, and the ways they build capital, will be altered. Possibly such a change may lead to marginalization of Hadendowa women.

This may strike someone coming from a Western cultural sphere as contradictory. We have a strong normative tradition of equal rights and possibilities for women and men. This is reflected in our development ideologies, which currently revolve around ideas of empowerment of women in the Global South (Nederveen Pieterse 2010). Empowerment is in the mainstream of the majority of large development programmes. Seeing the gendered power balance from the viewpoint of Hadendowa women is a reminder of the need to be sensitive to local difference.

Social and cultural change is also taking place in the Hadendowa communities. In Sudan change occurs very much in the shape of Islam and education. One of the cultural traits attacked by Islamist and educationalist advocates is *zar,* which is regarded as backward and contrary to proper Islam. Furthermore, women are regarded as not very good Muslims due to some of their other practices, such as specific female hairstyles that some preachers argue prevent proper ritual washing before prayer (Vågenes 1998).

Male symbolic capital is dependent on many elements, and essential among these is the behaviour of a man's female dependents (wife, daughters, sisters and mother). He needs to control them. They may not want to be controlled, and may use such measures as absence from the husband (for example staying for long periods with his mother-in-law) or *zar* problems (although *zar* is also seen as a 'real' problem caused by spirits). In cases like these, women's actions cause discussion and rumours. Women's claims are often indirect in the sense that women rarely address their husbands directly. Unhappy, poorly dressed or otherwise disturbed women – operating in female spaces – transmit damage to the symbolic capital of men. Because men are excluded from female spaces, they may find themselves in a difficult situation when family members and neighbours talk about their marital problems or their lack of ability as responsible men. Because men have little access to female spaces, they remain vulnerable and subject to female muted power. Gender segregation may appear as backward and against modern development, but there are some repercussions of this system that benefit the situation of women. Gender segregation needs to be carefully considered in each case.

References

Attanapola, C. 2008. Women's Empowerment for Promoting Health: Stories of Migrant Women Workers in Katunayake Export-Processing Zone, Sri Lanka. *Norsk Geografisk Tidsskrift–Norwegian Journal of Geography*, 62, 1–8.

Bjørge, E. 2009. Michel Foucaults bidrag til forståelsen av 'makt'. *Jussens venner*, 05/2009, 302–14.

Bourdieu, P. 1977. *Outline of a Theory of Practice*. Cambridge: Cambridge University Press.

Bourdieu, P. and Thompson, J.B. 1991. *Language and Symbolic Power*. Cambridge: Polity Press.

Foucault, M. 1972. *The Archaeology of Knowledge and The Discourse on Language*. New York: Pantheon Books.

Gramsci, A., Hoare, Q. and Smith, G.N. 1971. *Selections from the Prison Notebooks*. London: Lawrence and Wishart.

Holt, P.M. and Daly, M.W. 1979. *The History of the Sudan: From the Coming of Islam to the Present Day*. London: Weidelfeld and Nicholson.

Holy, L. 1989. *Kinship, Honour and Solidarity: Cousin Marriage in the Middle East*. Manchester: Manchester University Press.

Katz, C. 2009. Social Systems: Thinking About Society, Identity, Power and Resistance, in *Key Concepts in Geography*. Second Edition, edited by N.J. Clifford. Los Angeles: Sage.

Kenyon, S.M. 1991. *Five Women of Sennar: Culture and Change in Central Sudan*. Oxford: Clarendon Press.

Lund, R. 2002. Gendered Spaces – Socio-Spatial Relations of Self-Employed Women in Craft Production, Orissa, India. *Norsk Geografisk Tidsskrift– Norwegian Journal of Geography*, 56, 207–18.

Macleod, A.E. 1991. *Accommodating Protest: Working Women, the New Veiling and Change in Cairo*. New York: Colombia University Press.

Manger, L. (ed.) 1996. *Survival on Meagre Resources: Hadendowa Pastoralism in the Red Sea Hills*. Uppsala: Nordiska Afrikainstitutet.

Mazawi, A.E. 2006 [2002]. Educational Expansion and the Mediation of Discontent: The Cultural Politics of Schooling in the Arab States, in *Education, Globalization and Social Change*, edited by H. Lauder, P. Brown, J.-A. Dillabaugh and A.H. Halsey. Oxford: Oxford University Press, 980–90.

Nederveen Pieterse, J. 2010. *Development Theory: Deconstructions/Reconstructions*. Los Angeles, Calif.: Sage.

Syed, J. 2010. Reconstructing Gender Empowerment. *Womens Studies International Forum*, 33, 283–94.

Vantini, F. G. 1975. *Oriental Sources Concerning the Nubia.* Heidelberg and Warsaw: Polish Academy of Sciences and Heidelberger Akademie der Wisschenschaften.

Vågenes, V. 1998. *Women of Interior, Men of Exterior: The Gender Order of Hadendowa Nomads, Red Sea Hills, Sudan.* Dissertation for the Dr.Polit. degree. Bergen: University of Bergen.

Chapter 8

Gender, Politics and Development in Accra, Ghana

Samuel Agyei-Mensah and Charlotte Wrigley-Asante

Introduction

African towns and cities are in the midst of social and economic change, which has attracted scholarly attention from a diverse range of disciplines. Some of these studies have focused on particular cities, as seen in the works of Stewart (1999) on the changing urban morphology of Cairo in Egypt, Gaule (2005) on the changing spatial patterns of Johannesburg in South Africa, and Grant (2009) on the impact of globalization on Accra in Ghana. Simone (2004) has studied strategies deployed to control, shape and make sense of urban life in several African cities: Pikine, a suburb of Dakar in Senegal; Winterveld, a neighborhood on the edge of Pretoria in South Africa; Douala in Cameroon; as well as the strategies of Africans seeking work in Jeddah, Saudi Arabia.

While such works have greatly expanded our knowledge of the dynamics of urban change in African towns and elsewhere, few have specifically addressed the linkage between gender relations, politics and development. This chapter provides an overview of gender relations, power structures and development issues in Accra's history since the precolonial period. Accra has a long history of growth and change dating back to this time. A wealth of information on Accra's past is found in a number of studies. This chapter contributes to existing scholarship by highlighting issues of gender, politics and development in a chronological narrative. The chapter seeks to identify the relationship between different political ideologies and how these have had an impact on different facets of development over time. We examine how global and local forces interact to shape the socioeconomic and urban landscape. The chapter resonates with some of the historical scholarship of Ragnhild Lund on gender and development, particularly in China and Malaysia (Lie and Lund 2005).

As we move through the urban socioeconomic history of Accra, we need to take cognizance of the divergent political economic regimes that have had an impact on the socioeconomic and urban landscape. Four periods can be recognized: precolonial, colonial, postindependence and postadjustment. The main focal point of the precolonial period was the contact with Europeans and how this affected the fortunes of Accra, especially through trade. The colonial period, which started in 1874, was a critical phase in the history of Accra. This gave legitimacy to European rule and had a significant impact culturally in administration and on the way of life of the people of Accra. In the immediate postindependence period after 1957, the leadership of the country embarked on a socialist philosophy of development. The postadjustment period starting in 1983 marked a dramatic shift to a neoliberal economic regime with a withdrawal of state control over resources and the involvement of the private sector in economic development. The impact of globalization and middle class ideologies has become more profound on the urban landscape.

Precolonial and Colonial Accra

There are differing historical accounts of when Accra was first settled. The city developed not from one centre but from a series of contiguous settlements formed at different times by different peoples, beginning in the fifteenth century. The Ga people settled there at this time, coming from Ayawaso, 16 km to the north. Asere, Abola and Gbese are thought to be the oldest and most Ga-influenced of the quarters. Otublohum was originally settled by Akwamu and Denkyera from the eastern region of Ghana. These four together make up what is called Ussher Town, the area the Dutch claimed jurisdiction over in the seventeenth century. Three other quarters are Alata or Nleshi, Sempe and Akanmadze. These three are commonly referred to as James Town, which was the area of British jurisdiction in Accra. Alata was settled by Nigerian workers imported to build a European fort. Of all the seven quarters, Asere is by far the largest in population and area. All seven make up what is now known as central Accra and all have chiefs called *mantsemei* (plural of *mantse*, meaning 'father or chief of the area') (Robertson 1984).

The growth of Accra was stimulated by the arrival of the Europeans. The Portuguese built a small fort at Accra in 1482. In the seventeenth century, the British, Dutch, Swedes and Danes followed suit and established their spheres of interest. Initially, Accra was not the most prominent trading centre but rather the ports at Ada and Prampram along with the inland centres of Dodowa and Akuse. Accra, however, became an important centre for the slave trade and trade in other commodities with the Europeans, who had built the James and Ussher forts nearby. Thus Accra became an important facilitator for European commercial activities. The Portuguese, Swedes, Dutch, French, British and Danes had built forts in Accra by the seventeenth century. In the 1850s, Denmark sold Christianborg castle and others to the British. After a long period of hostility from the Asante people of central Ghana, the British attacked the Asante capital of Kumasi and officially declared Ghana a crown colony. Even though the original settlements have grown into a continuous whole, many indigenous settlements in the city preserve their customs and traces of their different origins.

Socioeconomic Change in Colonial Times

Before the arrival of the Portuguese, Accra was probably participating in lateral coastal trade. The coastal plain around Accra, which is much drier than most of the West African coast, was a suitable environment for raising livestock. Fish and salt extracted from sea water were traded inland. The coming of the Europeans added more commodities to the repertoire of the Ga traders. By the middle of the seventeenth century, Accra was the major centre for the gold trade on the West African coast. The Gas were raising cattle for sale and taking European goods, salt and fish to the interior to exchange for gold and slaves. Thus until the second half of the nineteenth century, most men were farmers or fishermen and traders. Then a number of wealthy Ga merchants became involved in agricultural enterprises and owned coffee plantations.

Men rather than women most often possessed high educational qualifications, which helped them to get ahead in the European trade. The activities of the Basel Mission Society in 1857, which taught skills such as carpentry, masonry and shoemaking, further added to the pool of European-type skilled labourers. By the 1870s, Accra had become the principal source of skilled labour on the West African coast. This also gave men better access to government jobs. However, while educational and occupational options changed for men, women did not receive the same opportunities. Between 1860 and 1952, there occurred a shift for many Ga men from a corporate kin to a capitalist mode of production, in which they played the role of wage earners working for expatriate administrators and businessmen. Their necessity for and capability of recruiting labour through the lineage system diminished greatly. Meanwhile the female hierarchy remained the chief mechanism for women to secure labour because women were still involved in labour-intensive activities such as fish-processing (Robertson 1984: 15). The beginning of colonial rule in 1874 reduced the power of the male elders by stripping them of much of their political power, while the women elders turned their attention to the increased trading opportunities for middle-women.

In 1877, Accra became the administrative capital of the British Gold Coast, replacing Cape Coast. Until this time, the settlement of Accra was confined between the Ussher fort to the East and the Korle lagoon to the West. One of the most influential decisions in the history of the city was of the building of the Accra–Kumasi railway in 1908. This connected Accra, the major port at that time, with Ghana's main cocoa producing regions. The railway was completed in 1923, and by 1924 cocoa was Ghana's largest export. Accra was the main exporter of cocoa until 1928 and this was one of the main reasons for its rapid growth.

Because of their position economically and politically, the Gas had strong contacts with many other peoples over a long period of time and this resulted in many intermarriages. In the process of lateral migration along the West African coast, some families of Brazilian, Sierra Leonean and Nigerian origin settled in Accra and became Ga. By the twentieth century many of them had long been exposed to strong European influence in politics and were open to outside influences, having intermarried for generations with other peoples. They maintained, however, a strong and distinct cultural identity and language (Robertson 1984).

Population Growth, Education and Health

In the late nineteenth century, the British took rudimentary sanitary measures in Accra such as building public latrines and sewerage ditches, and providing potable water. The construction of the Accra waterworks began at Weija in 1904 and was completed in 1914. In the period before World War I, other infrastructural development included the installation of streetlights, construction of cemeteries and laying-out of streets. By 1930, electricity was common in many homes in Accra. The significant improvements in the provision of amenities raised the standard of living and of healthcare.

   This socioeconomic and infrastructural development led to increased immigration to Accra from other parts of the Gold Coast and from elsewhere in West Africa, leading to a change in the population structure from an indigenous Ga-dominated town to an urban complex with many ethnic groups and foreign migrants (Agyei-Mensah and de-Graft Aikins 2010). In 1911, the population of Accra was 18,574, increasing to 61,558 in 1931 and 135,926 by 1948 (Acquah 1972: 31). Population growth was aided by the increased provision of European-type health care facilities, as well as by an awakened concern of the colonial officials for sanitary conditions, especially in Central Accra. The first hospital, Korle Bu, was opened in 1923, and the Accra Maternity Hospital and Princess Marie Louise Hospital for children followed shortly after. By 1938, some 35,000 outpatients were treated yearly at these three hospitals. Three other hospitals were the Achimota and Ridge Hospitals, and the Accra Mental Hospital. However, traditional medicine was also practised widely (Acquah 1972).

   In the colonial and postcolonial eras, men and women have occupied distinct positions in the labour force. The 1948 census gave the number of employed females in Accra as 18,672 as against 40,476 males. Women provided 6 per cent of all persons in the public services and 73 per cent of those in commerce. Of the 18,672 employed women, 16,526 (89 per cent) were traders. They formed the greatest number of persons in any occupational group and illustrate the importance of buying and selling over all other economic activities in the town. Educated women were employed as teachers, nurses, telephonists, clerks, typists and shop assistants.

   Housing development in Accra during the colonial period benefited mainly British civil servants. Little was done to improve the housing conditions of the majority of the population except for a few estates built in response to an earthquake which occurred in Accra in 1939. A few thousand dwellings were built for veterans and junior civil servants in the late 1940s and early 1950s. Since independence in 1957, the State Housing Corporation has provided houses on behalf of the Ghana government (Tipple and Korboe 1998).

Postindependence Period 1957 to 1983

Industry and Education

The period between 1960 and 1970 saw rapid industrialization and expansion in the manufacturing and commercial sectors in some major areas within the metropolis in accordance with the socialist ideology of the then president, Kwame Nkrumah. The rapid industrialization contributed to high immigration to Accra and a resultant high population growth rate between the 1960 and 1970 censuses. Industrial and educational policies initiated after independence improved the socioeconomic position of the people, especially women. Educational expansion involved the establishment of new institutions and the absorption of private schools into the public system. In 1961, the Nkrumah government initiated fee-free compulsory education, regardless of gender, for primary and middle school levels (Pellow 1977: 112). These policies, combined with a change in attitude on the part of women, greatly improved female school attendance. Improvements in educational levels can be discerned from the changes in illiteracy rates over the years. The overall illiteracy rate in Accra fell from 47.3 per cent in 1960 to 31.6 per cent in 1970. However, the illiteracy rate for females was 60.6 per cent in 1960. By 1970 it had fallen to 41.9 and by 1984 to 24.9 per cent. The corresponding figures for males were 35.9 per cent in 1960, 21.7 in 1970 and 12.2 per cent in 1984 (Agyei-Mensah 1997).

With the development of Tema as a new port and industrial hub of the country, the Accra and Tema region became the industrial heartland of Ghana. Thus, Accra became not only the political centre of the country but also its cultural, educational, financial and business centre. There was a strong presence of small-scale industries along the lower Odaw Ring Road West. Concentration on capital-intensive development led to the neglect of agriculture and small business (Agyei-Mensah 1997).

Socioeconomic Change

Developments in education and industry greatly affected the socioeconomic position of women who began to be drawn more into the cash economy by trading in imported goods made available by the industrial policies. These socioeconomic changes affected the institution of marriage, which substantially changed in form. Some of these changes included increasing infidelity, broken marriages, female-headed households and a relatively large number of unmarried women (Robertson 1984, Songsore and McGranahan 1996). According to Robertson (1984), most of the changes in marital forms can be traced to the interwar years and show the declining power of the male elders of the lineages in controlling marriage, a result of their reduced political power. Another common explanation for the rise in marital instability was the economic independence of women. Thus marital instability often arose in situations where women had independent access to money and males found themselves in marginal positions. The fragility of marriages was not only a result of the economic changes, but also a symptom of the shift away from conjugal economic cooperation due to the increasing differentiation of men's and women's jobs. During the period when fishing and farming were the most important occupations of Ga men, economic cooperation between spouses, or male and female relatives, was more the rule than the exception.

The postindependence era marked a period of major transformation in the health delivery system, particularly the growth of public and private clinics. Several government polyclinics were established, and there was growth in the private medical industry. Records from the office of the Medical and Dental Council show that only ten private clinics and hospitals were registered between 1960 and 1969 in Accra. By the end of the 1970s 29 per cent of the clinics and hospitals were private. Accra benefited from these improvements because of its position as the seat of government. Increasing numbers of health workers were trained, most importantly through the establishment of the University of Ghana Medical School at Korle Bu in 1978. The postindependence era also saw increased efforts to eradicate endemic and epidemic diseases as well as to improve infrastructural services. Major water supply schemes were established through the Kpong waterworks (80 km east of Accra) and Weija waterworks (15 km west of Accra), providing long-term health benefits in the city.

Housing Development

In the late 1950s and early 1960s, the housing industry in Ghana officially commenced in Accra, the national capital. Provision of housing was central in the 1960–65 National Development Plan. Two main state bodies, the State Housing Corporation (SHC) and the Tema Development Corporation (TDC), were formed to address housing issues. As part of a major industrialization drive, the TDC was created with the specific purpose of establishing residential units in the rapidly growing Tema area, where a second sea port had been constructed to serve Accra and the eastern part of the country (Bank of Ghana 2007). The SHC worked in Accra and other areas across the country.

The 1970s saw a period of very poor economic performance for Ghana and much of the developing world due largely to the energy crisis and oil price shocks. This recession brought price controls and sporadic building by worried developers. At this point various individuals in Accra built houses around areas like Kaneshie, where they could access markets and other amenities. Such individualized, unplanned buildings collectively created new communities like Kwashieman and Abeka in Accra. Old Fadama was settled by people from northern Ghana, giving birth to the new phenomenon of slums (*The Statesman* 2007).

Population Growth

Because of Accra's growing position as an industrial, administrative and commercial centre, the town attracted large numbers of people who came to search for jobs and experience urban life. The population of Accra was 364,719 in 1960 and increased to 617,415 in 1970 and 956,157 in 1984. In 1994, the population was estimated at 1,198,116. Accra's population grew at an annual rate of 5.1 per cent during 1960–1970 and 3.1 per cent during 1970–1984. The increase of population created severe housing problems (Austin-Tetteh 1972, Owusu 2008). According to the 1960 census, 60 per cent of the urban population of Ghana was living with three or more persons per room while 40 per cent were living with four persons or more. In the case of Accra, about 60 per cent of the population was living at a density of four persons or more per room.

Slower population growth during the period 1970–1984 and hence slower urban growth reflected the general economic decline. This was a reflection of the depressed state of the international economy as evidenced by an oil glut and declining earnings from commodity export markets as a result of falling prices of major export goods (Potts 1996).

The age structure of the population of Accra corresponded with the pyramid structure often seen in developing countries, with a broad base at age 0–14 compared to older age groups. A significant increase in the percentage of the population aged 45–64 could be seen in 1984 compared to the 1960 and 1970 censuses. A slight increase in the percentage of those aged 65+ suggested an ageing of the population. Significant changes were also observed with respect to fertility trends. In 1960 the child to woman ratio was 777 per 1,000 women. In 1970 it declined to 705 per 1,000 women, and in 1984 it went further down to 554 per 1,000 women (Agyei-Mensah 1997).

Thus by the early 1980s, Accra was experiencing a general economic decline similar to other African cities such as Lagos and Freetown, attributed to the depressed state of the international economy and a decline in earnings from the commodity and export markets (Konadu-Agyemang 2000). This led to serious developmental crises as export earnings fell by 52 per cent while domestic savings and investments dropped from 12 per cent of GDP to almost zero. An unprecedented number of both Ghanaian professionals from the educational, health and public sector as well as non-professionals left the country to seek 'greener pastures' elsewhere.

1983 to Present

Structural Adjustment Programmes, Globalization Processes, Gender and Development

The economic difficulties the country faced in the late 1970s and early 1980s led the then ruling government to invite the International Monetary Fund (IMF) and the

World Bank to provide solutions to the crisis. The IMF and World Bank proposed Structural Adjustment Programmes (SAPs). These programmes were in many ways similar to those that had been prescribed for other Third World countries. The package included: currency devaluation; reducing inflation; downsizing the public service; major cutbacks in government expenditure on education, health and welfare; financial reforms; privatization of public enterprises; export promotion; and other policies to enhance economic growth (Agyei-Mensah and de-Graft Aikins 2010).

The initial years of adjustment in Ghana involving macroeconomic stabilization achieved some success. Output began to revive, with an annual growth of about 5 per cent between 1984 and 1989. Export volume increased as a result of the rehabilitation of the mineral sector, cocoa and timber and the construction of roads. From 1986 the budget started showing surpluses, although these were partly illusory since they included increased foreign aid. Expansion of domestic credit was severely curtailed and, although external inflows kept the growth of aggregate money supply high, inflation was brought down to an annual average of about 25 per cent (Asenso-Okyere et al. 1992). Thus at the macrolevel the economic recovery programme (ERP) and structural adjustment programme (SAP) are said to have had a positive impact on the economy.

On the other hand, the impact of the adjustment programme was not positive at the microlevel. The SAP came with social costs, which hit vulnerable groups, particularly women and children, among villagers, food crop farmers and urban informal sector workers, who together constitute the majority of Ghanaians. The introduction of fees in schools and hospitals and the withdrawal of subsidies prevented many in these groups from receiving these services (Ardayfio-Schandorf et al. 1995, Baden 1997). The elimination of subsidies on foodstuffs and essential services also affected vulnerable groups, especially women. Cuts in government subsidies led to rises in transportation costs and deterioration of the road system due to declining public investment (Baden 1993, Young 1993). The economic reforms led to a series of changes in the household structure, thereby intensifying alternative livelihood strategies. Ghanaian female traders working under adjustment policies in Accra and Kumasi reported a crowding of the sector with new entrants (Baden 1993, Overå 2007). These newcomers included men, some of whom had begun to move into areas of trade traditionally associated with females. Young men took up roles traditionally performed by women, thus further displacing them and increasing the burden of the already overburdened women, many of whom had become the breadwinners of their families (Baden 1993, Overå 2007).

One visible manifestation of postadjustment Accra is migration to the city from the country's three northern regions and impoverished regions of the south, such as the Central and Volta regions. Subsequently, there appears to be a growing trend of men moving into the informal economy and into traditionally female domains in the city as a result of growing youth unemployment, lack of employment opportunities in rural areas, low educational levels and migration from the rural areas (Overå 2007, Wrigley-Asante 2010). Following the economic reforms, street vending has become a highly visible feature within the landscape of Accra, with the majority of vendors being males who have migrated from other parts of the country (Asiedu and Agyei-Mensah 2008). With lack of employment opportunities and low educational levels, many of these men have limited options. This has serious consequences for their socioeconomic lives. A large number of young females have also migrated from the three northern regions of the country to the city of Accra to work as 'headporters' (*kayayie*) in order to acquire minimum assets for either better marriage prospects or greater economic stability (Awumbila and Ardayfio-Schandorf 2008).

With the introduction of the government's economic recovery programme, the divestiture policy and the liberal market economic policies initiated after 1983, Accra has been transformed into a much more modern city. Globalization has contributed significantly to the modernization process. Globalization, seen as complex economic, political, cultural and geographical processes involving mobility of capital, organization, ideas, discourses and people, is associated with changing experiences of time and space. It is also associated with the development of new communications technologies and the rise of the information society (Beall 2002, Heine and Thakur 2011). While globalization brings opportunities to improve and transform the wellbeing of people, there is increasing evidence that the process is creating an increasingly unequal world and having serious impacts on cultural values and traditions (Ghai 1997, Heine and Thakur 2011). One key challenge of globalization is its influence on Ghanaian cultural practices and norms, including changing fashions, eating habits and, recently, homosexual practices. With rapid globalization there have been increasingly rapid changes in the social, economic and political spheres of life. According to data from the Ghana Broadcasting Corporation, the use of mobile telephony and internet services has witnessed an astronomical increase with mobile phone subscriptions increasing from 90,000 in 2000 to almost 15 million by the end of 2009, the majority of subscribers being in the city of Accra.

Over the years, Accra has expanded its residential boundaries but maintained its central business district. This has increased distances between work and home for a growing number of people. At the same time preferences for Western products and lifestyles such as food, technology and language have increased. With the increasing number of both international and local restaurants, especially the American-style fast-food industry that has taken root in the city, the traditional patterns of family life such as sharing of meals at home have changed with increasing consumption of out-of-home meals. This provides evidence of a structurally mediated lifestyle shift. This is implicated in the rising prevalence of lifestyle diseases and risk factors such as hypertension, diabetes and obesity (Agyei-Mensah and de-Graft Aikins, 2010).

Accra also boasts of a number of three- and five-star hotels, which have improved the city's status for tourism. There have also been developments in the investment and financial sector with a number of new foreign-owned banks being established. The banking industry in Accra has entered a new era with the introduction of automated or autocash teller machines as well as internet banking. A stock exchange has been established, and controls on foreign exchange have been eased. These changes in the financial sector have opened up employment ventures for many people as well as attracting both local and foreign investors. There has also been a proliferation of credit and loan schemes in Accra to cater for the needs of both women and men, particularly in the informal sector (Wrigley-Asante 2008).

These socioeconomic and infrastructural developments have led to a change in the ethnic composition of Accra. According to the 2000 census, the Akans formed the largest group in the city, estimated at 42 per cent. They were followed by the Ga-Dangme (29 per cent), Ewe (14.8 per cent), Mole-Dagbon (5.6 per cent), and other minority ethnic groups (8.6 per cent). The proportion of the Ga-Dangme ethnic group in Accra is on the decline and contrasts sharply with earlier figures. For instance, in 1948 and 1960 the Ga-Dangme made up 51.6 per cent and 37.5 per cent respectively of Accra's population. The decline of the proportion of the indigenous Ga–Dangme ethnic group is not due to a decrease in their numbers, but is mainly a result of the large influx of other ethnic groups into the city (Agyei-Mensah and Owusu 2010).

This demographic change has had an impact on the quality of housing due to the development of squatter settlements and overcrowded housing conditions. With in-migration from rural areas, the city has expanded with no regard to zoning, giving rise to urban sprawl. The growth of the city has outstripped the rate of provision of services such as waste collection, potable water and electricity, resulting in slums. Some of the more severe problems of poor households in the city of Accra include inadequate potable water supply, unsanitary conditions, insect infestations, poor waste disposal and collection, and crowding. These environmental problems have a seriously negative health impact due to infectious and communicable diseases such as malaria, upper respiratory tract infection and diarrhea (Songsore et al. 2009).

Housing and the Physical Landscape

Another aspect of post-1980 Accra is the rapid transformation of the urban landscape. Shopping malls have been constructed at major locations within the city, the newest being the Accra Shopping Mall. The first shopping malls such as Koala were opened in the early 1980s. The prices these outlets charge are aligned with Western standards and lead to an exclusive selection of customers, clearly indicating the social groups for which these shopping malls are designed. The malls symbolize a fast-growing consumerist culture.

With the SAP and the new market-oriented approach, housing built by the TDC and SHC during the postindependence era was converted into limited liability corporate entities. Mortgages that had been state-administered were bought up, under housing finance reform, by the new Home Finance Company (HFC), essentially a secondary mortgage generator (Bank of Ghana 2007). With the establishment of the HFC, many privately-owned real estate developers emerged to provide residential houses, mainly for middle and high income Ghanaians in Accra.

Rapid urbanization and increasing globalization have led to increasing concentration of economic activities by multinational companies, expatriates and other foreign nationals. This has led to pressure on land resources and shelter within and at the peripheries of the city. As a result, Accra is unable to meet the demand for housing. Many residents of Accra are forced to live in areas of the city where much of the housing stock is characterized by absence of basic services such as in-house water supply, toilets and bathroom facilities. Consequently, slums and squatter neighbourhoods are becoming very common. With rapid urbanization and urban growth, housing has become one of the critical challenges facing residents in Accra. This situation is attributed to deficiencies and weaknesses of national and city level policies and strategies on housing development.[1]

A new phenomenon is sporadic and uncontrolled land and residential development in the suburban areas of Accra. Most buildings are developed without prior approval from planning agencies. The key agencies responsible for initiating, approving and executing plans, the Department of Town and Country Planning, the Greater Accra Metropolitan Area and the Accra City District Council, have not instituted growth management ordinances to control urban sprawl (Attoh 2010). As a result of the exponential growth of the informal sector, traders – particularly hawkers – continue to experience harassment and face ejection by city authorities. Many, however, resist relocation and return to the sidewalks after being ejected. This suggests that there is no fundamental shift in the state's or the city authority's attitude to the informal sector (Tsikata 2007). The high income suburb of East Legon, for example, is located in a low-lying area prone to flooding and is directly underneath the flight path to the airport. Some houses in East Legon are located dangerously close to high-tension electric cables that run through the area. The extension of utilities to these outlying areas is an expensive process. Excess demand on the existing water system in Accra makes the water pressure in these outlying regions so low that some residents have resorted to well water (Attoh 2010). Another visible manifestation of the postadjustment urban challenges of Accra is increased traffic congestion, making it difficult to move through the city, especially at peak hours.

Socioeconomic Change and Gender Relations

The proclamation of the Decade for Women (1975–1985) and the International Women's Year in 1975 by the United Nations led to the establishment of the National Council on Women and Development (NCWD) in Accra. The Council became one of the most important instruments for the integration of women in the

---

1   Owusu, G. 'Urban Growth, Globalization and Access to Housing in Ghana's Largest Metropolitan Area, Accra'. Unpublished paper presented at the 4th European Conference on African Studies (ECAS 4), Uppsala, Sweden, 15–18 June, 2011.

development of the country. One of its major functions was to develop and promote action programmes to integrate women in all sectors of national development (Nikoi 1998). The first major task of the NCWD was consciousness-raising, whereby educational programmes were launched to create awareness among the Ghanaian public of the plight of women. One of the results of this educational programme was a debate in schools and newspapers about women's capabilities and their roles in society (Dolphyne 1995). These activities contributed to changes in sociocultural attitudes towards females, particularly regarding the education of girls, especially in mathematics, technology, sciences and vocational skills.

By the 1990s, significant efforts were made by many international and national NGOs to achieve poverty reduction, particularly in rural communities in Ghana. Many operated from Accra and provided human-centred integrated rural development programmes including the provision of educational facilities, health facilities, business management training, and counselling and legal aid services particularly to women in other parts of the country. With changes in Ghana's political system, the Ministry of Women and Children's Affairs (MOWAC) was established in 2000 in Accra to promote the welfare of women and children. Accra has since become the centre of gender activism in Ghana with a number of NGOs, women's networks and coalitions formed to take actions and address women's rights and concerns or to improve services particularly for women (Manuh 2007, Tsikata 2007).

Having grown over the years into a cosmopolitan city, Accra offers a wide range of educational opportunities. Although female employment remains concentrated mainly in low-skill sectors, the gap in education between men and women is narrowing, with younger women desiring and attaining higher levels of education. Data from the University of Ghana show that in 1961 only 9.1 per cent of undergraduate students were females. By 1981, female undergraduates accounted for 16.4 per cent. This increased to 34.4 per cent females in 2001 and 45.4 per cent females in 2010. With increasing gender activism in the city, more women are involved in various occupations and professions within the service sector, with women making impressive inroads into professional services such as law, banking, accounting and academia. Women are increasingly taking leadership roles, particularly in politics and local governance in Accra, and as members of parliament (MPs), district assemblies and unit committees.

Changes in the labour market have influenced unemployment, job security and the type of labour, affecting the lives of both men and women, particularly those of low socioeconomic status. For many young people, especially men, this can pose a major obstacle to the timing of marriage and other life choices. Oppong's studies among senior civil servants in Accra show that the process of decision-making in the home regarding how domestic tasks and resources should be allocated is a complex sequence of events taking place between spouses as well as between them and their kin. She reveals that a couple's power relationship may vary among the urban elite. While some couples share major decisions, others make their decisions separately. There are also husband-dominated relationships where husbands take major decisions affecting both his and his wife's use of resources. However the relative power of husbands and wives is being influenced by the comparative resources which husbands and wives bring to the marriage (Oppong 1974).

Recent studies in Accra have shown that the economic emancipation of women has given them an opportunity to play an active role in the major decision-making processes within homes and communities. On the other hand, economic hardship has resulted in lack of economic support from husbands, mistrust and conflict, lack of respect and lack of dialogue between spouses. This is particularly pronounced in low socioeconomic communities, where men are constantly confronted with their inability to provide for the household (Songsore and McGranahan 2003, Wrigley-Asante 2010). Songsore and McGranahan (2003) further note that the growing economic hardships and the increasing role of women in informal economic activities have enhanced women's influence in the household. This is being activated by the growing power of the women's movement and associations, which is creating a consciousness of the need for greater gender equality.

Unemployment and increasing poverty in Accra, and men's inability to fulfill their social roles and expectations, challenge the social value and traditional position of men as heads of the household, threatening the man's honour, reputation and masculinity (Wrigley-Asante 2010). Silberschmidt notes that male identity and self-esteem have become increasingly linked to sexuality and sexual manifestations, creating tension and domestic violence.[2] Wrigley-Asante's (2010) study in Accra supports this. Structural changes in the economy resulting from macroeconomic policies, trade liberalization and the process of globalization are influencing the lives of men in both low-income indigenous areas and other residential areas in Accra. Many men are struggling to fulfill their expected role at the household level due to job insecurity, thus creating a feeling of anxiety and loss of sense of power and self-esteem, which may lead to gender antagonism and extramarital affairs. Conflicts of interest embedded in gender relations are becoming more pronounced and visible.

Socioeconomic change within Accra has reduced the value of children and increased their cost. In the last four decades, the likelihood of assistance from the extended family has dwindled and the burden falls almost entirely on the nuclear family. In the elite suburbs of Accra, parents are concerned about the quality of their children and the quality of their proposed spouses. As a result, actual and preferred family size ideals have converged to an average of three children, with successful family limitation achieved by easily accessible contraception. Marriage is occurring at a later age and the first birth after marriage may also be postponed (Agyei-Mensah et al. 2003).

---

2   Silberschmidt, M. 'Are Men Interested in Engaging in the Struggle for Gender Justice and Broader Social Change – Or What Would Make Them Interested?' Unpublished paper presented at an International Symposium Linking Lessons from HIV, Sexuality and Reproductive Health with Other Areas for Rethinking AIDS, Gender and Development, Dakar, Senegal, 15–18 October, 2007.

Conclusions

The study of urban change in Accra provides a framework for understanding the interplay of gender, politics and development. Since its origins in the nineteenth century, Accra has become a modern metropolis. The city's architecture reflects this history, ranging from nineteenth-century British colonial buildings to middle and upper class residential houses. As Accra has transformed into a modern city with improved hospitality, housing, banking and telecommunication services, its position as an industrial, administrative and industrial centre has attracted a large number of immigrants. Migration into the city has posed major challenges in terms of housing, environmental and health-related issues. Moreover, the traffic situation makes movement across and within the city quite chaotic. The ethnic profile is also changing as the Akans have replaced the indigenous Gas as the largest population group.

These transformations are not only reflected in the physical landscape but also in socioeconomic and gender relations. More and more women are becoming educated and empowered, and are postponing entry into marriage. Some of these changes have brought new challenges such as marital disruption leading to divorce, infidelity and female-headed households.

The evolution of Accra from a colonial town to a modern global city and the socioeconomic changes that have occurred through intricately connected economic, sociocultural and geopolitical processes illustrate the combined influence of gender relations, globalization and politics on the transformation of African cities.

References

Acquah, I. 1972. *Accra Survey.* Accra: Ghana Universities Press.

Agyei-Mensah, S. 1997. *Fertility Change in a Time and Space Perspective: Lessons from Three Ghanaian Settlements.* PhD thesis. Trondheim: Norwegian University of Science and Technology (NTNU).

Agyei-Mensah, S. and de-Graft Aikins, A. 2010. Epidemiological Transition and the Double Burden of Disease in Accra, Ghana. *Journal of Urban Health*, 87, 879–97.

Agyei-Mensah, S. and Owusu, G. 2010. Segregated by Neighbourhoods? A Portrait of Ethnic Diversity in the Accra Metropolitan Area. *Population Space and Place*, 16, 499–516.

Agyei-Mensah, S., Aase, A. and Awusabo-Asare, K. 2003. Social Setting, Birth Timing and Subsequent Fertility in the Ghanaian South, in *Reproduction and Social Context in Sub-Saharan Africa: A Collection of Micro-Demographic Studies*, edited by S. Agyei-Mensah and J.B. Casterlin. Westport CT: Greenwood Press, 89–108.

Ardayfio-Schandorf, E., Brown, C.K. and Aglobitse, P.B. 1995. *The Impact of PAMSCAD on the Family: A Study of the ENOWID Intervention in the Western Region of Ghana.* FADEP Technical Series 6. Family and Development Programme. Accra: University of Ghana.

Asiedu, A. and Agyei-Mensah, S. 2008. Traders on the Run: Activities of Street Vendors in the Accra Metropolitan Area, Ghana. *Norsk Geografisk Tidsskrift–Norwegian Journal of Geography*, 62, 191–202.

Asenso-Okyere, W., Asante, F.A. and Oware Gyekye, L. 1992. *Case Studies on Rural Poverty Alleviation in the Commonwealth, Ghana.* London: Food Production and Rural Development Division, Commonwealth Secretariat.

Attoh, S.A. 2010. Urban Geography of Sub-Saharan Africa, in *Geography of Sub-Saharan Africa*, edited by S.A. Attoh, 3rd edition. Upper Saddle River, NJ: Pearson Prentice Hall, 265–304.

Austin-Tetteh, P. 1972. Housing and Population, in *Interdisciplinary Approaches to Population Studies*, edited by A.S. David, E. Laing and N.O. Addo. University of Ghana Population Studies No. 4. Accra: University of Ghana, 98–112.

Awumbila, M. and Ardayfio-Schandorf, E. 2008. Gendered Poverty, Migration and Livelihood Strategies of Female Porters in Accra, Ghana. *Norsk Geografisk Tidsskrift–Norwegian Journal of Geography*, 62, 171–9.

Baden, S. 1993. *Gender and Adjustment in Sub-Saharan Africa.* Report Commissioned by the Commission for European Communities. Brighton: Institute of Development Studies.

Baden, S. 1997. *Economic Reform and Poverty: A Gender Analysis.* Report prepared for the Gender Equality Unit, Swedish International Development Cooperation Agency (Sida). Bridge Report No. 50. Brighton: Institute of Development Studies.

Bank of Ghana 2007. *The Housing Market in Ghana.* Prepared by the Research Department. Accra: Bank of Ghana.

Beall, J. 2002. Globalization and Social Exclusion in Cities: Framing the Debate with Lessons from Africa and Asia. *Environment and Urbanization*, 14, 41–51.

Dolphyne, F.A. 1995. *The Emancipation of Women: An African Perspective.* Accra: Ghana Universities Press.

Gaule, S. 2005. Alternating Currents of Power: From Colonial to Post-Apartheid Spatial Patterns in New-Town, Johannesburg. *Urban Studies*, 42, 2335–61.

Ghai, D. 1997. *Economic Globalization, Institutional Change and Human Security.* DP 91, Geneva: United Nations Research Institute for Social Development.

Grant, R. 2009. *Globalizing City: The Urban and Economic Transformation of Accra.* Syracuse, NY: Syracuse University Press.

Heine, J. and Thakur, R. 2011. Introduction: Globalization and Transnational Uncivil Society, in *The Dark Side of Globalization*, edited by J. Heine and R. Thakur. Tokyo: United Nations University Press, 1–16.

Konadu-Agyemang, K. 2000. The Best of Times and the Worst of Times: Structural Adjustment Programmes and Uneven Development in Africa: The Case of Ghana. *Professional Geographer*, 52, 233–6.

Lie, M. and Lund, R. 2005. From NDL to Globalization: Studying Women Workers in an Increasingly Globalized Economy. *Gender, Technology and Development*, 9, 1–30.

Manuh, T. 2007. Doing Gender Work in Ghana, in *Africa After Gender*, edited by C.M. Cole, T. Manuh and S.F. Miescher. Bloomington, IN: Indiana University Press, 125–49.

Nikoi, G. 1998. *Gender and Development*. Accra: Ghana Universities Press.

Oppong, C. 1974. *Marriage Among a Matrilineal Elite: A Family Study of Ghanaian Senior Civil Servants*. Cambridge: Cambridge University Press.

Overå, R. 2007. When Men Do Women's Work: Structural Adjustment, Unemployment and the Changing Gender Relations in the Informal Economy of Accra, Ghana. *Journal of Modern African Studies*, 45, 539–63.

Owusu, G. 2008. Indigenes' and Migrants' Access to Land in Peri-Urban Areas of Ghana's Largest City of Accra. *International Development Planning Review (IDPR)*, 30, 177–98.

Pellow, D. 1977. *Women in Accra: Options for Autonomy*. Algonac, MI: Reference Publications Inc.

Potts, D. 1996. Shall We Go Home? Increasing Urban Poverty in African Cities and Migration Processes. *Geographical Journal*, 161, 246–64.

Robertson, C. 1984. *Sharing the Same Bowl: A Socio-Economic History of Women and Class in Accra, Ghana*. Bloomington, IN: Indiana University Press.

Simone, A. 2004. *For the City Yet to Come: Changing African Life in Four Cities*. Durham, NC and London: Duke University Press.

Songsore, J. and McGranahan, G. 1996. *Women and Household Environmental Care in the Greater Accra Metropolitan Area, Ghana*. Urban Environmental Series, Report No. 2. Stockholm: Stockholm Environmental Institute.

Songsore, J. and McGranahan, G. 2003. Women's Household Environmental Caring Roles in the Greater Accra Metropolitan Area: A Qualitative Appraisal. *Institute of African Studies Research Review*, New Series 19, 67–83.

Songsore, J., Nabila, J.S., Yangyuoru, Y., Avle, S., Bosque-Hamilton, E.K., Amposah, P.E. and Alhassan, O. 2009. *Environmental Health Watch and Disaster Monitoring in the Greater Accra Metropolitan Area (GAMA), 2005*. Accra: Ghana Universities Press.

Stewart, D.J. 1999. Changing Cairo: The Political Economy of Urban Form. *International Journal of Urban and Regional Research*, 23, 128–46.

The Statesman 2007. A Brief History of Housing in Ghana [Online, 27 January]. Available at: http://www.ghanaweb.com/GhanaHomePage/NewsArchive/artikel. php?ID=117756 [accessed: 20 January 2010].

Tipple, G.A. and Korboe, D. 1998. Housing Policy in Ghana: Towards a Supply-Oriented Future. *Habitat International*, 22, 245–57.

Tsikata, D. 2007. Women in Ghana at 50: Still Struggling to Achieve Full Citizenship? *Ghana Studies*, 10, 163–206.

Wrigley-Asante, C. 2008. Men are Poor but Women are Poorer: Gendered Poverty and Survival Strategies in the Dangme West District of Ghana. *Norsk Geografisk Tidsskrift–Norwegian Journal of Geography*, 62, 161–70.

Wrigley-Asante, C. 2010. Rethinking Gender: Socioeconomic Change and Men in Some Selected Communities in the Greater Accra Region of Ghana. *Ghana Social Science Journal*, 7, 52–71.

Young, K. 1993. *Planning Development with Women: Making a World of Difference*. London: Macmillan Press.

Chapter 9

Ignored Voices of Globalization: Women's Agency in Coping with Human Rights Violations in an Export Processing Zone in Sri Lanka

Chamila T. Attanapola

Introduction

Employers in a lay-off stunt on pretext of global economic crisis
(*Shramika* 2009a)

Let us rise against violence against women (*Shramika* 2009b)

Due to floods we lost our salary (*Shramika* 2010)

These newspaper headlines from 2009 and 2010[1] indicate that the situation of women factory workers in the export processing zones (EPZs)[2] of Sri Lanka had not changed significantly since my previous encounter with EPZ women in 2004. The Sri Lankan government has ratified all eight of the International Labour Organization's (ILO) conventions on labour rights (Compa 2003, Egels-Zandén and Hyllman 2007). Also, a large number of awareness-raising campaigns have been organized by international and local nongovernmental organizations (NGOs) (Hale 2004), and throughout the years voice has been given to women EPZ workers in academic publications and the media. Yet women workers still experience various forms of violation of their rights as workers, as women and as human beings in everyday life within factories and in local society.

In this chapter I explore how women factory workers respond to human rights violations in the EPZ sector in Sri Lanka. I ask how rights of women factory workers are violated in their daily life and how they respond to these violations, and I ask in what ways women's actions demonstrate their agency and empowerment and further female activism. Through the stories of three women, I investigate the

---

1 From *Shramika*, Newspaper Published by the Women's Centre Ja-Ela, Sri Lanka.

2 Sri Lankans refer to free trade zones (FTZ) but in this chapter I use the more common international appellation export processing zones (EPZ).

strategies of women factory workers for coping with the challenges of this 'glocal' place. Ritzer (2003: 194) defines *glocalization* as the interpenetration of the global and the local, resulting in unique outcomes in different geographical areas. Hence a place where global and local processes interact is identified as a 'glocal' place. An export processing zone is a place where global and local processes meet with the global economy as well as general gender standards and feminist ideas (Davids and van Driel 2007). I gathered information during three sessions of fieldwork in Sri Lanka's largest export processing zone in Katunayake in 2002, 2004 and 2010. The Women's Centre has served as my gatekeeper to the women workers, women activists and trade union activists with whom interviews and observations were conducted. The Women's Centre is an organization formed by and working for women workers in export processing zones in Sri Lanka. It was established in response to the various forms of harassment faced by EPZ women at the workplace, home and boardinghouse and in the streets. The Women's Centre aims to empower and organize women in order to overcome the problems they face as women (*Shamrika* 2009c).

Globalization and Gender: From Exploitation to Activism

Since the late 1970s, a new international division of labour and neoliberal economic policies with structural adjustment programmes have accelerated economic globalization by recommending the establishment of EPZs in developing countries (Davids and van Driel 2007). EPZs demand primarily female labour for several reasons. First, in the era of globalization, industrial capitalists have increasingly become interested in investing in consumer items such as textiles, garments, jewellery, food and electronics. Such industries do not look for skilled labour that demands high wages but for semiskilled or unskilled labour to work on assembly lines. Second, it is believed that manual dexterity, patience and other 'gender-specific attributes' such as tidiness render women more suitable than men for carrying out assembly line tasks that are repetitive and demand painstaking attention to detail. Third, female workers are preferred because they are considered docile, obedient and willing to accept tough work discipline, are less likely to press demands for better pay and working conditions, and are less inclined to join trade unions than men. Fourth, low wages are rationalized on the basis that women do not have the primary responsibility for earning the main income for the family. Industrial capitalists are able to make higher profits through employing women rather than men due to the high level of productivity and low level of costs involved (Elson and Pearson 1981, Pearson 1992, 2000). Thus feminization of the global workforce has been inevitable. Particularly young and unmarried women from rural parts of developing countries are attracted to EPZs in urban areas because unemployment has been higher among young educated women in rural areas due to mechanization and modernization of the rural agricultural sector (Lund 1993, Standing 1999, Villarreal and Yu 2007).

Literature on globalization and women or gender focuses on four major issues. The first set of literature focuses on how women working at EPZ factories are exploited by multinational enterprises and therefore can be seen as victims of globalization and economic liberalization policies. Wages (Braun and Gearhart 2004), safety and health conditions in working environments (Thorborg 1991, Theobald 1996, Attanapola 2005, Root 2009), and poor labour relations or lack of rights (Gunatilake 1999, Ngai 2004, Gopalakrishnan 2007, Pearson 2007, International Labour Office 2008, Wick 2010) are the three areas that have received the most criticism with regard to the exploitative situation in EPZs. EPZs have been referred to as 'zones of oppression' (Romero 1995), 'danger zones' (Frumkin 1999) and 'zones of exploitation' (Glassman 2001).

Lim (1985) pioneered the second set of studies on gender and globalization, which focuses on factory work as liberation for women, giving them an opportunity to participate in income-generating activities and liberate themselves from patriarchal dependency. While agreeing that multinational enterprises practise several forms of exploitation (e.g. low-paid female labour and relocation), Lim (1985, 1990) argues that they do so only to survive within the global market system and that the literature has not examined the challenges in the sector and its positive influence on women. According to Lim (1990), it is not the operations of the multinationals that create exploitation of women and gender discrimination but the sociocultural environments of the countries in question. Lim (1997) further argues that in the long run demand for female labour will lead to a reduction of the imperialist and patriarchal components of capitalist exploitation of Third World female workers. Studies carried out in Malaysia (Lie and Lund 1994), Indonesia (Wolf 1992) and China (Zhang 1999) identified that work in EPZ factories is a positive experience for women, involving poverty reduction, change in gender roles from dependent daughters to breadwinners of the family, and social and economic freedom while living away from their families.

The third set of studies analyses the dynamics of women's employment in the EPZ sector. Pioneered by Elson and Pearson (1997), this literature identifies the EPZ women as empowered women who make strategic choices and act upon their needs within their sociocultural and economic contexts. For example, studies in India (Chakravarty 2004, Soni-Sinha 2004), Sri Lanka (Attanapola 2004) and the Philippines (McKay 2006) show that, regardless of the subordination position and oppression of women in general, women workers are empowered because they are engaged in income-generating activities, have control over their income, have increased status within families by contributing to the family economy – some have even become breadwinners – and have decision-making power over the matters that concern them most, such as marriage.

The fourth set of literature studies how women workers actively participate in organizational work such as NGOs and trade unions, and raise awareness of their rights and mobilize fellow women to exercise their rights as EPZ workers, women and humans (Mills 2005, Attanapola 2008, Dominguez et al. 2010). Further studies unravel how women workers make alliances with global institutions to act against not only unjust employers and practices at the workplace but also international labour standards and trade agreements that are violating workers' rights (Prieto and Quinteros 2004, Egels-Zandén and Hyllman 2007, Anner 2009). These studies illustrate women's activism within the EPZ sector as a form of globalization from below (Rigg 2007).

Coping with Violations of Rights through Women's Agency

My analysis of women's choices and actions relating to human rights violations is based on theories of women's empowerment. Gender and development (GAD) identifies women as active agents, who use their agency to overcome different barriers they encounter in their life paths, make strategic choices and take actions within their sociocultural, economic, legal and political contexts. Women who use their agency to gain control over the matters that are important to their lives are described as empowered women (Moser 1993, Afshar 1998, Rowlands 1998, Kabeer 2002). Kabeer (2002: 19) defines empowerment as 'expansion in people's ability to make strategic life choices in a context where this ability was previously denied to them'. Accordingly, empowerment is a process whereby women gain the ability to make choices that they could not previously and subsequently achieve a better condition or position. To measure empowerment, there must be a positive change. A coping strategy that does not lead to a better condition or position is not an achievement according to Kabeer's definition of empowerment. However, a successful coping strategy reflects the capacity of women to make the best decisions within their sociocultural, economic and legal context.

In order to explore actions taken by women to cope with and survive human rights violations, and to examine whether the chosen actions reflect empowerment, I use Kabeer's (2002) three interrelated concepts of change that make up choice:

1. Resources: sociocultural, economic and legal contexts that enhance or constrain women's ability to make choices.
2. Women's agency: the process of using women's 'powers' to make choices and take actions within their sociocultural, economic and legal context.
3. Achievement or positive change: the outcome of choices.

Within the gender and empowerment discourse, power is disaggregated into four categories (Rowlands 1998):

1. *Power over:* the ability of the actions of relatively powerful actors to affect the actions and thoughts of the relatively powerless.
2. *Power to:* the capacity to act.
3. *Power within:* gaining the sense of self-identity, confidence and awareness in the situation that is a precondition for action.

4. *Power with:* the synergy that can emerge through partnerships and collaboration with others or through processes of collective action and alliance building.

Violations of women's rights are results of acts of 'power over'. Studying women's agency relating to human rights violation clearly concerns how women use their 'power within', 'power to' and 'power with' in reaction to 'power over' actions. The conceptual framework is shown in Figure 9.1.

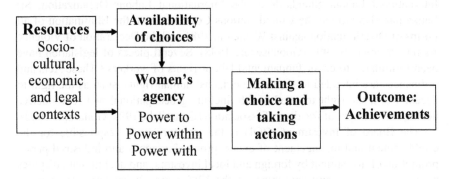

Figure 9.1   **A conceptual framework for studying women's empowerment in actions to cope with human rights violations**

Institutions play important roles as either enablers or supporters or as barriers or constrainers for achieving workers' rights. Institutions determine what kinds of resources are available for EPZ women, including availability of a legal framework, service providers (lawyers, NGO workers and trade unions) and information regarding human rights and labour rights. If the resources are available, women can choose according to their preference. When women are informed, assisted and mobilized they use their agency to take actions as individuals, as a community or as a part of an organization. These actions may ultimately lead to achievement, in this case coping with human rights violations.

The EPZ Sector in Sri Lanka: Supportive Mechanisms and Networks of Women Workers

This section provides a description of the sociocultural, economic and legal context of resources available for women EPZ workers in Sri Lanka, whereby they make choices and take actions against human rights violations.

In 1977, the Sri Lankan government adopted a Structural Adjustment Programme, mainly at the request of the International Monetary Fund and the World Bank, in order to reduce unemployment and increase economic growth (Winslow and Woost 2004). To accommodate export-oriented industrialization through foreign direct investment (FDI), nine export processing zones (EPZ) and two industrial parks (IP) with sophisticated infrastructure were established during 1978–2000 in different parts of the country (Board of Investment of Sri Lanka 2013).

Sri Lanka is among the countries which have formally subscribed to several hundreds of human rights instruments and is committed to the realization of corresponding standards at home. The legal framework of the country covers all aspects of the Universal Declaration of Human Rights and the Fundamental International Labour Standards of the International Labour Organization. Sri Lanka has also ratified the United Nations Convention on the Elimination of all Forms of Discrimination against Women in 1981 and the Declaration of Violence Against Women in 1993 (Goonasekere 1998). Several pieces of legislation have been formulated to cover fundamental labour rights of workers.[3] These Acts and ordinances are intended to secure the citizens' human rights, such as the right to decent work, the right to good health, including occupational and reproductive health, and the right of freedom of association (Gunatilake 1999, Gunaratne 2002).

The Board of Investment (BOI) is the statutory body responsible for the establishment and management of export processing zones and industrial parks, promotion of investment by foreign and local investors, and dealing with disputes between employers and employees in the EPZ sector. It has formulated rules, regulations and guidelines for investors and for employers in factory building and maintenance. These concern the environment, occupational safety and health, and labour standards and relations, based on the country's labour laws (Board of Investment of Sri Lanka 2004). The BOI determines the minimum wages in the EPZ sector, which are higher than in other sectors in the country. Further, there are regulations regarding working hours, payments and leave for workers in the export processing sector in order to protect them from exploitation as well as to encourage them to work efficiently.

Since the EPZ was first established in Katunayake, a number of organizations have set up centres for female workers. Although the organizations may have different approaches and ideologies, their support is important to the women and they have devised ways to work together on specific issues. Organizations such as the Women's Centre, *Kalape Api* (We in the Zone), *Mithuru Sevena* (Friendship House), and *Dabindu* Collective (Drops of Sweat) are the most popular NGOs working for female workers in the EPZ. Further, many religious organizations have worked with the women since the beginning of the EPZ. They have established centres and organized activities in attempts to integrate women into the village community. These NGOs encourage EPZ women to participate in nationwide

---

3   Minimum Wage/Wage Board Ordinance (1941), Factory Ordinance (1942, amended 2002), Trade Union Ordinance (1935), Night Work Act (1984), Termination of Employment of Workmen Act (1970), Employees Provident Fund Act (1958) and Employees Trust Fund Act (1980) (Gunatilake 1999, Gunaratne 2002).

actions, for example Women's Day and Labour Day celebrations. Further, the NGOs working for the EPZ women collaborate with other women's organizations in Sri Lanka and sign petitions to obtain legal protection for women against sexual harassment at the workplace and violence. NGOs raise EPZ women's awareness of worker's and women's rights and health problems, and mobilize workers to take action against employers who violate their rights.

Until 2000, the workers right to organize themselves within the EPZs was systematically banned in Sri Lanka through 'employees' councils' and gate pass systems, which forbade trade unions from entering into the zones. An employees' council consists of one representative from the managerial staff and three representatives from the factory workers. Recognition of the Free Trade Zones & General Services Employees Union (FTZ&GSEU) in 2000 as the EPZ workers' trade union by the government, BOI and multinational enterprises (MNEs) was one of the great achievements of collaboration between NGOs and EPZ workers. However, there are still only seven factories in Katunayake EPZ that allow workers to join the union and recognize unions at factory level. Regardless of the law, MNEs discourage workers' unionization inside and outside the factories (interview with women activists, 2010).

Not Unheard but Ignored: Stories of Globalization

The voices of women EPZ workers are well documented in local and national newspapers, in television dramas and films, in research articles, in ILO reports from tripartite meetings where governments, NGOs and trade unions have participated, and in internet campaigns. Nonetheless, appalling stories continue to appear in newspapers published by the Women's Centre or are told by women in person. While many have heard their stories, unfortunately their voices are ignored. Hence, I present here some women's stories to remind responsible actors that, even though EPZ women's rights still in practice remain unprotected, the women are not entirely powerless victims, but active agents who make positive changes.

Choose Not to be a Victim: The Story of Suba

I met Suba in 2004 at her boardinghouse. She was happy and packing her bags to return home permanently. She said that, since she was 27 years old (at the time of the interview), her parents were arranging a marriage partner, who was a farmer. Although she had worked for eight years as a factory worker, she had not saved enough money to start a new life as she had expected. She said that, since her parents did not expect money from her, she spent most of her salary on fashionable clothes and jewelry, which are important items for a young woman, as a dowry. She enjoyed the new found 'power to' make economic choices and her position as an independent woman able to satisfy her own needs through engagement in productive work. However, she explained:

> Everything began to change after four years. I was severely sick and admitted to
> a government hospital and treated for two weeks. My family worried about me
> and they wanted to take me home. Factory managers allowed me to take sick
> leave and they told me to come back when I got well.

There was no written permission regarding the amount of sick leave Suba could
have. Her lack of knowledge of workers' rights in the EPZ system prevented
her getting sick leave in a legal manner. From the factory's point of view, it is
profitable to reduce an inefficient workforce; due to her sickness Suba had become
an 'inefficient worker'. Factories can replace workers since there are always many
women waiting outside to fill vacancies (interview with NGO workers, 2004).
Trustful Suba believed the manager's words and came back to the factory after
three months. What happened afterwards made her angry and disappointed:

> They [the factory managers] told me that I could start as a trainee. They would
> 'allow me to start working as a favour since I was such a good worker'. They
> wanted to give me a new number [identification number] and a new card and
> they would pay me trainee's salary.

Suba's employers wanted to register her as a new worker for three reasons. First,
as a new worker, Suba would have to work as a trainee for six months and would
receive a trainee's salary, and she would not have been entitled to any benefit
from the employment during that period. Second, although she was employed as
a trainee, she could work as a machine operator since she had the necessary skills.
The company would gain in both respects, but Suba would lose money. Third,
if Suba was registered as a new worker, the company would not be responsible
for her social security benefits from the Employees' Provident Fund (EPF) and
Employees' Trust Fund (ETF), which she had being paying into for four years.
The funds would have been automatically cancelled and the company would have
saved this amount, while Suba would have lost her life savings (interviews with
NGO worker 2004). The employers could not trap Suba easily in their deception.
She was aware that the employers wanted to offer a trainee job, not as a favour to
her, but to save money:

> I refused their offer and told them I wanted to resign. I did not receive a letter of
> service termination, which proves my working experience in that factory for four
> years. They were angry with me because I didn't want to work for them anymore.

When Suba left she understood that she would not be able to get a better job than
as a trainee. She made a rational decision by choosing between leaving the job at
hand as a trainee and finding the same type of job in another factory. She foresaw
the possibility of a hostile working environment since the managers were angry
over her arguments with them regarding her job. At a new factory she could
make a fresh start.

Getting a new job was not difficult since at the interview Suba declared that she had experience even though she could not prove it. Most employers would know claims like this were true because avoiding issuing 'Service Terminate Certificates' is their strategy. Suba started working as a 'trainee' and received a very low salary. She continued:

> Workers noticed that we should be paid more than we received as our wage. Some months I expected a higher wage since I did many overtime hours and worked on public holidays. Supervisors informed us that our salary was deducted for unfinished targets, unpaid leave, damage to production etc. We did not have any control over how they calculated our wages. Nobody wanted to complain about this. The supervisors and managers did not like the workers who confronted them about wages and working conditions. They would notice those workers and fire them as soon as they got the opportunity. Therefore, everybody wanted to protect their jobs even though they felt that something was not just.

The production target system, which is practised in EPZ factories, is not laid down by law. Workers must reach a certain target per day in order to receive the minimum salary. Otherwise, wages are deducted to compensate for the unfinished target, or workers must work unpaid overtime to achieve their targets. In this situation, it is not enough to raise workers' awareness concerning the situation, but there is need for collective action. In the case of Suba's factory, lack of 'power with' led workers to accept rights violations as normal.

After three and a half years of work in the second job, Suba decided to get married since her parents put pressure on her by telling her that she would be too old to marry and have children if she waited any longer. She explained:

> I don't think about the money which I should have received. I only think that I never again need to listen to the harassing words of supervisors and from the society. I am satisfied with my situation, since I am going to be married. He is a good man and we are planning to grow paddy and other crops in our fields. I can take free if I am sick without being yelled at and there is my family to help us in our work. I worked in paddy fields before I started work in the factory. I know that life in the village is much better than the life in the zone area.

Too Old to Work in the EPZ: The Story of Kumari

I met Kumari in July 2010 when I revisited the Women's Centre. Apart from an extremely worn-out, thin and pale body that demonstrated the evidence of years of work at the EPZ factory, Kumari was not typical of most of the factory workers at Katunayake EPZ. She told me that she was 48 years old and had worked at a factory for 21 years. Kumari was at the Women's Centre with several other colleagues to gain legal assistance regarding her dismissal in 2009. Her story is typical of the stories of several thousand women who have experienced sudden loss of jobs due to the 'global economic crisis':

> I have experienced several forms of rights violations such as being unable to get sick leave, deduction of wages for not completing the quota, and working continuously three shifts without rest and many more. It is the normality for EPZ jobs. However what happened to ten factory workers including myself since December 2008 was absurd. One day the personal manager called us and asked us to sign a document. It was a letter of resignation. We were also asked to submit the papers for obtaining EPZ benefits.

In a situation like this Kumari and her colleagues did not have time to discuss the consequences of signing the document among themselves. They were almost forced by the manager to sign. Kumari thought that at least she would receive the EPF money and be able to find another factory job. A few hours later, they were offered jobs at the same factory, under the condition that they registered as new workers:

> Even though this does not affect me badly I was concerned about those who had worked less than five years. There were seven of them. They will not receive the EPF since they signed the resignation document. We tried to talk to the managers but they did not listen. We [three of workers who had been working for more than five years] were afraid to complain further due to the fear of losing our EPF money too.

Even though Kumari and those of her colleagues who were similarly affected wanted to confront their employers, she could not get the support of the Workers' Council. Workers' lack of solidarity and empathy leads to absence of 'power with' actions by workers when it comes to individual rights. Women workers must bear the individual losses either because there are no resources, such as information on the legal framework, legal support and moral support, or because women workers do not know about the existing resources, or available resources are inefficient. In this case, even though the resources should have been available through the Workers' Council, it was inefficient or ignorant in giving the individual workers support when their rights were violated.

According to Kumari, their situation worsened after two months:

> The manager called us again and told that they must terminate our jobs. The reason given by the manager was the global economic crisis – 'the factory does not get any quota, and the profits have lowered and are unable to cover the costs of wages'.

Kumari and her colleagues become 'victims' of the global economic crisis, according to the factory manager. However, what Kumari heard next made her furious:

I got to know that since we 'old' workers lost the jobs due to economic crisis, several young workers were hired by the factory. If there is real crisis they cannot hire new workers, can they? I come to the Women's Centre to see what our options are to charge against the factory managers. The Women's Centre advises us and helps us writing letters to the Department of Labour and other responsible institutions.

Kumari's story represents the ugly truth of employment in the EPZ. She found out that she lost her job not due to the global economic crisis but due to her age. She has decided to bring a charge against her employers and works to get every support she can. She has aroused awareness of the injustice that she and other colleagues experienced and allied herself with the Women's Centre to take action.

The Victim Who Became an Activist: The Story of Mali

Mali is a 35-year-old unmarried woman who had been working at Katunayake EPZ for 14 years. When I met her at the Women's Centre in 2002, she was preparing legal action against her employers. She had just been dismissed from her second job. Coming from an extremely poor family in rural Sri Lanka, where educated young women do not have access to paid work except as helpers in the agricultural sector (Attanapola 2005), Mali did not have a better choice than factory work. According to Mali, economic status and affiliation to the powerful political party have become the determinant factors in achieving 'decent' jobs in Sri Lanka. She was subjected to human rights violations through discrimination in getting a job even before she started to work in the EPZ.

Mali believed her choice to sacrifice her young life working in the EPZ was not wasted since her family survived extreme poverty and now all of them live better lives. Although the work was hard and harassment took place everywhere in the workplace and society, Mali continued to work at the first factory for 12 years. She said:

> At that factory we even received a New Year bonus, which was not common in some factories. I tried to attend every day so I could get the monthly attendance allowance and annual attendance bonus too.

According to an NGO activist:

> Bonuses are lucrative incentives for the workers. Factories introduce different strategies to maximize the efficiency and flexibility of workers, regardless of the impact of longer working hours and continuous work throughout the year, especially on the health status. Factories know that those employees who cannot tolerate this type of intensive work will exhaust themselves easily with the result that they give up the work quickly, while only those who are strong enough to do the demanding work will remain. New and younger workers replace those who

have left. Employers gain more than they pay out to the efficient workforce in the form of bonus and allowances. First, they manage to maintain an efficient and productive workforce. Second, the workforce is maintained in accordance with the demand from the global market. (Interview with the secretary, the Women's Centre, 2010)

In this context, Mali was only concerned with the income she received through hard work. However, Mali and her colleagues became 'victims of globalization' in January 2001. She described this incident:

Before the holiday [in December 2000] our boss informed us that the factory didn't make enough profit and they could not pay us the bonus for that Christmas. He wished us happy New Year and we went on holiday. When we came back there was a banner hanging on the gate stating that the factory was closed and the owners had moved away from the country. We were helpless. There were about 6,000 workers in that factory and most of them were women. We didn't know what to do. So I went to my village.

Like most of her colleagues, Mali did not see any other option than returning home. Later, however, when her colleagues started mobilizing ex-workers to fight for their rights, she returned to Katunayake:

With the help of the Association of Free Trade Zone Workers, a few active members of our Workers' Council took the initiative to press charges against the company. I wanted to know what would happen to the workers' EPF and ETF. It is our life savings. I also joined them. Further, through the activities of the Women's Centre, I raised awareness of our rights as workers at the EPZ. At the same time, I managed to get a job at a factory.

Mali demonstrated her agency to 'power with' actions. She was motivated by her fellow colleagues who initiated collective action against unjust employers. Her participation in NGO activities also demonstrated her further awareness of available resources enhancing her knowledge of human rights and workers' rights. She decided to continue working at a factory since:

Considering my age (35 years), without a job and a considerable dowry, no man would want to marry a poor woman like me. I need money to look after my old parents and myself.

In her second job she was hired as a trainee since she did not have any certificate to prove her 12-year working experience as a machine operator. She tolerated the situation, hoping that she would be promoted after six months. But things did not turn out as she expected:

I kept my mouth shut during the first six months because I hoped I would be promoted after the 'training period'. I know the law says that a worker must be given permanent status after the training period. Although eight months passed I was not promoted to permanent machine operator. My salary was so low, and it was hardly enough for me to satisfy my own needs. As a temporary worker I did not receive benefits from the factory except the salary. So I began to talk with supervisors and managers about being a permanent worker. Three months passed after I first talked to my managers and nothing special happened except that I lost faith in the managers and always tried to confront them with my situation. The supervisor in my section was sympathetic towards me and encouraged me to talk with the managers. But one day, after eleven months of work as a temporary worker, instead of giving me permanent status, as the law states, they fired me without any reason. With me, three other girls who were on a temporary basis (they have not been working as long as me) were fired. The only reason the managers gave us was that they had to cut down the workforce of the factory.

For the second time, Mali became unemployed, leaving her once again with an insecure future, without savings or a service certificate. EPZ workers become the victims of a global process of flexible labour. For the managers of companies, only their survival within the global market matters and they are unable (or unwilling) to see the impacts on individual workers (interview with the treasurer at the Women's Centre 2010). However, Mali sees her situation as revenge taken by the managers since she argued with them about her right to be promoted after six months:

In that factory we don't have the right to talk about our rights and injustices happen to us. They just fired me because I talked about my right to be a permanent worker in the factory. I think they are afraid of those girls who want to fight for their rights. They are afraid that if one worker achieves her rights, then others might follow her. So it is easier to get rid of those who make trouble for the managers.

The Workers Council at the factory did not come forward to help her because there is a group of workers who are supporters of managers and wanted to achieve their personal goals. As an active member of the Association of Free Trade Zone Workers' (later the FTZ&GSEU), she started again a legal battle against her employers. Both the Women's Centre and the FTZ workers' association assisted her in this matter.

Women's Agency in Coping with Violations of Human Rights

Drawing upon the stories, I have gained insights into various forms of human rights violations of EPZ workers in Sri Lanka. The women's stories demonstrate how the workers' rights are being violated due to the strategies taken by the MNEs and EPZ employers. Further, the stories allow exploration of women workers' strategies to overcome the violations and their ability to make use of knowledge of human rights, laws and the role of supportive institutions.

All three workers acknowledged that they were willing to work under tough conditions, such as longer working hours and supervisor pressure, being motivated by their target of earning a better income. They tolerated verbal harassment as a normal part of EPZ life. Further, they did not question the production target system, which is not imposed by law, and its consequences for workers who are unable to keep up with it despite wage deductions and unpaid overtime work. Moreover, the women tolerated the avoidance by managerial staff of written documents such as employment contracts, sick leave permits or letters of service termination, which could have been useful for the women in claiming their rights. Hence, EPZ women are systematically denied their rights on a day-by-day basis. Subsequently women workers have internalized this kind of violation of rights and oppression and do not confront their oppressors.

However, the stories also demonstrate how the women have made choices according to their lived experiences, preferences and priorities and acted to cope with the negative impacts of being incorporated into the global economy. For example, Suba did not allow herself to be exploited and harassed by the same employer twice. She was aware that she had little to lose if she resigned, while the employer would lose a well-trained machine operator. She chose 'not to choose' the factory manager's offer to register her as a new employee, instead choosing a job at another factory. In that way she avoided possible harassment by her employer. In the end she made a rational choice as an obedient traditional daughter in Sri Lankan society, namely marriage at a suitable age, which she preferred to the harassment of working life in the EPZ.

The story of Kumari is both unique and general at the same time. According to the Secretary of the FTZ&GSEU:

> Multinational corporations use the global economic crisis to maximize their profits. They do not hesitate to cut down the number of workers. This is a move to use the global economic crisis as a pretext to remove an excessive number of workers, curb their rights, and impinge on labour laws in order to maximize productivity through getting the most out of least possible number of workers. (Interview, August 2010)

Kumari's employer strategically terminated her employment, blaming the global economic crisis. After learning that the real reason behind her termination was her age, Kumari and her colleagues allied themselves with the Women's Centre and the FTZ&GSEU. They are taking action to bring the unjust employers to the labour court. Their goal is to prevent this kind of unjust treatment happening to other workers.

Mali, on the other hand, was motivated to get to know her rights as an EPZ worker after she lost her first job as a result of factory relocation. She joined a trade union, participated in campaigns and sued her former employers as she is determined that justice be done for the sudden loss of jobs. She participated in activities arranged by the Women's Centre and helped arouse awareness of the various forms of rights violations at EPZ workplaces. Her actions clearly demonstrate her agency, both 'power within' and 'power with'. However, her dismissal from the second job demonstrates that EPZ employers are more abrasive towards the employees who confront them and demand their rights. Hence women are inclined to remain silent and avoid challenging their employers. This can be identified as a conscious and considered survival mechanism rather than lack of agency.

Further, Mali found that the Workers' Council at the second factory was ineffective in assisting individual workers and thus she sought help from the Women's Centre and the Trade Union instead. She has chosen to continue working at an EPZ factory since she realized that there is no better alternative if she is to continue to be an economically independent woman as well as to help her poor, elderly parents. Her only other option would have been to move to the village and marry but, according to her, this was not an option.

Even though all women encounter various forms of rights violations and have access to similar kinds of resources, the stories provide evidence that women's actions may either be enabled or hindered by their knowledge of workers' rights and human rights.

Table 9.1 summarizes women's agency in the form of the three powers that demonstrate their choices of actions against human rights violations in export processing zones in Sri Lanka. A further dimension is added to the analysis, the scale from local to global, whereby women scale up their actions relating to violations of human rights. On one hand, the examples demonstrate women's agency at the local level such as factories, NGOs and Workers' Councils. On the other hand, they demonstrate women's agency at the national level, for example by bringing their case to the labour court and obtaining assistance from the national trade unions such as the FTZ&GSEU. The stories presented here do not demonstrate women's agency at the global level. According to the Secretary of the Women's Centre, global level actions are already beginning to happen through alliances with several international nongovernmental organizations such as War On Want, Clean Cloth Campaign and TieAsia.

Concluding Remarks

Globalization has a complex impact on women. While providing economic opportunities, it is also accompanied by the violation of women's rights. Despite the large amount of research giving voice to women's accounts of human rights violations in EPZ industries, and despite protective mechanisms created by the

**Table 9.1   Women's agency demonstrated in choices and actions taken
against human rights violations in export processing zones in
Sri Lanka**

Type of power	Actions at factory/ local level	Actions at national level	Actions at global level
Power within	Understand the different forms of rights violations within and outside EPZ factories. Identify the roots and consequences of violations of rights of women workers. Identify the supportive agents.	Obtain knowledge of national labour rights and rights as EPZ factory workers. Identify the supportive agents.	Understand the UN and ILO Conventions on Human Rights and their mechanisms to protect fundamental human rights. Identify supportive agents and institutions.
Power with	Maintain collective feeling and empathy with fellow workers. Build effective Workers' Council at the factory.	Build alliances with NGOs, e.g. the Women's Centre, trade unions, as well as with the Ministry of Labour Relations and Manpower.	Build alliances with international NGOs, global institutions and clients/customer organizations.
Power to	Confront employers. Complain to higher authorities. Resign from job, look for alternatives, find another job, or get married.	Complain to the Department of Labour with assistance from the trade unions and the Women's Centre. Participate in strikes and other campaigns. Sign petitions to change the legal framework.	Sign petitions to international organizations (ILO, EU).

ILO and national governments as well as the work of local and international NGOs and trade unions, the situation has not changed significantly. MNEs and employers continue to ignore women's voices and resistance and find opportunities to maximize their profits.

The voices, experiences and actions documented and analysed in this chapter indicate that change – achieving rights – is possible. I argue that, even though some women EPZ workers accept the situation as it is and internalize oppression, they are not powerless victims. Rather I identify them as women with agency who make choices and act upon their powers to cope with the situation and minimize the disadvantages with the limited resources they have within their sociocultural, economic, legal and political context. Moreover, making alliances with resourceful institutions, these women take action to minimize the negative impacts of globalization in attempts to overpower their oppressors.

This provides us with hope for these women who are heading towards female activism in grassroots globalization.

References

Afshar H. (ed.) 1998. *Women and Empowerment: Illustrations from the Third World.* London: Macmillan.

Anner, M. 2009. Two Logics of Labor Organizing in the Global Apparel Industry. *International Studies Quarterly*, 53, 545–70.

Attanapola, C.T. 2004. Changing Gender Roles and Health Impacts Among Female Workers in Export-Processing Industries in Sri Lanka. *Social Science & Medicine*, 58, 2301–12.

Attanapola, C.T. 2005. Experiences of Globalization and Health in the Narratives of Women Industrial Workers in Sri Lanka. *Gender, Technology and Development*, 9, 81–102.

Attanapola, C.T. 2008. Women's Empowerment for Promoting Health: Stories of Migrant Women Workers in Katunayake Export-Processing Zone, Sri Lanka. *Norsk Geografisk Tidsskrift–Norwegian Journal of Geography*, 62, 1–8.

Board of Investment of Sri Lanka 2004. *Labour Standards and Employment Relations Manual.* Colombo: Board of Investment of Sri Lanka.

Board of Investment of Sri Lanka 2013. *Setting Up in Sri Lanka – Where to Set Up? BOI Administered Zone* [Online: Invest in Sri Lanka]. Available at: http://www.investsrilanka.com/setting_up_in_srilanka/free_trade_zones_industrial_parks.html [accessed: 18 January 2013].

Braun, R. and Gearhart, J. 2004. Who Should Code Your Conduct? Trade Union and NGO Differences in the Fight for Workers' Rights. *Development in Practice*, 14, 183–96.

Chakravarty, D. 2004. Expansion of Markets and Women Workers: Case Study of Garment Manufacturing in India. *Economic and Political Weekly*, 39(45), 4910–16.

Compa, L.A. 2003. *Justice for All: The Struggle for Worker Rights in Sri Lanka* [Online: American Centre for International Labor Solidarity, Washington, DC/ Cornell University, School of Industrial and Labor Relations]. Available at: http://digitalcommons.ilr.cornell.edu/reports/32/ [accessed: 18 January 2012].

Davids, T. and van Driel, F. 2007. Changing Perspectives, in *The Gender Question in Globalization: Changing Perspectives and Practices*, edited by T. David and F. van Driel. Aldershot: Ashgate, 3–22.

Dominguez, E., Icaza, R., Quintero, C., Lopez, S. and Steman, Å. 2010. Women Workers in the Maquiladoras and the Debate on the Global Labor Standards. *Feminist Economics*, 16, 185–209.

Egels-Zandén, N. and Hyllman, P. 2007. Evaluating Strategies for Negotiating Workers' Rights in Transnational Corporations: The Effects of Codes of Conducts and Global Agreements on Workplace Democracy. *Journal of Business Ethics*, 76, 207–23.

Elson, D. and Pearson, R. 1981. Nimble Fingers Make Cheap Workers: An Analysis of Women's Employment in Third World Export Manufacturing. *Feminist Review*, 7, 87–107.

Elson, D. and Pearson, R. 1997. The Subordination of Women and the Internationalization of Factory Production, in *The Women, Gender & Development Reader*, edited by N. Visvanathan, L. Duggan, L. Nisonoff and N. Wiegersma. London: Zed Books, 191–203.

Frumkin, H. 1999. Across the Water and Down the Ladder: Occupational Health in the Global Economy. *Occupational Medicine (Philadelphia, Pa.)*, 14, 637–63.

Glassman, J. 2001. Women Workers and the Regulation of Health and Safety on the Industrial Periphery: The Case of Northern Thailand, in *Geographies of Women's Health*, edited by I. Dyke, N.D. Lewis and S. McLafferty. London & New York: Routledge, 61–87.

Goonesekare, S. 1998. *A Rights Based Approach to Realizing Gender Equality* [Online: UN Division on the Advancement of Women -DAW]. Available at: http://www.un.org/womenwatch/daw/news/rights.htm [accessed: 28 January 2011].

Gopalakrishnan, R. 2007. *Freedom of Association and Collective Bargaining in Export Processing Zones: Role of the ILO Supervisory Mechanisms*. Working Paper no. 1. Geneva: International Labour Standard Department, International Labour Office.

Gunaratne, C. 2002. *International Labour Standards and the Employment of Women in Sri Lanka*. Study Series No. 24. Colombo: Centre for Women's Research (CENWOR).

Gunatilake, R. 1999. *Labour Legislation and Female Employment in Sri Lanka's Manufacturing Sector*. Colombo: Institute of Policy Studies.

Hale, A. 2004. Beyond the Barriers: New Forms of Labour Internationalism. *Development in Practice*, 14, 158–62.

International Labour Office 2008. *Freedom of Association in Practice: Lessons Learned*. Report I (B) [Online: International Labour Office, Geneva]. Available at: http://www.ilo.org/wcmsp5/groups/public/---dgreports/---dcomm/documents/publication/wcms_096122.pdf [accessed: 27 January 2011].

Kabeer, N. 2002. *Resources, Agency, Achievements: Reflections on the Measurement of Women's Empowerment*. SIDA Studies 3. Stockholm: Novum Grafiska, 17–57.

Lie, M. and Lund R. 1994. *Renegotiating Local Values: Working Women and Foreign Industry in Malaysia*. Richmond: Curzon Press.

Lim, L.Y.C. 1985. *Women Workers in Multinational Enterprises in Developing Countries*. Geneva: International Labour Office.

Lim, L.Y.C. 1990. Women's Work in Export Factories: The Politics of a Cause, in *Persistent Inequalities*, edited by I. Tinker. Oxford: Oxford University Press, 101–21.

Lim, L.Y.C. 1997. Capitalism, Imperialism and Patriarchy: The Dilemma of Third World Women Workers in Multinational Factories, in *The Women, Gender & Development Reader*, edited by N. Visvanathen, L. Duggan, L. Nisonoff and N. Wiegersma. London: Zed Books, 216–39.

Lund, R. 1993. *Gender and Place*, Volume 1: *Towards a Geography Sensitive to Gender, Place and Social Change*; Volume 2: *Gender and Place: Examples from Two Case Studies*. Trondheim: Department of Geography, University of Trondheim.

McKay, S.C. 2006. Hard Drivers and Glass Ceilings: Gender Stratification in High-Tech Production. *Gender and Society*, 20, 207–35.

Mills, M.B. 2005. From Nimble Fingers to Raised Fists: Women and Labour Activism in Globalizing Thailand. *Signs: Journal of Women in Culture and Society*, 31, 117–44.

Moser, C.O.N. 1993. *Gender Planning and Development: Theory, Practice and Training*. London: Routledge.

Ngai, P. 2004. Women Workers and Precarious Employment in Shenzhen Special Economic Zone, China. *Gender and Development*, 12(2), 29–36.

Pearson, R. 1992. Gender Issues in Industrialization, in *Industrialization and Development*, edited by T. Hewitt, H. Johnson and D. Wield. Oxford: Oxford University Press, 222–47.

Pearson, R. 2000. Moving the Goalposts: Gender and Globalization in the Twenty-First Century. *Gender and Development*, 8(1), 10–19.

Pearson, R. 2007. Beyond Women Workers: Gendering CSR. *Third World Quarterly*, 28, 731–49.

Prieto, M. and Quinteros, C. 2004. Never the Twain Shall Meet? Women's Organizations and Trade Unions in the *Maquila* Industry in Central America. *Development in Practice*, 14, 149–57.

Rigg, J. 2007. *An Everyday Geography of the Global South*. London: Routledge.

Ritzer, G. 2003. Rethinking Globalization: Glocalization/Globalization and Something/Nothing. *Sociological Theory*, 21, 193–209.

Romero, A.T. 1995. Labour Standards and Export Processing Zones: Situation and Pressures for Change. *Development Policy Review*, 13, 247–76.

Root, R. 2009. Hazarding Health: Experiences of Body, Work, and Risk Among Factory Women in Malaysia. *Health Care for Women International*, 30, 903–18.

Rowlands, J. 1998. A Word of the Time, but What Does it Mean? Empowerment in the Discourse and Practice of Development, in *Women and Empowerment: Illustrations from the Third World*, edited by H. Afshar. London: Macmillan, 1–34.

Shamrika 2009a. Employers in a Lay-Off Stunt on Pretext of Global Economic Crisis. *Shamrika*, 4(13), July/August, 1–3.

Shamrika 2009b. Let Us Rise Against Violence Against Women. *Shamrika*, 4/14), November/December, 1–3.

Shamrika 2009c. Women's Centre – Women Workers' Shelter. *Shamrika*, 4(14), November/December, 6.

Shamrika 2010. Due to Floods We Lost Our Salaries. *Shamrika*, 5(2), July/August, 1, 3.

Soni-Sinha, U. 2006. Where are the Women? Gender, Labour and Discourse in the Noida Export Processing Zone and Delhi. *Feminist Economics*, 12, 335–65.

Standing, G. 1999. Global Feminization Through Flexible Labor: A Theme Revisited. *World Development*, 27, 583–602.

Theobald, S. 1996. Employment and Environmental Hazard: Women Workers and Strategies of Resistance in Northern Thailand. *Gender and Development*, 4(3), 16–21.

Thorborg, M. 1991. Environmental and Occupational Hazards in Export Processing Zones in East and South Asia, with Special Reference to Taiwan, China and Sri Lanka. *Toxicology and Industrial Health*, 7, 549–61.

Villarreal, A. and Yu, W. 2007. Economic Globalization and Women's Employment: The Case of Manufacturing in Mexico. *American Sociological Review*, 72, 365–89.

Wick, I. 2010. *Women Working in the Shadows: The Informal Economy and Export Processing Zones*. Siegburg: SUDWIND Institut för Ökonomie und Ökumene, and Munich: Evangelical Lutheran Church in Bavaria.

Winslow, D. and Woost, M.D. (eds) 2004. *Economy, Culture, and Civil War in Sri Lanka*. Bloomington, IN and Indianapolis: Indiana University Press.

Wolf, D.L. 1992. *Factory Daughters: Gender, Household Dynamics, and Rural Industrialization in Java*. Berkeley, CA: University of California Press.

Zhang, H.X. 1999. Female Migration and Urban Labour Markets in Tianjin. *Development and Change*, 30, 21–41.

Chapter 10
'No More Tears Sister': Feminist Politics in Sri Lanka

Jennifer Hyndman

Introduction

This chapter pays tribute to the extensive body of scholarship created over more than three decades by Ragnhild Lund. The gender-and-development literature, pioneered by scholars such as Lund (1993), has been transformed in places such as Sri Lanka, where it helps to account for the context of conflict that has characterized much of the last three decades. Lund's own published research traces this trajectory, first in the context of macrodevelopment in the Mahaweli Dam and Maduru Oya Projects (Lund 1979, 2000), and more recently in the light of the ongoing conflict and the aftermath of the 2004 tsunami (Brun and Lund 2008, Blaikie and Lund 2010). In arguing for an explicitly feminist approach to complex humanitarian crises, I am not saying anything entirely new. In many ways, Lund's work foreshadowed this argument. In 1988, she presented a feminist study of women in the Mahaweli area at the Centre for Women's Research (CENWOR) Women's Convention in Colombo.[1] At that time, feminist analysis had not yet come to replace 'women' or even a singular gender analysis. Such an assessment was a radical contribution to gender and development.

In this chapter, I draw on the story and works of Dr Rajani Thiranagama, a highly politicized Tamil Sri Lankan woman who sought social justice for many groups and under many guises. I illustrate how those struggling for social justice are themselves changed by the conflict.

Social relations are destabilized by conflict. Development programming also changes social relations, and ideally is changed by them. Speaking of violence and trauma, Jenny Edkins (2003: 4) writes: 'Our existence relies not only on our personal survival as individual beings but also, in a very profound sense, on the continuance of the social order that gives our existence meaning and dignity: family, friends, political community, beliefs.' Edkins is not referring to any fixed notions of identity or home here. She continues, 'Events of the sort we call traumatic are overwhelming but they are also a revelation. They strip away the diverse commonly accepted meanings by which we lead our lives in our

---

1 Lund, R. 'Women in the Mahaweli Area – A Feminist Assessment.' Paper for the CENWOR Women's Convention, Colombo, March 1988.

various communities. They reveal the contingency of the social order ...' (Edkins 2003: 5). Social change created by conflict and its attendant disordering of norms produces both new vulnerabilities and new possibilities for those situated at the margins of society.

As Butler et al. (2000: 17) remind us, '"identity" itself is never fully constituted; identification is not reducible to identity.' In Sri Lanka, all citizens may be required to carry identification cards, but such markings of identity are only ever partial. They tend, however, to 'fix' people in place and culture in particular ways (Hyndman and de Alwis 2004).

In the first part of the chapter, I extend and illustrate an argument that has been made before: that analysis of gender identities and relations alone is insufficient to understand women's (and men's) social and political struggles during war. A feminist approach is needed, one that takes into account multiple bases of identity and histories of belonging, including but not limited to ethnonational affiliation.

In the second part, I pay tribute to the many people who have died trying to stop the violence and human rights atrocities that have been part and parcel of the conflict in Sri Lanka over the past 26 years. In particular, I examine the struggles of one woman, Dr Rajani Thiranagama, who worked tirelessly for social justice – in its multiple meanings – and to document abuses against Tamil people in Jaffna, particularly women. Twenty years after her death, after thousands more Sri Lankans have been killed in their efforts to protest against state-based and rebel-induced violence, exclusion and discrimination, her life and political passion remain an inspiration and example.

Her short lifetime of activism embodied a startling range of political affiliations and politicized identities. During her early student days, Rajani Thiranagama was an antinationalist supporter of Marxist policies in the student movement to create greater social justice. Later, as a doctoral student in London, England, she joined the militant nationalist rebel group, the Liberation Tigers of Tamil Eelam (LTTE), to fight human rights atrocities against Sri Lankan Tamil people. After returning to the Jaffna war zone in the 1980s, Rajani worked as a professor of medicine at Jaffna University. Back in Sri Lanka, Rajani renounced her LTTE membership, and instead worked doggedly with women (and men) affected by conflict in the north into the late-1980s. This unusual political pedigree in a single lifetime speaks to the shifting grounds for antiwar politics in a country besieged by war.

Many Sri Lankans have taken up critical antiwar roles since the pogroms of 1983 in which Tamils were ethnically cleansed from the capital, Colombo. These acts, and those that instigated them – namely the killing of Sinhalese army officers in northern Sri Lanka by the rebel Tamil Tigers – ushered in repeated cycles of war and violence in which peace activists and antichauvinists lost their lives.

Feminists have been front and centre in struggles against the conflict. From the Sri Lanka feminist pioneer Kumari Jayawardena (1986) to the scholar–activists who have contributed to the feminist collective commentary of 'Cat's Eye' in the Colombo newspapers throughout the late 1990s and early 2000s and to the women's collectives, grassroots nongovernmental organizations (NGOs) and human rights activists, all have worked to highlight injustice and promote change just as Rajani Thiranagama, a founder of University Teachers for Human Rights, Jaffna (UTHR(J)), did two decades ago. Brian Seneviratne, in a foreword to *The Broken Palmyra* (1987), wrote:

> No worthwhile contribution to solving the mess in Sri Lanka can be made by those such as the writer of this Foreword and the thousands of other Sri Lankans who have not had the courage to stay in Sri Lanka. It can only be made by those such as the authors of "The Broken Palmyra" who have the courage, determination and patriotism to stay where they are needed and say what is right rather than what is convenient or acceptable to some power base.

In what follows, I briefly contextualize the conflict in Sri Lanka and elaborate a feminist analysis of conflict that contends 'gender is not enough'. I extend that analysis by fleshing out the difference that ethnonational identity makes in defining gender identities, such as 'woman', in Sri Lanka. I then turn to the story of Rajani, to show that not only gender and ethnonational identity position people in specific ways during conflict, but that one's identity shapes and is shaped by political identity and activity over the course of a lifetime. Her efforts to make change happen and seek justice for the people of Jaffna illustrate how politicized identities change as the grounds of conflict shift. Social, economic, political and cultural identity cannot be essentialized nor separated from one another.

Context and Background

The 26-year-long war in Sri Lanka that technically ended in May 2009 was characterized by violent and competing nationalisms between the Liberation Tigers of Tamil Eelam (LTTE) and the Government of Sri Lanka's armed forces, beginning with the pogroms of 1983 (Jayawardena and de Alwis 1996). Sri Lanka has a long history and geography of struggle, well-documented by scholars (Abeysekera and Gunasinghe 1987, Spencer 1990, Jeganathan and Ismail 1995, Thiruchelvam 1996). The conflict spawned large-scale displacement within the country and well beyond its borders, where a significant Sri Lankan Tamil diaspora has emerged from this country of just under 20 million people (Hensman 1993, Daniel 1997, Fuglerud 1999).

People opposing the terms and tactics of the conflict come from all ethnonational groups. Many have disappeared or been killed for their efforts. Rajani Thiranagama is but one such person, shot dead for her role in exposing violence and human rights atrocities visited upon Tamil people in Jaffna. Too many Sri Lankans have been murdered for their efforts to stop violence since then, their conciliatory, sometimes federalist, non-violent proposals for change punished with murder. In 2000, an esteemed human rights lawyer with an international reputation, Neelan Thiruchelvam, an elected member of parliament and a Tamil political party leader, was killed by a suicide bomber in Colombo while driving to work in gridlock traffic. On 13 August 2005, unknown gunmen shot and killed Sri Lanka's Foreign Minister, Lakshman Kadirgamar, at his home in Colombo. Kadirgamar, a Tamil member of the ruling Sri Lanka Freedom Party, had led the campaign to have the LTTE labelled a terrorist organization by several countries. It was widely acknowledged that he topped the LTTE's list of political targets, despite being Tamil himself. In August 2006, Kethesh Loganathan, Deputy Secretary General of the Sri Lanka Government's Secretariat for Coordinating the Peace Process (SCOPP), was shot dead by an unknown gunman on a Saturday evening in Colombo. Although very different in their political positionings, these are just a few people who paid the highest price for their efforts to effect political change.

In February 2002, a ceasefire was signed and with it a glimmer of hope that it might provide a political space for reconciliation and an end to hostilities. Relative calm, which included the dismantling of many army checkpoints that dotted the country, held throughout the post-tsunami period until 2006. Despite further hopes that the 2004 tsunami might bring peace or at least extend the ceasefire, intense fighting between government forces and the Tamil Tigers resumed. More than a thousand people were killed in 2006, including 17 NGO staff working for an international relief organization, who were murdered in an unprecedented attack in August (Apps 2006). In February 2007, an apparent assassination attempt on the US and Italian ambassadors in Batticaloa signalled an escalation of the conflict. The renewed fighting forced thousands of tsunami survivors in the Eastern Province to flee their homes yet again (Pathoni 2006).

In May 2009, military conflict in Sri Lanka ended with the defeat of the LTTE and the death of its leader, Prabakharan. Hundreds of thousands of Tamil Sri Lankans caught in the crossfire in the final confrontation in northern Sri Lanka were put in internment camps after the government's victory, ostensibly because they posed a security risk. A state of emergency was maintained in order to evade constitutional protection, which would have made it impossible to detain one's own citizens, and Sri Lankan Tamils languished in these camps until late 2009. President Mahinda Rajapakse announced the release of thousands before elections were held early in 2010, when he was reelected with an overwhelming majority. However, peace and the reconciliation that is a precondition for it remain elusive.

In the following, I reiterate my conception of gender and show how its meaning is co-constituted with, through and against ethnonational identities.

Gender is Not Enough

Gender relations, as a dimension of all human relations, and the constitution of gender identities are central concerns of feminist politics and thought. However, the primacy of gender as an analytical category over other historical and geographical locations is not fixed across time and place (Hyndman and de Alwis

2003). As Gerry Pratt (2004: 24) notes, Western feminism has been criticized for its own exclusions:

It is not simply that women's experiences differ depending on other aspects of their social locations; feminism as a political movement takes a different trajectory in different places, depending on how it articulates with other political struggles.

Daiva Stasiulis (1999: 194) elaborates on the importance of relationality, positionality and 'relational positionality' to feminist politics: 'They refer to the multiple relations of power that intersect in complex ways to position individuals and collectivities in shifting and often contradictory locations within geopolitical spaces, historical narratives, and movement politics.' Positionality situates the producer of knowledge in a web of historically specific, culturally inflected and geographically unique although globalized power relations. Relationality builds upon this *place* of knowledge-making, not as a fixed point in space or time, but as a grid of power/knowledge characterized by connectivity, linked political projects, and transnational political and social movements. I agree with Stasiulus that materiality is vital, but argue that poststructuralist analyses do not categorically deny the material bases of power relations. A poststructuralist analysis can reveal the very processes by which particular constellations of power are effaced or naturalized (Butler 1990). As Mala de Alwis and I argued in our critique of 'gender' within humanitarian and development work in Sri Lanka:

We contend that a thoroughly feminist analysis must incorporate *multiple* bases of identity and social relations, not exclusively gender. A feminist analysis provides a more powerful lens with which to examine the place of women and men, and a more compelling position from which to transform relations that provoke or perpetuate violence, hate, and inequality. The intersection of one's class, caste, religion, sexuality, nationality and/or race, and membership in social groups *produces* different gender relations across time and space. The contingencies of place and time also shape the meaning of 'woman' and 'man' in different contexts of development and humanitarian work. (Hyndman and de Alwis, 2003: 215 emphasis added)

In short, a gender identity or analysis that prefigures the specific context of conflict risks creating a reductionist reading of 'women in war'.

I extend this argument throughout the chapter, noting that the use of feminist thinking is deliberate. To assume that gender is somehow the primary category of analysis or most disparate axis of difference within a context of neoliberal imperatives, competing nationalisms and conflict is too simplistic. Adopting an approach to power relations that includes but does not privilege asymmetrical gender relations is vital. While patterns of patriarchal power can be identified across world regions, social classes and diverse cultural contexts, it does not operate evenly across geographical space or in trans-historical ways.

As Chandra Mohanty (2003) has argued, the category 'woman' is inherently unstable. Its meaning also varies across ethnonational identity and geographical location. Gender cannot be analysed in isolation from these co-constitutive factors. Ultimately, these differences intersect in ways that position people very specifically in relation to each other. While 'gender and development' (GAD) analyses examine power relations and are generally committed to political change where necessary, this approach assumes that the change will need to be made specifically in the realm of gender relations (Rathgeber 1990). Gender identities – what it means to be a man or woman in a particular society – are co-constituted through historical, geographical and social locations. A GAD approach does not account for caste bigotry, ethnonational differences and the economic geographies of displacement due to war.

Gill Valentine (2007) analyses intersectionality in feminist scholarship, arguing that not only are identities co-constituted across multiple axes of difference (including but not limited to gender), but also that expressions of identity may come to the fore at different times. Valentine uses the metaphor of 'geometries of oppression' to trace how identity is 'done', by whom and when. 'The identity of particular spaces ... are in turn produced and stabilized through the repetition of the intersectional identities of the dominant groups that occupy them' (Valentine 2007: 19), yet the foregrounding of particular identities can vary across a single lifetime. Valentine, however, fails to grasp Pratt's (2004) important observation that women's different experiences depend not only on aspects of their social locations, but also on the different trajectories of history in different places, and how feminism articulates with other political struggles in situ.

With respect to Sri Lankan widows, for example, Ruwanpura and Humphries (2004: 187) reason that a young widow with children is likely to face greater financial responsibilities than an older one with grown children: 'The needs of a young widow with several dependent children whose husband has been killed in ethnic violence may be very different from the needs of a middle-aged widow with several children old enough to work ...'. Furthermore, widows are caste as more deserving female heads of households than their non-widow counterparts. These authors question homogenizing accounts of Sri Lankan women's lives and explore ethnicity as a source that produces difference among female-headed households in eastern Sri Lanka.

Significantly different experiences of war were noted across female-headed households despite 'oppressive gender standards within both Sinhala and Tamil ethnic groups ... emphasizing motherhood and sacrifice as the archetypal feminine path .' (Ruwanpura and Humphries 2004: 179). Jayawardena and de Alwis (1996) have written about 'moral motherhood' in the context of the war in Sri Lanka, and the ways in which women are used in nationalist projects as reproducers of the nation and bearers of cultural identity. Gender and ethnicity, then, *co-constitute* and *differentiate* the meaning of 'woman', and by definition 'widow', in Sri Lanka.

The differential exposure of these women from different ethnonational backgrounds to the conflict in Sri Lanka is remarkable. Ethnonational belonging, socioeconomic class, gender and geographical location *produce* identities in ways that change over the life course and across the political terrain as it evolves. I analyse selected writings by Dr Rajani Thiranagama to illustrate how political identities, like social ones, are complex composites that are always in process.

University Teachers for Human Rights, Jaffna

Rajani's sisters describe their childhood, including Rajani's, as a privileged, Christian, middle-class one. She read novels like *Little Women,* choices very much shaped by the British colonialism that officially ended in Sri Lanka in 1948. Leaving her home in the predominantly Tamil area of Jaffna in northern Sri Lanka, Rajani attended Colombo University's medical school where she joined the student movement. She married a Sinhalese man, Dayapala, from the same JVP (Janatha Vimukthi Peramuna, People's Liberation Front) movement in which she worked hard for nonviolent change (Klodawsky 2005).

She was also influenced by her sister, Nirmala, who earned a scholarship to study in the United States and attended university during the Vietnam War era. Nirmala was imprisoned soon after her return to Sri Lanka for harbouring suspected rebels from the Tamil Tigers. Rajani took a doctoral fellowship in London and campaigned for her sister's release while working on her dissertation. There she joined the LTTE, a decision that her by then estranged husband loathed, and worked to strengthen its international presence while campaigning for her sister's release. As he pointed out, there was no room for Dayapala, a Sinhalese activist, in a nationally homogenous (Tamil) militant group (Klodawsky 2005).

As a highly educated Tamil woman, Rajani was a coveted spokesperson for the LTTE in London, but as the bloodletting continued in Sri Lanka, Rajani became disillusioned with the movement. Once Nirmala had escaped from prison with the help of the LTTE and Rajani had completed her studies, both sisters decided to leave the LTTE, a dangerous step away from a movement that – as Nirmala put it – did not allow members to leave (Klodawsy 2005). Both Rajani and Nirmala were accused of being Western feminists by the LTTE as both had studied abroad, questioned the tactics of the LTTE and left the movement. In her own words from *The Broken Palmyra*, Rajani was acutely aware of her own colonization, especially upon her return to postcolonial Sri Lanka:

> Imperial psychology has over the centuries, developed increasing subtle and sophisticated means to subjugate and oppress people. But with regard to women, it still employs the most barbaric forms of control and repression – arrogance, dominance, men in battle garb, whether they come with swords or guns on a horse or in armoured cars. The price of conquest seems heightened by the violation of the women. (UTHR(J) 1987, 5.1)

Returning to Jaffna, Rajani aligned herself with no party or political stripe, instead committing herself to her medical students – mostly men – and the social development of the people of Jaffna, particularly women who were so profoundly affected by the conflict. In her search for justice, beyond the ideologically-driven and nationalistic social movements, she worked hard – as so many others have since then – with and for those harmed by war, protesting through the UTHR(J) at the atrocities committed by both government forces and Tamil Tigers in the ongoing conflict. She paid a high price for this commitment and affiliation in 1989, namely her life. Today the human rights activists that Rajini worked with in UTHR(J) live abroad or are not involved in human rights work in Sri Lanka, given the dangers it still presents.

The founders of UTHR(J) wrote *The Broken Palmyra* (see also Hoole et al. 1992). One of the chapters was authored solely by Rajani and titled, 'No more tears sister.' It recounts women's stories of the conflict told to Rajani, focusing on the extraordinary adversity ordinary women faced.

No More Tears Sister

Helene Klodawsky's (2005) film with the same title, *No More Tears Sister*, documents the life of Rajani through interviews with her sisters (Nirmala, Vasuki and Sumathy), parents, husband (Dayapala) and two daughters. The film is remarkable for the candour of its interviews and multiple 'takes' – not always favourable – of Rajani's decisions, her rage at the violence she witnessed, and her parenting, stubbornness and tenacity.

The clarity of Rajani's reporting in 'No more tears sister' brought the plight of Tamil women in Jaffna to the attention of a wider public. This work, like so many publications by UTHR(J), does not allow violence by any side against civilians to go undocumented. Much of Rajani's chapter narrates the terror invoked by the Indian Peacekeeping Forces, who were supposed to disarm the LTTE but ended up fighting against them. UTHR(J) is not a large human rights organization, but it is a powerful voice that refuses to be silenced by the threat of violence. Here is what one woman told Rajani:

> On 12 November, in the morning three Indian soldiers came to our house at about 8 o'clock. My mother was in the kitchen. Only my daughter and I met them. They merely said: "Checking," and started pushing my daughter into a room. I dragged her and shouted "Amma, Amma, checking, checking". Then the soldiers who were at the sentry point very near our house came running to our house ... We were scared. I then took my daughter and hid her in a small box-room at the rear of the house and at about 9:30 we saw the same three soldiers ... but instead [they] came through another vacant house, jumping over the common parapet wall. Then they locked my parents in one room, took me to a room, showed the gun and raped me, one after the other, all three of them. I did not scream. What if they shot my parents? I can still recollect those beady

eyes. I could not handle it. I left the village, and Jaffna, as soon as the first buses started running to Colombo. I started having nightmares. I started seeing their faces and hearing voices. I took my daughter and we went abroad. I even went to a psychiatrist. I could talk to him because he was a total stranger. He gave me drugs. It quietened me, but it has not taken the trauma away. I am becoming worse, much worse. At least I saved my daughter. (UTHR(J) 1987, 5.4)

Rajani recorded the trauma of this testimony, though her writings also speak to her own feelings of anger and impotence. Many more cases of rape are documented, one of which ends in suicide. One woman activist noted a geography and temporality of these sexual attacks:

Rape occurred mainly in November and December, when the families were trickling from the refugee camps to their old homes. Many women were quite isolated with few neighbours being around. It provided ample opportunity for the soldiers to rape. Many of the women were beaten before being raped. (UTHR(J) 1987, 5.4)

Hearing these stories from survivors of the violence, Rajani was herself clearly traumatized by the brutality visited upon the women of Jaffna. A male government official told Rajani, "'I agree that rape is a heinous crime. But my dear, all wars have them. There are psychological reasons for them such as battle fatigue.'" She writes, 'A screaming rocket burst into my head. I thought to myself, yes it is part of all wars; but still, we women cannot swallow it. Our bodies are ours. You cannot relieve yourselves on us' (UTHR(J) 1987, 5.4).

Rajani did not live long enough to learn that rape is now considered a weapon of war, punishable like any other war crime after 1996. Even this news, however, would have done little to assuage the anger and violation she expressed, living in the war zone.

The Sri Lankan Army targeted Tamil men between 14 and 40 as the most likely to be recruited as Tigers and terrorists, so many young men left their communities, leaving behind women, children and the elderly to carry on. One woman explained that 'with all these incidents of molestation, and rape, we cannot go anywhere without a male escort and most of the time we are forced to remain inside' with so many men gone. Rajani comments on this testimony: '… the ongoing incidence of sexual violence against women gives the impetus to the return of narrow values' (UTHR(J) 1987, 5.4).

Ranjani laments the cumulative effect of the fear generated by this violence, prompting women to retreat to their homes, hide behind closed doors and windows, and refrain from riding bicycles and moving freely without male escorts. However, this shift to more 'traditional' ways was also condoned by the Indian Peacekeeping Forces. One teacher read a note to her class provided to her by the Indian Army; it read: 'Girls should wear saris, and should not go around on cycles' (UTHR(J)

1987, 5.4). Leaving sexual violence in their wake, the Indian Peacekeeping Forces had the temerity to leave morality tales alongside the damage they engendered.

Class and Conflict

Dr Rajani Thiranagama was very critical of the political inertia of middle-class privilege from whence she came. Speaking of the fear created by rape, she noted:

> ... without a strong women's leadership in the movements, or grass roots organizations, the community of women had no path to organize along and come together so as to raise their voices against such gross acts as sexual violence. And we find two clear streams of action emanating. (UTHR(J) 1987, 5.4)

On the one hand, the middle-class families whose members experienced rape tried to hide the violence. They were averse to exposing perpetrators because they feared the censure of the society around them. They sidestepped the problem, thinking of themselves as individuals, not as part of community being literally raped and pillaged. Rajani wrote: 'This portrays the anaemic character of our middle classes in whom the community has reposed its power but who continually fail the people' (UTHR(J) 1987, 5.4). Rajani's analysis of class and gender echo those of many socialist feminist commentators of that time.

On the other hand, 'the victimisation of women,' Rajani argued, 'individualised the burden of the act carried by the woman, thereby internalizing the pain and trauma and creating far reaching damage to the inner life of a woman. The society stands apart and the most it does is to indulge in sympathetic gossip.' She cites a 20-year old woman on this troubling topic: 'Why can they not treat it as a wound sister and let it heal? The soldiers destroy once. But the village destroys us a thousand times' (UTHR(J) 1987, 5.4).

The Greek chorus of 'community', according to Rajani, could ruin women's reputations as much as the perpetrators of sexual violence. As Iris Marion Young (1990) has argued, 'community' is as much about exclusion as belonging. The shame and individualization of rape were clear issues in this particular period.

Rajani was disappointed by herself and many of those around her: 'One had seen the shattered interior of a wounded woman. However much one consoled and advised, it seemed so stupid. When I went home, I felt exhausted, impotent and angry at ourselves, our class, our men and our whole passive, stupid society' (UTHR(J) 1987, 5.4).

Rajani was operating largely on her own in her efforts to document atrocities in Jaffna. She was murdered while cycling to work in 1989.

The Broken Palmyra also documents the torture and degradation that Tamil women faced at the hands of the Sri Lankan army when taken into custody. The story of Suseela, a girl who was only 14 when she was married to a mentally disabled man from a rich family, rounds this out by revealing the injustice that fuelled Rajani's politics:

'Sister, life – happiness – had no meaning to me. Of course his family were kind to me. They gave me jewellery and built a small house for us. I lent the jewellery to my family to make a living. I had no life of my own. My husband would leave me and go to his parents' house to sleep … I want to fill my life. I want to do social service. I want to be of service to others. That is how I started helping the people in the village and then the Tigers. I want to do something. Sister, after all these happenings, though people are grateful, they are scared to associate with me, especially to send their daughters with me … Even when I come to see you, they say that I have gone to meet the Tigers. After all this, the barren life and the pain, *I have no more tears sister*'. (UTHR(J) 1987, 5.5)

Conclusion

From daughter and Marxist student protestor to sister, nationalist rebel and human rights campaigner to protofeminist champion of Tamil women in Jaffna, Rajani Thiranagama took on a range of political positions to promote social justice for Tamil people in Sri Lanka.

Ethnonational identity produces meanings of woman and man in a development setting and in a conflict setting. 'Any understanding of women's experiences based on a narrow conception of gender would simply be incapable of fully addressing the homogenizing and hierarchizing effects of economic and cultural process' (Alexander and Mohanty 1997: xvi); political processes can be added to this mix. Just as class and gender relations cannot be separated out from one another – they are co-constituted – so too do class and ethnonational identity produce one another in a recursive fashion. Identities are never fixed, neither socially nor politically.

Few scholars have contributed to feminist research during times of peace, war and disaster in Sri Lanka, but Ragnhild Lund is one who has, and I dedicate this work to her.

References

Abeysekera, C. and Gunasinghe, N. (eds) 1987. *Facets of Ethnicity in Sri Lanka* Colombo: Social Scientists' Association.
Alexander, M.J. and Mohanty, C.T. 1997. Introduction: Genealogies, Legacies, Movements, in *Feminist Genealogies, Colonial Legacies, Democratic Futures*, edited by M.J. Alexander and C.T. Mohanty. New York: Routledge, xiii–xlii.
Apps, P. 2006. Shelling a Call to War, Tamil Rebels Say. *The Globe and Mail*, 7 August.
Blaikie, P. and Lund, R. (eds) 2010. *The Tsunami of 2004 in Sri Lanka: Impacts and Policy in the Shadow of Civil War*. London: Routledge.

Brun, C. and Lund, R. 2008. Making Home During Crisis: Post-Tsunami Recovery in the Context of War, Sri Lanka. *Singapore Journal of Tropical Geography*, 29, 274–87.

Butler, J. 1990. *Gender Trouble: Feminism and the Subversion of Identity*. New York: Routledge.

Butler, J., Laclau, E. and Žižek, S. 2000. *Contingency, Hegemony, Universality: Contemporary Dialogues on the Left*. London: Verso

Daniel, V. 1997. Suffering Nation and Alienation, in *Social Suffering*, edited by A. Kleinman, V. Das and M. Lock. Berkeley: University of California Press, 309–58.

Edkins, J. 2003. *Trauma and the Memory of Politics*. Cambridge and New York: Cambridge University Press.

Fuglerud, Ø. 1999. *Life on the Outside: The Tamil Diaspora and Long Distance Nationalism*. London: Pluto Press.

Hensman, R. 1993. *Journey Without a Destination*. Colombo: Centre for Society and Religion.

Hoole, R., Somasundaram, D., Srtiharan, K. and Thiranagama, R. 1992. *The Broken Palmyra: The Tamil Crisis in Sri Lanka – An Inside Account.* 2nd edition. Claremont: The Sri Lanka Studies Institute.

Hyndman, J. and de Alwis, M. 2004. Bodies, Shrines, and Roads: Violence, (Im)mobility, and Displacement in Sri Lanka. *Gender, Place and Culture*, 11, 535–57.

Hyndman, J. and de Alwis, M._2003. Beyond Gender: Towards a Feminist Analysis of Humanitarianism and Development in Sri Lanka. *Women's Studies Quarterly*, 31, 212–26.

Jayawardena, K. 1986. *Feminism and Nationalism in the Third World*. London: Zed Books.

Jayawardena, K. and de Alwis, M. 1996. Introduction, in *Embodied Violence: Communalising Women's Sexuality in South Asia*, edited by K. Jayawardena and M. de Alwis. New Delhi: Kali for Women, ix–xxiv.

Jeganathan, P. and Ismail, Q. 1995. Introduction: Unmaking the Nation, in *Unmaking the Nation: The Politics of Identity and History in Modern Sri Lanka*, edited by P. Jeganathan and Q. Ismail. Colombo: Social Scientists' Association, 2–9.

Klodawsky, H. 2005. *No More Tears Sister: Anatomy of Hope and Betrayal* [video recording]. Montreal, QC: National Film Board of Canada.

Lund, R. 1979. *Prosperity through Mahaweli – Women's Living Conditions in a Settlement Area*. Master's dissertation in geography. Bergen: University of Bergen.

Lund, R. 1993. *Gender and Place*, Volume 1: *Towards a Geography Sensitive to Gender, Place and Social Change*; Volume 2: *Examples from Two Case Studies*. Trondheim: Department of Geography, University of Trondheim.

Lund, R. 2000. Geographies of Eviction, Expulsion and Marginalization: Stories and Coping Capacities of the *Veddhas*, Sri Lanka. *Norsk Geografisk Tidsskrift– Norwegian Journal of Geography,* 54, 102–9.

Mohanty, C. T. 2003. *Feminism Without Borders: Decolonizing Theory, Practicing Solidarity.* Durham, NC and London: Duke University Press.

Pathoni, A. 2006. Still Shaking, Asia Marks Tsunami, *The Globe and Mail,* 27 December, A11.

Pratt, G. 2004. *Working Feminism.* Philadelphia, PA: Temple University Press.

Rathgeber, E. 1990. WID, WAD, GAD: Trends in Research and Practice. *Journal of Developing Areas,* 24, 489–502.

Ruwanpura, K. and Humphries, J. 2004. Mundane Heroines: Conflict, Ethnicity, Gender and Female Headship in Eastern Sri Lanka. *Feminist Economics,* 10, 173–205.

Seneviratne, B. 1987. A Foreword, in *The Broken Palmyra* [Online: UTHR(J) University Teachers for Human Rights (Jaffna), Sri Lanka]. Available at: http://www.uthr.org/BP/A%20FOREWORD.htm [accessed: 4 September 2007].

Spencer, J. (ed.) 1990. *Sri Lanka: History and the Roots of Conflict.* London and New York: Routledge.

Stasiulis, D. 1999. Relational Positionalities of Nationalisms, Racisms, and Feminisms, in *Between Woman and Nation: Nationalisms, Transnational Feminisms, and the State,* edited by C. Kaplan, N. Alarcon and M. Moallem. Durham, NC and London: Duke University Press, 182–218.

Thiruchelvam, N. 1996. Sri Lanka's Ethnic Conflict and Preventive Action: The Role of NGOs, in *Vigilance and Vengeance: NGOs Preventing Ethnic Conflict in Divided Societies,* edited by R. Rotberg. Washington DC: Brookings Institution, 147–64.

UTHR(J) 1987. 'No More Tears Sister': The Experience of Women: War of October 1987, in *The Broken Palmyra,* volume 2, chapter 5 [Online: UTHR(J) University Teachers for Human Rights (Jaffna), Sri Lanka]., Available at: http://www.uthr.org/BP/volume2/Chapter5.htm [accessed: 13 September 2007].

Valentine, G. 2007. Theorizing and Researching Intersectionality: A Challenge for Feminist Geography. *The Professional Geographer,* 59, 10–21.

Young, I.M. 1990. The Ideal of Community and the Politics of Difference, in *Feminism/Postmodernism,* edited by L. Nicholson. New York and London: Routledge, 300–23.

PART III
Human–Environment Relations, Environmental Discourses and Development

PART III

Chapter 11

Renegotiating Local Values: The Case of Fanjingshan Reserve, China

Stuart C. Aitken, Li An, Sarah Wandersee and Yeqin Yang

Introduction

In the early 1990s, Ragnhild Lund (1993) argued for a place- and people-centred perspective in development theory. In doing so, she was one of the first geographers to recognize the devaluation of place in conventional development studies. Lund points out that guidelines for development during the latter half of the twentieth century were heavily focused on economic rather than social, cultural or ecological issues. Drawing from a range of theories from Marxism to liberalism, she notes that few of these perspectives focused on place as a dynamic factor in societal change. Rather, place was characterized as a relatively passive stage for social and political action and interaction. In a broad-ranging critique of 'modernist' Rostowian and Marxist perspectives, Lund argues against both the popular regionalizing systems theories based on Wallerstein and humanist perspectives that advocated endogenous self-reliance. The former tend towards purely economic solutions and the latter suggest a focus on local practices that are hugely valuable and tie in with perspectives that date back to Ernst Schumacher's famous *Small is Beautiful* (1973), but are also easily conscripted into the service of neoliberal policies that foist way too much responsibility for economic advancement on to the shoulders of people who are least able to bear it. Instead, Lund favours an 'alternative development' that takes account of gender and is influenced by social movements focused on ecology, peace and women (see also Nederveen Pieterse 1998). In so doing, she recognizes:

> … that women and men encounter a variety of external policies and interventions in a given place, and modify and adapt to external influences in accordance to norms, conventions and practices prevalent in the local society. Both internal and external factors are historically and geographically specific. Consequently, it is necessary to understand the relationship between gender and place to realize change. (Lund 1993: 197)

Lund's perspective highlights the need for indigenous women to renegotiate continuously local values in the light of broad structural economic and social transformations. Her poststructural feminism is pragmatically grounded in the realities of shifting economic and social conditions and how they play out in local landscapes. In a study of working women in Malaysia, for example, Lie and Lund (1995: 10) argue that studies must focus on local values and, from a feminist perspective, '... women's views on the changes taking place in their own lives as well as in their families and the local surroundings.' Lund's focus on women's perceptions and values ties in with 1990s literature on feminist, postmodern and populist approaches to development and postdevelopment (Nederveen Pieterse 1998, Blaikie 2000, Momsen 2004). It also resonates with Arturo Escobar's (2008) interest in the ability of indigenous peoples to create 'figured worlds' in which local practices, culture and identities are deployed effectively enough to create a visible (spontaneous, emotional and corporeal) space for authoring that contest external, hegemonic representations of that place.

With this chapter,[1] we bring Lund's feminism and Escobar's poststructuralism to bear on a participatory mapping project in Fanjingshan National Nature Reserve (FNNR), China (Figure 11.1). A critical issue in FNNR relates to the resource-use relations between local farmers and an endangered snub-nosed monkey species, *Rhinopithecus brelichi*, known as the grey snub-nosed monkey or the Ghizhou snub-nose golden monkey. The United Nations International Fund for Agricultural Development (IFAD) argues that participatory mapping is particularly important when dealing with indigenous peoples and forest dwellers that find their lives disproportionately threatened by reduced access to land and natural resources. The 21,000 farmers within FNNR have an intimate knowledge of their local environment, and participatory mapping is an important tool for accessing this knowledge in the complex gender and child/adult contexts of human–environment dynamics. Based on first impressions and pilot work, we talk about the efficacy of this technique, particularly in terms of how it embraces a place-based sensibility that empowers female, male, child and adult participants. We then discuss local women's participation in interviews and highlight, as examples of Escobar's figured worlds, their roles in a changing world of short-term work and boarding schools for children. We then speculate on ways the FNNR example demonstrates Lund's imperative for renegotiating local values and Escobar's (2008) concern for building on identity, territory and autonomy where they may exist locally.

Post-structural Feminism: A Failure of Heart

Lund's perspective on renegotiating local values in the face of economic and social restructuring presages contemporary poststructural feminism, which is not only critical of patriarchy but forefronts the far more radical idea that development

1   Some of the ideas in this chapter originated in Aitken and An (2010). The research was made possible by generous funding from the Margot Marsh Biodiversity Foundation and the Zoological Society of San Diego, and through education and facilities support from San Diego State University, the University of California – Santa Barbara, and Fanjingshan National Nature Reserve.

Figure 11.1 Fanjingshan National Nature Reserve (FNNR), Guizhou Province, China

problems actually have their origins in the (male) reasoning of enlightenment thinkers influenced by René Descartes and Baruch Spinoza amongst others.[2] Descartes attributed clear thinking to men and emotion to women. While recognizing the importance of emotions in how we think, Spinoza nonetheless believed that emotions were transformed into intellect through a strong man's detached understanding of grand questions such as universality and transhistorical necessity (Peet and Hartwick 2009). With an understanding that emotions play an important part in creating the complexities of men's and women's lives, Lund (1993: 195) notes that '... gender roles and gender relations are not framed on the basis of patriarchy alone.' By recognizing the importance of emotion- and place-based contexts of development that are not inordinately (and apolitically) humanistic, her work joins with a strand of late twentieth-century feminism that elaborates a poststructural critique of reason and one of its problematic enlightenment products,

---

2   Chris Norris (1991) provides a useful elaboration of the numerous works of Spinoza and Descartes in terms of how they influence both enlightenment and modern thinkers. For those interested in the original works, Spinoza's *Ethics* (2008 [1677]) and Descartes' (1998 [1664]) *Le Monde et l'Homme* are amongst their most influential works respectively.

modern development. In an important sense, a poststructural feminism argues that modern development is the problem for women (and men), not the solution.

A focus on local values destabilizes the grand terms of enlightenment-based, universal development that is planned from the Global North and implemented in the Global South. Tropes such as development, modernization, self-reliance and revolution may speak to important parameters of change and transformation, but poststructuralists and feminists argue that they also speak to the dominant policies and practices of international institutions, nongovernmental organizations (NGOs) and revolutionary governments whose bases are predicated upon masculinist endeavours and a male-dominated public sphere (Scott 1995). The ensuing power struggle places rationality, efficiency and optimism at the forefront of a regime that may also characterize women's work as inferior, backward or invisible. In discourses of this kind, social struggles focus on productive activities that exclude gendered power relations and retain notions of a subordinate reproductive sphere and ideas of nature that are seen as feminine.

A problem arises, however, from switching the valences of the discourse by putting women and their work at the centre of development discourses. Too often, accounts of 'women in development' are written in policy language amenable to the ongoing practices of development agencies. Making women central to development practices is here often about changing women (e.g. requiring them to speak bureaucratic policy language) rather than changing institutional practices. Putting women at the heart of development in this way is about fostering development practices that continue to ignore difference, indigenous knowledge and local expertise while 'legitimating foreign "solutions" to women's problems in the South' (Parpart and Marchand 1995: 16). This, in turn, shifts development solutions from local areas to development agency headquarters in Washington, Oslo, Geneva or Ottawa.

Early on, as part of this critique, Lund (1983) argued that placing women at the centre of development efforts while not monitoring larger globalization processes (with a particular concern for the movement of women's power away from the local) did little to further change in the Global South. Globalization is not a homogenizing force, she notes, but is rather a force of differentiation, as some people are integrated into the global economy while others are marginalized, abused or rejected (Lund 2008).

Alternatively, focusing on women *and* development rather than women *in* development draws from dependency theory and neo-Marxist approaches to underdevelopment but, in so doing, it deemphasizes Marxist class relations in favour of social relations between women and men, and the relations between women and the material contexts of their lives (Peet and Hartwick 2009: 259). A basic materialist argument is that women perform most of the labour in many societies of the Global South and a reformulated theory must focus on that as its heart. With women's labour as a central focus, traditional areas of developmental concern are seen from a different orientation. Gender relations become central to understanding productive activities (Figure 11.2). The focus turns to women

Figure 11.2   Women prepare food in Fanjing Valley
     Photo: Stuart C. Aitken

workers as part of the turn to industrialization (Lie and Lund 1995). The informal and rural sectors of the economy are emphasized and, as suggested by Gibson-Graham (1996), the reproductive sphere becomes central to the creation of economic communities that foster sustainable forms of development. A central concern of feminist, poststructural and postcapitalist economics is retheorizing the significance of women's empowerment through their work and agency from a relational perspective (Escobar 1995, Gibson-Graham 2006).

Lund's feminism and Escobar's poststructuralism move alternative development theories forward by focusing on a relational understanding of change and transformation with a focus on creating 'locally situated, culturally constructed and socially organized' 'figured worlds' as the sort of spaces 'in which cultural politics are enacted that result in particular personal and collective identities' (Escobar 2008: 218). These kind of poststructural relations are best articulated through Deleuze's (1993) notion of *folding*, wherein people have the capacity to unfold and enfold the spaces and discourses they encounter through the myriad of microbehaviours that comprise everyday actions. Folding suggests important relations with space and other people's relations with space.[3]

---

3   It is beyond the purview of this chapter to elaborate the ontological basis of Deleuze's relationality (but see Hayden 1995). In short, identity politics may be thought of

It is a different conception of relational than that elaborated by neopopulist and constructivist development theorists in the 1980s, who tend to focus on dyads such as insider/outsider or core/periphery.[4] Escobar's relational focus is tied specifically to activism and everyday behaviours as bases for challenging inequality and neoliberal policies.

Changes in behaviour are often strategies to preserve basic elements of lifestyle and traditions: changes seldom occur in the form of dramatic events, and social change may be seen as something that is discursively imposed on people (Lie and Lund 1995: 7–8). However, it is important to understand that from Escobar's relational perspective the material and gender contexts of life change over time and that marginalized peoples can take advantage of these changes:

> ... when women enter new fields, such as taking up work in the modern sector,
> this necessarily implies changing relationships to fathers, brothers and husbands
> and may lead to new socio-cultural definitions of what belongs to the male and
> female spheres. (Lie and Lund 1995: 11)

This idea of complex relations being unfolded and enfolded is picked up by Escobar (2008) when he argues for *redes* as networks or assemblages that open up the possibility of transformative action in the face of blistering and relentless attacks by corporate and colonial capitalism.[5] Life and social change, he points out,

> ... are ineluctably produced in and through relations in a dynamic fashion ...
> Images of *redes* circulated widely ... in the 1990s [in the Global South] ...
> represented graphically as drawings of a variety of traditional fishing nets,
> lacking strict pattern regularity, shaped by use and user, and always being
> repaired, *redes* referred to a host of entities, including among others social
> movement organizations, local radio networks, women's associations, and
> action plans. (Escobar 2008: 26)

---

in terms of Deleuze's 'folding of forces', which concerns the ability of a body to act upon itself and to produce itself as a subject. This force enables a bending back of power – a self-governance – to emerge. In this way, Deleuze's folding of forces enables a subject to construct actively and shape its own gendered bodily relations to the world: it is a body that is produced, reproduced and naturalized through everyday behaviours. Alternatively, the 'folding of bodies' concerns the body's material relations with space and the 'fold of the line outside the fold' forms when a body connects with the creative potential of pure matter energy, which Deleuze (1988: 104) calls the virtual: a very real but not yet actualized potential. This is the indeterminate space of change.

4   Robert Chambers (1983), for example, uses contemporaneous methodologies such as personal construct theory and the repertory grid, which are epistemologically focused on the dichotomies that were later heavily criticized by feminist geographers amongst others.

5   The Spanish *redes* is most closely related to the English term *network,* but Escobar (2008: 36) uses it to convey the more powerful Deleuzian idea of *assemblages* that constitute folding and dynamic relations.

Escobar (2008: 65) goes on to argue that relational strategies for battling externally imposed structures 'should take as a point of departure an understanding of resisting, returning, and re-placing that is contextual with respect to local practices, building on movements for identity, territory, and autonomy wherever they may exist.' Both Escobar and Lund argue that it is precisely how women and their work are integrated into the global economy by core countries that determines marginalization and oppression. From this vantage point it is important to explore the intersection of various axes of power in relation to participation, vulnerability, class and gender (Lund 2008: 134).

Relational Knowledge and Power

It was not until the 1990s that alternative, antimodernist development theories critiqued development practices for their concerns with fulfillment of basic needs with a focus on indigenous values, neoliberal self-reliance and so forth. Lund (1993) was one of the first to argue that identifying ways to empower the poor and marginalized was more appropriate than focusing solely on fulfilling needs, which were frequently inappropriately identified by 'experts' from elsewhere. Part of this critique questioned how men and women could sensitize themselves and act against oppressive structural forces, including patriarchy (Lund 1993). The structures affecting women's lives – production, reproduction, socialization, motherhood, gender and sexuality – contain different contradictions and dynamics but they nonetheless contain a unity in women's experience. Women are contextualized by the shifting social relations they inhabit and the types of labour they perform.

A focus on gender *and* development argues that the sexual division of labour in a society is one of the relations in which men and women become dependent upon each other, and these relations must necessarily change. Gender power relations rather than 'women in development' are the needed focus of analysis. In addition, a focus on gender and development emphasizes that women are not a homogenous group but rather are divided by class, ethnicity, age and so forth. Women are seen as social actors within wider social contexts, and the state can be an important actor promoting women's emancipation (Peet and Hartwick 2009).

Escobar (2008: 32) approaches power over the production of locality as being tantamount to two conflicting yet at times mutually constitutive 'processes of localization.' On the one hand, there are the dominant forces of the state and capital, which attempt to 'shift the production of locality in their favor,' thus ultimately creating '... a delocalizing effect with respect to places,' and, second, what Escobar refers to as subaltern forms of localization: 'place-based strategies that rely on the attachment to territory and culture; and network strategies.' In the first instance capital and the state mobilize the politics of scale that valorize local endeavours (e.g. some ecotourism programmes are foisted on indigenous peoples and are advertised globally as authentic, traditional experiences that do not hurt the environment). To the extent that these strategies do not originate from local places (they may come from the state or the Global North), they inevitably induce a delocalizing effect in terms of an unfolding of social and ecological life. In the second instance are subaltern strategies, which follow the Deleuzian notion that 'the oppressed, if given the chance ... and on their way to solidarity through alliance politics ... *can speak and know their conditions*' (Spivak 1988: 25). Escobar advocates two strategies that focus on (1) attachment to place, and (2) attachment to *redes* that empower social networks to enact the politics of scale from below. These latter strategies, as suggested by some of our work in FNNR, engage 'local movements with biodiversity networks, on the one hand, and with other place-based actors and struggles, on the other' (Escobar 2008: 32).

In what follows we highlight the FNNR project in Guizhou Province, China, where complex social and biodiverse relations between the endangered snub-nosed monkey, local hillside farmers and their traditional agricultural practices, tourist policies, economic development and education interweave in ways that highlight Escobar's delocalizing effects and place-based strategies and suggest the importance of Lund's renegotiation of local values on a continual basis.

Lessons from Fanjingshan Reserve, China

Recent decades have witnessed considerable interdisciplinary research and conservation efforts, pointing to a fundamental question of how we can better understand the space–time complexities of humans, protected species and the environment (e.g. Ehrlich and Wilson 1991, Vitousek 1994, Jeffers 1997, Vitousek et al. 1997, Dirzo and Raven 2003, Smith et al. 2003, Turner et al. 2003, O'Connor and Crowe 2005). To address this question from a relational perspective, it is necessary to engage with the policy and practical complexities at FNNR as they relate to the Guizhou snub-nose golden monkey and the farmers who occupy the reserve (Figure 11.3). Humans and monkeys compete for resources and space within the reserve, which features various complex heterogeneities and nonlinear relationships. For example, although some local farmers live in valleys close to roadways, at certain times of the year they may venture into the mountainous snub-nose golden monkey habitat in search of herbs. Other local farmers who are ethnic Tujia are spending time engaged in state-sponsored tourist activities, which takes them away from farming. Still other farmers are involved in grassroots environmental activism (that is not state-sponsored) to protect, for example, the numbers and quality of fish in local rivers. As the farmers' activities flow and change in response to external and internal issues, the location and range of the snub-nose golden monkeys change, but the relationship between human and monkey behaviours is by no means clear.

What is clear is that the Guizhou snub-nose golden monkeys are endangered because of their small population size (700–800 monkeys (Yang et al. 2002)), high infant mortality (20 per cent) and other life-history traits (e.g. three years of weaning and at least three years interbirth interval). Living today in the increasingly higher reaches of FNNR, this species may be threatened by a rapidly increasing human population within their sole and last habitat. As a national treasure of China and natural heritage of the entire world, the Guizhou has been listed as 'endangered' by the Chinese government and the International Union for the Conservation of Nature.

Figure 11.3 Typical landscape of the FNNR
Photo: Stuart C. Aitken

The project from which the work in this chapter is derived aims to establish a complex relational approach to (1) educate local indigenous people and FNNR staff, and (2) empower farmers, tourists and policy-makers in their conservation knowledge and awareness. The c.21,000 local residents who dwell within or near the reserve boundary mostly live a subsistence lifestyle. These residents are allowed to enter non-key-habitat forests and collect resources (e.g. fuel wood, herbs and mushrooms) and herd oxen.[6] Forests also provide shelter, cover and food sources for snub-nose golden monkeys. Local households primarily use fuel wood for cooking and heating in the winter. Pilot surveys (18 local households in 2007, 69 in 2009) showed that the average fuel wood consumption per household

---

6   Report. on Participatory Rural Appraisal (PRA) Project, Global Environment Facility, 2003 (in Chinese).

Figure 11.4   Buddhist temples atop the twin peaks of Golden Mountain
Photo: Stuart C. Aitken

amounts to 12,600 kg/year, and local farmers spend 233 days on fuel wood collection. The same surveys showed that local residents collect fuel wood all year round, with greater harvesting rates from late autumn to early spring. Other resource collectors enter the reserve in seasons when specific resources are most available, e.g. collecting bamboo shoots in spring.

Since the reserve was established in 1978, poaching and unintentional killing of snub-nose golden monkeys have rarely occurred. Fanjingshan is a place of natural beauty, with many spectacular scenic views, and cultural heritage, for example many Buddhist temples (Figure 11.4). As of 2009, over 40,000 tourists visited the reserve annually, primarily between April and October with peaks in August and September (Yang, personal communication). The establishment of a cable car service from the east entrance to a place near the Golden Peak in spring 2009 has substantially increased the number of tourists. Some expeditionary tourists avoid the designated sightseeing trails, entering core snub-nose golden monkey habitat areas with aid from local residents as guides.

Human-induced habitat degradation may cause reserve managers to restrict human entry into certain areas of the reserve or at particular times. This may encourage monkeys to return. It is reported that both loss of canopy forests and disruption by human visits (Yang, personal communication) are threatening the species. However, little is known about when, where, for what purposes and how often local people enter habitat areas, or how snub-nose golden monkeys may change their habitat use patterns as a result. The issue for us was to develop ways to engage local people that empowered their participation in the changing contexts of the FNNR with regard to human occupancy and monkey preservation.

Community Mapping as a Source of Power

IFAD (2009: 2) argues that participatory community mapping is a particularly powerful empowerment and educational tool when working with 'indigenous peoples and forest dwellers that find their lives disproportionately threatened by climate change, environmental degradation and conflict related to access to land and natural resources'. The farmers who reside within or near FNNR have an intimate knowledge of their local environment, and participatory mapping is an important tool for accessing this knowledge in a complex context of human-environment relations.

Participatory mapping amongst indigenous peoples is a widely used technique that dates back several decades (Nahanni 1977, Stull and Schensul 1987, Belyea 1989, Herlihy and Leake 1997). Initially developed by geographers, anthropologists and ecologists, the technique engages ethnographic practice from a multidisciplinary perspective and is recognized as a method that values local expertise. With the advent of spatial technologies for processing, documenting and presenting local information (Craig et al. 2002), participatory mapping has increasingly made use of access to geographical information systems (GIS) and the global positioning system (GPS). Combined with some of the issues we raised earlier, this suggests important possibilities and prospects for local engagement and empowerment (Dana 1998, Craig et al. 2002). Participatory maps are more than traditional cartographic representations; they are powerful spatial visualization techniques that include input from, and ownership by, individuals and local communities. Outputs may include depictions of local resources (Meredith et al. 2002), cultural features (Al-Kodmany 2002), and pathways and an individual's daily rounds (Aitken et al. 2006). In terms of engagement and ownership, it is always best if community members create the maps and control their use, although often researchers, NGOs and development agencies will initially provide maps, technologies and training. To avoid exclusivity, sensitivity to the role of women, elders and children is important as part of the participatory mapping process. It is important to be sensitive to a variety of stakeholders and to avoid making information and people vulnerable to exploitation.

Participatory mapping includes basic mapping methods in which community members draw maps from memory on the ground or on paper. The creation of maps is based upon local cultural and aesthetic practices in an attempt to visualize relations between land use and local communities. Kesby (2005) details a form of community mapping based upon performance that is sensitive to the dance culture of an African village. The incorporation of both temporal and geographical scales enables a detailed appraisal of information about individual movements, individual and communal resources, and seasonal practices. Participatory 3-D modelling integrates local spatial knowledge that is sensitive to land elevation to produce scaled and georeferenced information. Geographical features are identified on the model using pushpins (points), string (paths) and paint (areas), and may be digitized into a GIS (IFAD 2009).

The larger project at FNNR provides an important opportunity for participatory mapping. The project explores factors that may affect a local community's support for conservation in FNNR. Residents live in villages that are located at various distances from the area's main tourist areas and from the monkey habitat. Village locations differ in their elevations on Fanjing Mountain. The highly variable terrain and educational background amongst residents of FNNR present challenges to participatory mapping but also opportunities for empowerment.

In a pilot survey conducted in spring 2010, two participatory mapping techniques were experimented with and produced varying degrees of success. The first technique involved black-and-white gridded paper maps of the area, using as a base existing land-use maps of the area, complete with landmarks such as rivers and mountains (Figure 11.5, map 1). The intention was to gain an understanding of the high-frequency use areas. Local reaction to the maps was at times one of interest, as most participants had not previously seen land-use maps of their area. However, often participants did not identify with or understand these maps. This could be an educational gap but may also simply be that the type of map did not connect with the way local residents relate to the world around them. Judging distance and location was especially challenging due to the highly variable local terrain that was not immediately identifiable on the maps for orientation purposes. Although the maps raised the issue of local relations to the region and the importance of local experiences and ideas, empowerment opportunity was minimal due to the limited ownership of a base map that was not created by participants. IFAD (2009) argues that maps created by local residents give a sense of ownership and in conjunction with their expertise may lead to empowerment in resource preservation.

The second community mapping technique was more effective, utilizing a 3-D computer model of FNNR visualized on a laptop (Figure 11.5, map 2). Although participants did not create the base of this map either, the visualization model included satellite imagery and a toggle facility that switched between this and a draped location map. Participants were excited by the interactive nature of a model whereby they could zoom into and rotate around local villages and mountains. Furthermore, the colour variation and visible terrain lent themselves well to intuitive interpretation since participants could actually see the extent of villages and recognize ridgelines. The farmers' appreciation of the 3-D computer maps counters wisdom from IFAD, which suggests that technology of this kind should be used with caution. However, it supports evidence from studies in the USA of young children, who have little problem understanding scale changes and map transformations when they are animated on a computer screen (Aitken

1999). Through this technique, educational limitations were minimized because the participants could see and understand the model without much explanation. In addition, since the technique uses computers and mapping technology, it creates a space for the younger, more technologically comfortable generation to get involved in teaching older generations. The 3-D technique seems to cross barriers and opens up the discussion of local knowledge to all community participants. Knigge and Cope (2006) demonstrate the empowerment value of this kind of participatory mapping visualization through utilization of both qualitative and quantitative information.

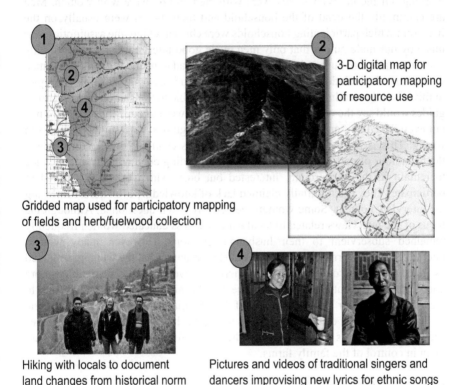

Gridded map used for participatory mapping
of fields and herb/fuelwood collection

3-D digital map for participatory mapping of resource use

Hiking with locals to document
land changes from historical norm

Pictures and videos of traditional singers and
dancers improvising new lyrics for ethnic songs

Figure 11.5   Multimedia participatory mapping in FNNR using qualitative information
Photos: Stuart C. Aitken. *Compilation*: Sarah Wandersee

The potential of computer-based multimedia 3-D participant mapping seems large but it is, as yet, untested in FNNR beyond the pilot work of spring 2010. Clearly the importance of inclusivity in any community mapping project relates to issues of gender, age and other power relations. In what follows, we look more closely at how the gender power relations in the FNNR relate to Lund's and to Escobar's work on development and transformation by considering the roles of local women in interviews and in the context of the FNNR.

Women in Interviews

A total of 263 interviews were conducted in spring 2010. Households were randomly chosen from lists of reforestation participants, with available and willing interviewees recruited on arrival since sometimes people had moved or were busy working. Of the interviews, 201 were with men and 62 were with women. Men are commonly the head of the household and their names were usually on the lists from which participating households were chosen. Often, the assumption was made by our male guides that only men made good interviewees or that we only wanted to talk with husbands and not wives, so it had to be made clear that women were also to be included. Although communication issues may have played a part in this misunderstanding, it is not entirely attributable to translation error. At times guides would say the wife would not be able to answer our questions when we had not even asked her, and when she was asked the guides were sometimes proven wrong. This is perhaps an indication of women's changing roles. Nonetheless, the behaviour of participants varied hugely depending on gender. Women tended to remain in the background, interested but busy with work unless asked to participate. They often initially claimed lack of knowledge, even before knowing the interview content. Some women seemed shy when questioned. Although the focus of the interviews related to local expertise, less educated women frequently remained subservient to their husband's opinions. However, many women interjected with information when their husbands gave what the women perceived as inaccurate or wrong answers. Furthermore, women were sometimes more knowledgeable about local resource use and farming patterns, partially due to the sharing of labour but also because of the changing role of women in the area. With the prevalence of temporary work taking husbands and older sons to cities and boarding schools taking young children away for periods of time, women are often in control of the family farms.

Women's and Children's Changing Roles

Women have strong roles in present-day China, both in terms of labour requiring physical strength and in terms of socioeconomic power. They take part in construction crews, and at times men readily admit that women are for the most part in charge of the household. There is social pressure for women to marry, but their roles extend beyond the traditional ones of wife and mother. Women are farmers, professionals in a broad range of fields, graduate students and valuable contributors of opinions. In FNNR, migration is additionally challenging past roles, along with education. As in other rural areas of China, the system of residence registration (*hukou*) demands that rural labour-oriented migrants – called a 'floating population' by many population researchers (Liang 2001) – only 'temporarily' stay (on a scale of weeks and sometimes years) in their migration destinations (often cities). Such 'temporary' migrants keep their *hukou* and possessions (e.g. farmland and houses) in their original villages and often come back to celebrate spring festivals at the Lunar New Year. Based upon interviews in spring 2010, we determined that an average of one third of family members in each household have done or are doing temporary work outside of FNNR, living in cities and interacting with a more cosmopolitan worldview. Sometimes women stay behind with children and run the farm, but it is also common to leave children with grandparents, especially as the children spend most of their time away at boarding schools. This provides women with more opportunities to participate in city work and invest in their own education and professional development. Even with older generations, women and men who participated in our pilot surveys had the same median education level of grade school. With newer generations, educational levels are increasing as girls and women are encouraged to continue their studies.

Children's education is a high current priority for families in FNNR compared with other rural areas in China. This is regardless of the interviewees' gender. Due to cost and infrastructure limitations, older generations had lower educational levels and recognized the importance of a good education for their children's and grandchildren's futures. Future priorities differ for men and women. Women show higher concerns for family issues than educational issues. This does not mean that women do not value education but that parents see their children's future success as linked with education. This is evidenced by the large numbers of fathers and mothers who, in order to get better education for their children, migrate away from remote areas with them for years at a time (interviews, spring 2010). In some areas, it was difficult to find households to interview since families had moved into the city for, we were told, better jobs and better education for children. Focusing on children's education not only improves their lives but also empowers families by equalizing gender and generational standing and exposing remote areas to new ideas. In FNNR, women may have little knowledge of climate change, but children can teach their parents about issues that they learn in school and on television. Education is not the only value imparted by parents and grandparents in FNNR. By staying alternately with grandparents and in boarding schools for much of the year and reuniting with parents on family holidays such as the Lunar New Year, children learn the importance of nuclear and extended family while also gaining autonomy. Migration and boarding schools expand and equalize women's roles through broadening horizons, redefining duties and increasing education.

The Gendered and Generational Politics of Human–Environment Complexities

Both Lund's discussion on local values and Escobar's views on identity, territory and autonomy connect with dynamics in FNNR. Strong values of education, self-sufficiency and family are present in the communities with which we are working. Families share views on protecting species, but women also pass on to their children a different approach to environment values. For example, when asked why snub-nose golden monkeys should be protected, more women than men said they did not know, but when giving specific answers, women focused less upon law and rarity (common answers for the men) and more on the harmlessness of the species and upon human-ecological connections, such as protecting the monkey being good for people. These values are being incorporated into women's evolving relationships within a shifting society in a way that suggests an appreciation of habitat loss and need for preservation, and how these issues may relate to the practicalities of education and resource use. With the predominance of ethnic minorities such as the Tujia and Miao in FNNR, local identity includes a history of marginalization and strong government involvement but also a growing dedication to preserving traditional songs, dances and costumes (Figure 11.6).

Figure 11.6   Tujia (an ethnic minority in China) performing a traditional dance in FNNR – the sticks historically held money at each end, creating a rattle from the coins
Photo: Stuart C. Aitken

Locally, women are at the heart of this change as coordinators, teachers and the majority of performers in cultural groups. Tourism development in the area encourages the preservation of traditional cultures, and migrant families have an opportunity to learn from examples elsewhere and apply new ideas in their traditional homes. Thus migration, in conjunction with education, is contributing to developing women's and children's identities in new directions, enabling them to realize expanding opportunities while their values preserve their family connections. Migration also increases the sense of family self-sufficiency already present from surviving in a remote area such as FNNR. Under the circumstances of migration, a family functions as a team by sharing household and family duties but with periods of separation that highlight personal growth and create a foundation for future development in FNNR by meshing tradition with innovation. Escobar's (2008) idea of *redes* as a traditional net that is constantly repaired and rearranged finds particular force with women's and children's work in FNNR

Conclusion

In this chapter, we speculate on the ways the FNNR example demonstrates Lund's imperative for renegotiating local values and Escobar's (2008) concern for building on identity, territory and autonomy where they may exist locally. Although our work is in its infancy, it nonetheless seems clear that short- and long-term migration patterns and changing educational opportunities are important for the place of women's and children in FNNR. As the project develops, we will test this supposition locally using interactive multimedia 3-D community mapping technologies because they seem appropriate tools for empowering residents, especially children. The technique was particularly attractive to young men and women who had little difficulty with computer-generated geovisualizations.

The project team is buoyed by Lund's feminism and Escobar's post-structuralism because each moves alternative development theories forward by focusing on a relational understanding of change and transformation that accounts for local contexts while also embracing less parochial changes. We are still developing the theoretical and empirical bases of the project, but the pilot work documented in this chapter gives some important insights. From feminist geography, Lund's ideas that renegotiation of local values are often inspired by changing gender politics are borne out in large part by women's work in the FNNR. The impact of webs of grassroots activism petitioned by Escobar is also suggested by some of the local contexts of environmental education. At the same time, state and multinational enterprises are beginning to develop traditional tourist and ecotourist facilities. While acknowledging that changes in behaviour are often strategies to preserve basic elements of lifestyle and traditions, we note that FNNR has seen an increase in national and global attention while continuing to embrace traditional ethnic cultural practices, albeit in different ways, and often in connection with tourism. Regarding the benefit and empowerment of women, women are taking on more community management roles and becoming more likely to develop education strategies for their children and for wider ecological understanding. Given that changes seldom occur in the form of dramatic events and almost always move forward as an amalgam of the old and the new, it is important to remember that from a relational perspective the material and gender contexts of life change over time and marginalized peoples can take advantage of these changes in unexpected ways.

References

Aitken, S.C. 1999. Scaling the Light Fantastic: Geographies of Scale and the Web. *Journal of Geography*, 98, 118–27.

Aitken, S.C. and An, L. 2010. Figured Worlds: Environmental Complexity and Affective Ecologies in Fanjingshan, China. *Ecological Modeling: An International Journal on Ecological Modeling and Systems Ecology*, 229, 5–16.

Aitken, S.C., Estrada, S.L. Jennings, J. and Aguirre, L. 2006. Reproducing Life and Labor: Global Processes and Working Children in Tijuana. *Childhood*, 13, 365–7.

Al-Kodmany, K. 2002. GIS and the Artist: Shaping the Image of a Neighborhood though Participatory Environmental Design, in *Community Participation and Geographic Information Systems*, edited by W. Craig, T. Harris and D. Weiner. London and New York: Taylor and Francis, 320–29.

Belyea, B. 1998. Inland Journeys, Native Maps, in *Cartographic Encounters: Perspectives on Native American Mapmaking and Map Use*, edited by M. Lewis. Chicago: University of Chicago Press, 135–55.

Blaikie, P. 2000. Development, Post-, Anti- and Populist: A Critical Review. *Environment and Planning A*, 32, 1033–50.

Chambers, R.1983. *Rural Development: Putting the Last First*. London and New York: Prentice Hall.

Craig, W.J., Harris, T.M. and Weiner, D. 2002. *Community Participatory Mapping and Geographic Information Systems*. London and New York: Taylor and Francis.

Dana, P.H. 1998. Nicaragua's 'gPSistas': Mapping their Lands on the Caribbean Coast. *GPS World*, 9(9), 32–42.

Deleuze, G. 1988. *Spinoza: Practical Philosophy*. San Francisco: City Light Books.

Deleuze, G. 1993. *The Fold: Leibniz and the Baroque*, translated by T. Conley. Minneapolis, MN: University of Minnesota Press.

Descartes, R. (1998 [1664]). The Treatise on Man, in *The World and Other Writings*, edited by S. Gaukroger. Cambridge: Cambridge University Press, 99–169.

Dirzo, R., and Raven, P.H. 2003. Global State of Biodiversity and Loss. *Annual Review of Environment and Resources*, 28, 137–67.

Ehrlich, P. R., and Wilson, E.O. 1991 Biodiversity Studies: Science and Policy. *Science*, 253, 758–62.

Escobar, A. 1995. *Encountering Development: The Making and Unmaking of the Third World*. Princeton NJ: Princeton University Press.

Escobar, A. 2008. *Territories of Difference: Place, Movements, Life, Redes*. Durham NC and London: Duke University Press.

Gibson-Graham, J.K. 1996. *The End of Capitalism (As We Knew It)*. Minneapolis: University of Minnesota Press.

Gibson-Graham, J.K. 2006. *A Postcapitalist Politics*. Minneapolis: University of Minnesota Press.

Hayden, P. 1995. From Relations to Practice in the Empiricism of Gilles Deleuze. *Man and the World*, 28, 283–302.

Herlihy, P.H. and Leake, A.P. 1997. Participatory Research Mapping of Indigenous Lands in the Honduran Mosquitia, in *Demographic Diversity and Change in Central America*, edited by A.R. Pebley and L. Rosero-Bixby. Santa Monica, CA: Rand Books, 707–37.

IFAD 2009. *Good Practices in Participatory Mapping* [Online: International Fund for Agricultural Development]. Available at: http://www.ifad.org/pub/map/PM_web.pdf [accessed: January 11, 2011].

Jeffers, J.N.R. 1997. Ecological Consequences of Biodiversity Loss. *International Journal of Sustainable Development and World Ecology*, 4, 77–8.

Kesby, M. 2005. Re-theorising Empowerment-Through-Participation as a Performance in Space: Beyond Tyranny to Transformation. *Signs: Journal of Women in Culture and Society*, 30, 2037–65.

Knigge, L. and Cope, M. 2006. Grounded Visualization: Integrating the Analysis of Quantitative and Qualitative Data Through Grounded Theory and Visualization. *Environment and Planning A*, 38, 2021–37.

Liang, Z. 2001. The Age of Migration in China. *Population and Development Review*, 27, 499–524.

Lie, M. and Lund, R. 1995. *Renegotiating Local Values: Working Women and Foreign Industry in Malaysia*. Richmond: Curzon Press.

Lund, R. 1983. The Need for Monitoring and Result Evaluation in a Development Project – Experiences from the Mahaweli Project. *Norsk Geografisk Tidsskrift*, 37, 169–86.

Lund, R. 1993. *Gender and Place*, Volume 1: *Towards a Geography Sensitive to Gender, Place and Social Change*. Trondheim: Department of Geography, University of Trondheim.

Lund, R. 2008. At the Interface of Development Studies and Child Research: Rethinking the Participating Child, in *Global Childhoods: Globalization, Development and Young People*, edited by S. Aitken, R. Lund and T. Kjørholt. London and New York: Routledge, 131–49.

Meredith, T., Yetman, G. and Frias, G. 2002. Mexican and Canadian Case Studies of Community-Based Spatial Information Management for Biodiversity

Conservation, in *Community Participation and Geographic Information Systems*, edited by W. Craig, T. Harris and D. Weiner. London and New York: Taylor and Francis, 205–18.

Momsen, J. 2004. *Gender and Development.* New York and London: Routledge.

Nahanni, P. 1977. The Mapping Project, in *Dene Nation – The Colony Within,* edited by M. Watkins. Toronto: University of Toronto Press, 21–7.

Nederveen Pieterse, J. 1998. My Paradigm or Yours? Alternative Development, Post-Development, Reflexive Development. *Development and Change,* 29, 343–73.

Norris, C.1991. *Spinoza and the Origins of Modern Critical Theory.* Oxford: Blackwell.

O'Connor, N.E., and Crowe, T.P. 2005. Biodiversity Loss and Ecosystem Functioning: Distinguishing Between Number and Identity of Species. *Ecology,* 86, 1783–96.

Parpart, J. and Marchand, M. 1995. Exploring the Canon, in *Feminism/Postmodernism/Development,* edited by M. Marchand and J. Parpart. London: Routledge, 1–22.

Peet, R. and Hartwick, E. 2009. *Theories of Development: Contentions, Arguments, Alternatives.* New York: Guildford Press.

Scott, C. 1995. *Gender and Development: Rethinking Modernization and Dependency Theory.* Boulder, CO: Lynne Riener.

Schumacher, E.F. 1973. *Small is Beautiful: A Study of Economics as if People Mattered.* New York: Blond and Briggs.

Smith, R.J., Muir, R.D.J., Walpole, M.J., Balmford, A. and Leader-Williams, N. 2003. Governance and the Loss of Biodiversity. *Nature,* 426, 67–70.

Spinoza, B. de. 2008 [1677]. *Ethics,* translated from the Latin by R.H.M. Elwes (1883) [Online: eBooksBrasil/Project Gutenberg]. Available at: http://www.ebooksbrasil.org/eLibris/spinoza.html [accessed: 22 January, 2013].

Spivak, G.C 1983. Can the Subaltern Speak? in *Marxism and the Interpretation of Culture,* edited by C. Nelson and L. Grossberg. London: Macmillan, 24–8.

Stull, D.D. and Schensul, J.J. (eds) 1987. *Collaborative Research and Social Change: Applied Anthropology in Action.* Boulder, CO: Westview Press.

Turner, B.L., Matson, P.A., McCarthy, J.J., Corell, R.W., Christensen, L., Eckley, N., Kasperson, J.X, Luerse, A., Martellog, M.L., Polskya, C., Pulsiphera, A. and Schiller, A. 2003. Illustrating the Coupled Human–Environment System for Vulnerability Analysis: Three Case Studies. *Proceedings of the National Academy of Sciences,* 100, 8080–85.

Vitousek, P.M. 1994. Beyond Global Warming: Ecology and Global Change. *Ecology,* 75, 1861–76.

Vitousek, P.M., Mooney, H.A., Lubchenco, J. and Melillo, J.M. 1997. Human Domination of Earth's Ecosystems. *Science,* 277, 494–9.

Yang, Y., Lei, X. and Yang, C. 2002. *Ecology of the Wild Guizhou Snub-nosed Monkey.* Guiyang: Guizhou Science and Technology Press.

Chapter 12

Right to Rights: *Adivasi* (Tribal) Women in the Context of a Not-So-Silent Revolution in Odisha, India

Smita Mishra Panda

Introduction

Why is the question of the 'right to rights' raised in India, one of the world's largest democracies? It has been observed that, with the growth in the economy following the reforms of the 1990s, the rights of underprivileged people are being eroded. Another outcome of economic growth is that inequality between rich and poor has increased. In the 1990s and 2000s, the wealthiest 10 per cent of the population held at least 50 per cent of total assets while the least wealthy 10 per cent held at most 0.4 per cent of total assets and 0.2 per cent of net worth (Jayadev et al. 2007, quoted in Weisskopf 2011). India witnessed widening income inequality during the postreform economic growth period 1993 to 2005. Further, despite the doubling of per capita consumption in the postreform decade, the decline in poverty lessened compared to the prereform decade. The prereform period witnessed a decline in poverty by 8.4 per cent and in the postreform period poverty declined by 6.7 per cent (Sarkar and Mehta 2010). When almost one third of the country's population lives below the poverty level there is something seriously wrong with the distribution of the benefits of growth – hence the question of rights strongly surfaces in a democracy like India. Some economists, however, would like to believe that increasing inequality is a necessary concomitant of economic growth (Weisskopf 2011).

The Constitution of India guarantees equal rights for all, regardless of caste, class, gender and ethnicity. However, there is widespread violation of rights of the marginalized sections of the population, a large majority among them being Indian tribal peoples, also termed *Adivasi*. The right to vote does not ensure other basic rights – food, shelter, education and a life of dignity among others. The constitutional provision of equal rights to all therefore has very little practical impact on the ground. Of the many urgent concerns linked with the economic growth of India, the present chapter addresses environmental degradation and erosion of livelihood rights of local communities (mostly the indigenous tribal communities) resulting from large-scale extraction-based industries in the Indian state of Odisha.[1] There are similar issues in the neighbouring states of Chhattisgarh and Jharkhand. These states are mineral-rich and also have large concentrations of tribal populations (World Bank 2007, Bhushan et al. 2008). There is rampant violation of laws that are intended for social protection and to prevent the alienation of livelihood resources of tribal communities (Ramdas 2009, Das 2010, Lund and Panda 2011). Despite several laws and targeted policies to protect tribal peoples and their habitats,[2] their provisions have been systematically violated and the habitats of tribal peoples have been encroached upon by large national companies and multinationals. Profits have been siphoned out of these areas. This indicates not only corruption at the highest level, but also that there is no strong mechanism at any level to check such actions. As a consequence, tribal peoples in Odisha continue to suffer land deprivation and dispossession of livelihood resources and cultural identity (Ambagudia 2010). Such violations over the years have led to innumerable protests and resistance by tribal communities and other civil society groups working for them (Mishra and Maitra 2006). Political parties have tried to maximize their advantage from such situations with total disregard for tribal communities, who are supposed to be on the receiving end of any development benefit. Only recently, the Supreme Court of India and the Ministry of Environment and Forests of the central government intervened to stop the UK-based aluminium company Vedanta expanding further its operations in the Niyamgiri Hills of Odisha, which have the highest concentration of bauxite in the country (*Indian Express* 2011). The estimated value of the bauxite found in the Niyamgiri Hills within the tribal areas is four trillion US dollars (at 2009 prices) (Roy 2009). The action by the government was only taken after much public outcry and sustained demonstrations against the state by civil society.

Since the advent of neoliberal policies in India in the early 1990s, tens of millions of people have been displaced from their lands by floods, droughts and desertification caused by indiscriminate environmental engineering and massive infrastructural projects such as dams, mines and Special Economic Zones (Roy 2009). The ruling party and the government take the view that such projects will encourage the right kind of development for Odisha and lift the conditions of the poor. What is germane to the right kind of development is equitable distribution of benefits and equal access to opportunities and resources for all, so that people living below poverty levels can hope to improve their lives. The chief minister of the state is often heard to say that the 'government wants

---

1   The name of the state was changed from Orissa to Odisha with effect from 1 November 2011 (Prasad 2011).

2   For example, Schedule V of the Indian Constitution (Chakravarty 2012); Panchayat Extension to Scheduled Areas 1996 (PESA) (Bijoy 2012); Forest Rights Act (FRA) 2006 (The Forest Rights Act n.d.); Land Alienation Act in the form of Orissa Scheduled Area Transfer of Immovable Property (by scheduled tribes) 1956 and Orissa Land Reforms Act 1960 (Ambagudia 2010); and more recently the proposals of a new Land Acquisition Bill (*The Hindu* 2012).

peaceful industrialization'. In reality however, the process of industrialization has not been peaceful. There is conflict over claims for adequate compensation in money and land between local communities (mostly tribal peoples) and the companies engaged in setting up industries. These conflicts are a direct outcome of the manner in which mining operations are conducted, encroaching on the land, water and forests of the tribal communities, threatening ecological sustainability and denying social justice. Yet, the principal chief conservator of forests (PCCF) of the government of Odisha claimed that mining activities are 'indispensable for growth'. Although minerals, tribal peoples and forests occupy the same physical space in the state, this should not be a reason to stop development according to the PCCF (*Indian Express* 2011). The benefits of 'development' have not flowed to the tribal peoples of the area but instead have accrued to the middle class in the urban areas with their accumulation of wealth and status while a vast majority of the underprivileged classes (mostly tribal and other marginalized communities) still languish in poverty (Roy 2009, Das 2011). Odisha State, which is one of the poorest states of the country, currently has a 10 per cent growth rate and yet 47 per cent of the people live below the poverty line, indicating the extremely inequitable distribution of the benefits of growth (Government of Orissa 2010). *Adivasi* rights have been severely violated. Some scholars have described it as genocide and others as 'ethnocide' as there is systematic cultural erosion of tribal ways of life and dispossession from their livelihood resources of land, water and forests (Das 2010, Lund and Panda 2011). Forests, mountains and water systems are taken over in large land grab projects by multinational companies backed by the state in the name of development. In Odisha, bauxite and iron-ore mining is destroying entire ecosystems and converting fertile lands into desert (Kumar and Choudhary 2005, Kumar 2006). Processing of iron ore for steel and bauxite for aluminium consumes large amounts of water and electricity and is driven by a cartel that is allied with mineral companies, investment bankers, government dealers, metal traders and arms manufacturers (Padel and Das 2010). The quest to maximize financial benefits by the corporate sector, with the support of the Indian state, is linked to new power structures of global financial capital, described as 'corporate imperialism' (Roy 2009). According to Padel and Das (2010: 10) 'Mining projects are fueled by an entrenched notion of development so powerful, that democracy and human rights often seem to wither in the face of it'.

The gender dimension of such 'development' is important to understand. *Adivasi* women have strongly responded and mobilized against these interventions that have dispossessed them of their livelihood sources and displaced them physically and culturally. The protest against this systematic 'ethnocide' can be seen in different tribal regions of the country. *Adivasi* women in the era of neoliberalist policies and globalization are subjected to a range of violations of economic and political rights basic to safety and security. The ensemble of social, political and economic rights is varied and has multiple implications, including livelihoods (Krishna 2007). The rights are basically to protect their resource base, hinder exploitation by the informal sector, prevent sexual abuse and secure freedom to express their grievances and seek legal recourse. *Adivasi* women are taking a leading role in demonstrations and protests against the authorities and companies responsible for draining their natural resources.

The mining sector is flourishing in the state of Odisha, while agriculture, which is the major determinant of per capita income for the vast majority of the population, is lagging behind (Mishra 2010). Although the ruling class is of the view that industrialization is best for Odisha's economy and is doing everything to make it possible, the question remains whether local communities, especially marginalized tribal communities, are benefited by these development efforts. The concentration of major minerals such as bauxite and iron ore coincides with the regions where there are extensive forests and tribal inhabitants. As much as 1019 sq. km of land has been leased out to companies for mining. They mine iron ore and manganese in Sundergarh and Keonjhar Districts, coal in Sundergarh, and bauxite in Kalahandi, Koraput and Rayagada Districts. The tribals affected are Dongaria Khonds in Lanjigarh, and Juangs and Paudi Bhuyan in Keonjhar and Sundergarh Districts (Kumar 2006). Rampant illegal mining is taking place whereby the area that is actually mined is much higher than the permissible limits set by the government. The state has no monitoring mechanism in place to check such illegal practice by companies.

The extraction of minerals has led to either displacement of local communities or drastic reduction in land, water and forests used by them, leaving the tribal peoples homeless, resourceless and pauperized. As they have low levels of literacy and lack specialized skills, there is little scope of employment in the mining sector. At best, they are recruited as unskilled temporary workers. In 2011 the state proposed a Mining Bill, which would provide for sharing profits from minerals with local communities to ensure a smoother process of land acquisition.[3] It is expected that, once local communities are convinced that they will share the profits with the industries, land acquisition will go smoothly for the companies. The impact of mining-based industrialization has a differential impact on women and men. It is observed that in the division of labour women are responsible for provision of household resources in the form of water, forest products, fodder and fuelwood (Shiva 1988, Agarwal 1992, Panda 1996). Therefore, in the event of environmental degradation and resource depletion, it is women and children who are adversely affected relative to the menfolk, as the latter tend to migrate to take wage employment in urban areas.

This chapter deals with *Adivasi* women's resistance struggles and movements in the state of Odisha to illustrate their mobilization around issues of forest rights, mining-led industrialization, nonpayment of wages by the state, dependence on food grains from the public distribution system, the proliferation of illicit alcohol and increasing domestic violence in tribal communities. The information sources

---

3   Ummar, S. and Patnaik, B.C. 'Mining Legislation and Right to Property: With Special Reference to Odisha, India.' Paper presented at the International Seminar on Mining Legislations, Bhubaneswar, 2–3 December, 2011.

are primarily civil society organizations and tribal women leaders with whom intensive discussions were conducted to obtain relevant information in 2010 and early 2011. Besides direct information on the struggles, further information was obtained from documentation available from nongovernmental organizations (NGOs). Civil society groups working in tribal areas are directly responsible for the mobilization of *Adivasi* women and their sustained activism.

Some Conceptual Underpinnings

Grassroots women's struggles and resistance have been conceptualized by several feminist scholars, significantly by Chandra Talpade Mohanty et al. (1991) in their book *Cartographies of Struggle*, where they raise pertinent questions of political consciousness and self-identity that are crucial to defining Third World women's engagement with feminism. The activism of *Adivasi* women has become more pronounced in the last 15 years, during which the outcomes of neoliberalist policies have made a visible impact on the local communities in the country. It is a situation where *Adivasi* women are fighting against capitalist intrusion that is the prime reason for their dispossession and displacement. A feminist reading of antiglobalization is aptly described by Mohanty (2003), who argues for a more intimate, closer alliance between women's movements, feminist pedagogy and crosscultural feminist theorizing, intertwining questions of subjectivity, agency and identity with political economy and the state. Thakar (2011), in her study of South Asian women, has raised the question of the interpretation of women's political activism (agency, activism and empowerment). She has pointed out that community politics can be regarded as an empowering process, particularly where women organize on the basis of collective identity.

Another important question is how *Adivasi* women conceptualize their struggles and resistance against the corporate sector or the hegemonic state. They are mothers, farmworkers and carers of families like any other rural women in India. They are conscious of their roles in the struggles vis-à-vis their menfolk. Their struggles are no longer confined to their habitats alone, but have spread beyond their own territories to other marginalized groups in the region (Bhanumathi n.d., Burra 2005). *Adivasi* women's agency has been a conduit to initiating and sustaining resistance against those structures responsible for their loss of livelihood resources. Simultaneous processes of Hinduization and 'patriarchalization' (implicit or explicit) of tribal social systems have directly affected the *Adivasi* women. Therefore, their struggles are also against the patriarchal social order that is slowly creeping into tribal societies. Insights into womens movements and resistance have been provided by poststructuralist researchers. Mahmood (2001) argues that a set of capacities possessed by a subject is a product of the operation of power (domination) that is manifested in the form of self-conscious identity and agency: 'such a conceptualisation of power and subject formation encourages us to understand agency not simply as synonym for resistance to relations of domination, but as a capacity for action that specific relations of subordination create and enable' (Mahmood 2001: 210). In the case of *Adivasi* women, who otherwise are known to be shy and docile, manifestation of their subject formation is based on their everyday lives of deprivation of livelihood resources by the corporate sector and the ruling class. Their subjectivity is embedded in lived and told stories that can be captured as narratives that are a combination of rational arguments, emotions and lived bodily experiences. Agency is strengthened by *Adivasi* women's subjectivity, which is a product of their lived experience and political consciousness.

Gender has limitations as a single analytical category. The concept of intersectionality has proved to be useful as it encompass multiple dimensions and modalities of social relations (McCall 2005). Feminist sociological theory suggests how various socially and culturally constructed categories such as gender, race, class, disability and other axes of identity interact on multiple and often simultaneous levels, contributing to systematic social inequality (Browne and Misra 2003, Knudsen 2006). Due to their ethnicity (*vis-à-vis* the caste society), tribal women in Odisha are marginalized as poor, pauperized women in their habitats or as dispossessed or displaced women struggling against a declining resource base, as well as against the patriarchal state and global forces (personal communications with grassroots women leaders working with tribal women, 2010–2011). Tribal women's activism and the rights question need to be situated and analysed from an intersectionality perspective. The ensuing sections will discuss how intersectionality plays out in *Adivasi* women's struggles.

Adivasi Women in Odisha

Odisha is home to 62 tribal communities, concentrated mostly in the southern and northern parts of the state in Koraput, Sundargarh and Mayurbhaj Districts. They comprise around 23 per cent of the total population of the state (Census of India 2011). The tribes are at various stages of socioeconomic development. At one extreme, there are tribes that are relatively isolated in remote hill regions, with their core culture intact and having little contact with the mainstream population. At the other extreme are those that have become Hinduized or Sanskritized by adopting caste traditions or have converted to Christianity and are indistinguishable from nontribals. Such a situation has often led to loss of tribal identity and tribal peoples are likely to be confused with the Hindu castes by researchers (SCSTRTI 2004). There are examples of tribal communities being wrongly classified as 'Other Backward Castes' (OBC) by government functionaries. Tribal people have a distinct cultural identity expressed through their social organization, language, dress, ornaments, rituals, festivals, housing, art and crafts. Except for some of the large tribes such as Santhals, Gonds, Saora and Khonds, most live in concentrated pockets and are often referred to as 'sons of the soil' or autochthonous population (Ota and Mohanty 2010). These include Bonda, Chenchu, Lodha, Mankidia, Kutia Khonds and Dongaria Khonds. Economic impoverishment and political marginalization of tribes go hand in hand in Odisha. According to 1999–2000 estimates, 73 per cent of the tribal population in the state lives below the poverty level (de Haan and Dubey 2004). Similarly, their representation in local governments and the state legislative assembly of Odisha is negligible.

As most tribes are patrilineal, tribal women traditionally have been accorded secondary status with respect to religion and political decision-making in the public domain. Women do not inherit land. *Adivasi* women are known for being faithful to their duties and responsibilities for ensuring the survival of their households and maintenance of the village and its surroundings. *Adivasi* women have a better place in the village community than their caste counterparts because of their economic contribution to the household and their community work for the well-being of the village. Typically a tribal man will pay a certain 'bride price' for a wife. If the marriage fails, she is free to leave and marry another, but she will not have any rights over the children, and he does not have any claim over the property. An *Adivasi* woman's labour is her only asset, which can be appropriated by her family as well as the community. In spite of such inequalities within tribal communities, they are considered to be comparatively egalitarian with respect to gender relations compared with the Hindu castes (Xaxa 2004, Chakrabarty and Bharati 2010, Mohindra and Labonte 2010). In addition, *Adivasi* women have more freedom than their caste counterparts and there is no discrimination between girls and boys. Strong son preference does not prevail among tribal communities. A clear indication is the sex ratio in the tribal-dominated districts of the state, which is much more equally balanced compared to the other areas. The sex ratio of the tribal population (for nine tribal-dominated districts) is 1024 females to 1000 males, which is in stark contrast to the rest of the state (978) (Census of India 2011). It is perhaps due to the aforementioned features that *Adivasi* women's activism has found a place in the public domain.

With the advent of development interventions such as policy advocacy and entrepreneurship through tribal women's collectives in tribal areas, facilitated by civil society organizations, there has been a visible change among *Adivasi* women. Examples are collective initiatives in the formation of groups at the village level, participation in economic activities and interaction with government and police functionaries. Civil society organizations include a series of local NGOs that have established their credibility in working towards the advancement of tribal women. Depending on the area and the involvement of the NGOs and Christian missionaries working with development among the tribals, *Adivasi* women have been increasingly exposed to the outside world. They may appear shy and silent, but they understand the injustices that surround them. Over the past ten years, they have come out of their passivity to assert their rights. They have become vocal and through collectives such as self-help groups and *mahila mandals* (women's groups) they have made their presence felt in the public domain discussing issues that have affected their livelihoods and rights.

The Not-So-Silent Revolution

Silent revolution refers to simmering discontent. The majority expressed their dissent, which was subdued and could be disregarded by the ruling class. However, over the years marginalized tribal peoples have mobilized and found channels and means for expressing their anger and frustration over the outcomes of development that has left them increasingly pauperized and helpless. The revolution that is seen particularly among the *Adivasi* women is no longer silent. In the past decade or so, it has become louder to the outside world. Historically, *Adivasi* movements in Odisha have been in existence since the early nineteenth century, protesting against British rule, Indian kings, landlords (*zamindars*) and upper castes, all of whom have encroached on their lands and economic resources, and attempted to control them. There were the Ghumsar risings, Mariah revolts, Bhuinya risings, Dharani Mela, Sambalpur revolution, Kalahandi uprising and Gangpur revolt (Pattanayak 2010). The objective of such *Adivasi* revolts was to protect their resources and cultural identity. *Adivasi* women have been in the forefront of most movements and struggles to protect their local resources and culture. Yet the involvement of men gets reported more prominently as they are more visible, more mobile and more articulate in the mainstream language and can express themselves better in the media. Further, the media and researchers who document such actions also have a patriarchal orientation and are not sensitized to notice women's actions in the public domain (personal communication with Vidhya Das of Agragamee, August 2010). There are, however, some examples from elsewhere in India where women have been in the forefront of struggles, e.g. the Chipko Movement in northern India, which received much publicity (Jain 1984).

Discontent has led to growing left-wing extremism in the area, where it is believed that *Adivasi* women have provided support to the Maoist and Naxal groups who are fighting against the state and the establishment, particularly in southern Odisha with its concentration of large-scale mining activities (Nayak 2006, Padhi and Panigrahi 2011). Maoists and Naxalites have overtly supported tribal peoples in their fight against the ruling class and injustice.

Multinational companies that have been the target of protests by *Adivasi* women include Vedanta in Lanjigarh area of Kalahandi District, Utkal Alumina in Kashipur in Rayagada District, Posco in Keonjhar District and Tata Steel in Kalinganagar. In all these districts where companies have set up plants for mineral extraction (bauxite and iron ore), *Adivasi* women have resisted the interventions for establishing processing units and have refused to give up their lands and homes. So far the administration has responded by putting the men in custody and declaring them as criminals and antisocial elements. *Adivasi* women (belonging to Dongaria Kondh, Jhodia Paraja and Kondh groups) who have blocked roads have been beaten, flogged and knocked down by the police. In Maikanch in Koraput District in 2001, police fired on *Adivasi* women who were in the forefront of resisting the setting up of an aluminium plant by Utkal Alumina (a conglomerate of Indian and foreign companies). *Adivasi* women appear to be more attached to their lands and households than their men, who can be lured by monetary compensation (discussion with tribal women leaders in workshop on Mobile Livelihoods, 12–13 March, 2010). *Adivasi* women's collectives began their protests as small groups in their villages and have over the years been successful in soliciting support from other marginalized sections of society such as *Dalits* (Scheduled Castes). In the Kashipur area of Rayagada District, such grassroots mobilization of *Adivasi* women has made its presence felt through actions against the state and the corporate sector.

The *Adivasi* women struggling in Rayagada, Sundergarh, Keonjhar and Sambalpur Districts (the leading areas of mining bauxite, iron ore and coal) are of the view that, if industries are established in the tribal areas, they will not only lose their natural resources but also their identity, as the companies will bring a different culture that would affect the social fabric (personal communication with tribal women leaders in a participatory workshop, March 2010). Sumani Jhodia, a tribal women leader from Kashipur, stated that drinking of illicit country liquor among men had increased due to inroads made by the 'company'. This implies that with the increasing presence of agencies engaged in mining and industrialization, there is an expansion in the sale of liquor (now packaged in sachets for convenience) and with it violence against women. She also remarked that men will 'get attracted by money given as bribes but women will never take money from the "company". They would rather beg in the streets than take money from "company men"' (personal communication in the field with tribal women leaders, March–April 2010). The liquor trade has proliferated in the last five years with a government policy to provide licenses to liquor shops in tribal areas. Liquor brewing and sale have made inroads to the remotest areas of the state. Increasing alcoholism among tribal men negatively affects household income and leads to domestic discord and violence against women. Further, the quality of the alcohol is questionable. Several cases have been reported of fatal diseases and deaths due to overconsumption of alcohol. The Panchayat Extension to Scheduled Areas (PESA) 1996, which gives full constitutional autonomy to the *gram sabha* (village council) in tribal areas to protect their resources and culture (Bijoy 2012), allows tribal people the right to brew their own liquor and engage in its trade. However, despite complaints from the *gram sabha,* the district administration has encouraged outside liquor breweries by providing them with licenses. Several illegal breweries also operate in the area, to which the administration turns a blind eye. The administration argues that the liquor trade generates substantial revenue, which is used to run schools and child-care centres and to construct roads and other infrastructure. The administration, instead of taking any action itself, merely asks women to counsel their menfolk to give up drinking. It is not surprising that *Adivasi* women have lost all faith in the administration and have taken matters into their own hands. The protests are referred to as *Mada Mukti Abhiyan* ('Fight Against Liquor') and have spread to several tribal areas. *Adivasi* women in all corners of the state are keen to fight against the growing liquor menace as the problems associated with it are common. The police are insensitive to the problem and have been routinely harassing the *Adivasi* women by falsely implicating their menfolk or threatening them with dire consequences if they protest against the administration (personal communication with NGO leaders in Rayagada and Khurda Districts in 2011).

Ama Sangathan, which means 'Our Group', was formed in 1993 in Kashipur in Rayagada District with the objective of empowering *Adivasi* women and providing them with a platform to raise their voice for minimum wages, rights over land, water and forests, and proper delivery of government programmes. *Ama Sangathan* is promoted by Agragamee, a well-regarded NGO that has worked in the region for more than 25 years. The origin of *Ama Sangathan* and its role as an apex organization can be traced to a successful struggle by the women of Mandibisi, a village in the Kashipur Block. They were successful in gaining control over the procurement, processing and marketing of minor forest produce (in this case, hill brooms) after a long battle with the government and contractors. *Ama Sangathan* is a block-level federation of *mahila mandals* (women's groups) and now has a wider outreach in the tribal areas of southern Odisha. It extends over 412 villages in 17 *gram panchayats*, with a growing membership. The women leaders of *Ama Sangathan* are well regarded by their group and community and have the support of a few men. Over the years, they have institutionalized their demands and the procedures to meet them. The major objectives of *Ama Sangathan* are:

- To increase the income of people through the promotion of small-scale collectives, enterprises and village cottage industries.
- To develop backward and underdeveloped villages.
- To take adequate and effective steps towards a needs-based training for village development.
- To ensure the proper utilization of government funds and see to the efficient delivery of services for the poorest and most marginalized sections of tribal society.
- To establish a networking system among local-level organizations.
- To ensure women's participation in developmental programmes within the community.
- To ensure free flow of information to the tribal community on the various funded programmes involving the region.
- To ensure access, control and conservation of land, water, forest and other natural resources in tribal regions by the tribal communities themselves.[4]

Ama Sangathan has a management structure with a president, vice president, secretary, joint secretary and treasurer, and an executive committee of seven members. All the positions are occupied by *Adivasi* women. Presently there is a network of more than 45 villages where *mahila mandals* (women's groups) are in place. They actively participate in the meetings, many of which involve protests against mining and the sale of illicit liquor.

---

4   Agragamee 2010. 'Ama Sangathan'. Kashipur, Odisha (Mimeo.).

Orissa Nari Samaj is another example of tribal women's activism in the state. It is a federation of 55 block-level tribal women's organizations. It covers 3,255 village-level groups or *sanghas* with a membership of 250,000 tribal women. A civil society organization called the Team for Human Resource Education and Action for Development (THREAD) was instrumental in forming this federation. It is based in Jatni in Khurda District and has been in the news for its role in protesting against genetically modified (GM) brinjal (aubergine) seeds, based on evidence of trials by Monsanto, a multinational company. Tribal women organized in protest against Monsanto and the government in 2009 by forming a massive rally in the capital city Bhubaneswar. They are the only pressure group fighting against GM crops. Janani Hasda, a tribal leader from Mayurbhanj District in northern Odisha, stated that because of illiteracy in the tribal villages women were hesitant to fight against the state. However, with training from THREAD, she now has the motivation to train other women, especially those in the *panchayati raj* system. She was one of the prime movers of the protest against GM crops in the state (THREAD 2009).

THREAD has promoted certain strategies to empower *Orissa Nari Samaj* through campaigns for advocacy on issues relating to atrocities against women. This resulted in the establishment of all-women police stations in remote areas. The organization also trained tribal women to contest *panchayat* (local government) elections. With a large collective base of 250,000 members, of which 1290 are elected members of the local governments, THREAD ensured that the distribution of food grain through the public distribution system was streamlined. They also ensured the distribution of BPL (below poverty line) cards to people, maintained proper functioning of the integrated child development services in the village and called for regular meetings of the traditional village councils. However, *Orissa Nari Samaj*'s most important achievement was the declaration of villages as free from GM crops (THREAD 2009). THREAD also focus on tribal women in other areas of development, such as income security, health, renewal of core tribal values and culture, establishing community-based ecology schools, promotion of organic farming and a host of capacity development programmes for economic and political empowerment.

Mati Ma Mahila Morcha (MMMM) is another civil society organization working since 1995 among *Adivasi* women in over 200 villages in three blocks in Nayagarh District on the issues of livelihoods, access to natural resources and governance. The major tribal people in the area are the Sabara, who depend on natural resources for their livelihoods. Sabara women have come together with the support of the MMMM to fight against the district administration mainly for food security and forest rights and against the sale of country liquor. Around 10,000 tribal women have been mobilized by MMMM's '*Lok Sangathan*' (people's collective). By means of the government's Right to Information Act, women's groups could obtain information about the food grains they are entitled to through the public distribution system. After the *Adivasi* women found out that they had received much less food grain than their due, they organized a massive rally and protest against the district collector. This led to further demonstrations and the administration found itself under pressure to meet their demands (Jha 2009). Similarly, they protested against wage irregularities under the Mahatma Gandhi National Rural Employment Guarantee Scheme (MGNREGS), started by the government all over the country in 2006. Under the programme, villagers are entitled to 100 days of paid work a year on road and other government construction work. According to the NGO activist leader Pravata Sahoo, the struggles of the tribal women prompted the district administration to make timely payment of their dues. There are other microlevel struggles and the NGO feels that after the initial mobilization process, *Adivasi* women have been in the forefront of strengthening collective resistance against the state (Sahoo 2011).

There are many more examples in Odisha where *Adivasi* women's activism in the form of micromovements against alcohol, mining industries, violence against women, moneylenders and the ruling class in general has emerged in different locations within the tribal districts. What is striking is that most struggles by *Adivasi* women are peaceful but loud enough for the state to hear and act on. Efforts at sustained activism by *Adivasi* women show how marginalized and excluded sections of society can collectively mobilize against denial of their rights.

Concluding Remarks

Grassroots mobilization has led to the formation of community-level people's institutions with a clear mandate to resist the expansion of mining-based industrialization, the liquor trade, violence against women and other governance-related irregularities affecting their daily lives and livelihoods. The cases demonstrate that the groups and networks of *Adivasi* women do not have a single agenda. The main objective is to restore their livelihood rights and dignity through struggles on various fronts.

The range of struggles discussed in this chapter signifies more than just protest against outsiders and the establishment. It is collective agency that has found a political voice for the assertion of rights. *Adivasi* women's everyday experiences and values provide a strong impetus for them to take to action in the public domain. *Adivasi* women's agency has been most effective as collective action. The actions are becoming more sophisticated and effective with time as the women are increasingly able to lobby and confront the external forces that are responsible for their plight and marginalization. *Adivasi* women's identity is strengthened through the facilitation of their actions by the local NGOs. Adivasi women have become less shy and docile and, with a transformation in their world view, they are poised for further action to pursue their rights as citizens of India. The analysis reveals that the intertwining of subjectivity, agency and identity has a positive influence in helping to restore rights to *Adivasi* women and their communities in Odisha. Thus the revolution that one sees emerging is the state of Odisha is not silent anymore and cannot be ignored.

References

Agarwal, B. 1992. Gender and Environment Debate: Lessons from India. *Feminist Studies*, 18, 119–58.

Ambagudia, J. 2010. Tribal Rights, Dispossession and the State in Orissa. *Economic & Political Weekly*, XLV(33), 60–67.

Bhanumathi, K. n.d. *Tribal Women's Struggles Against Political Violence in the Eastern Ghats*. Andra Pradesh: Samata-Centre for Advocacy and Support.

Bhushan, C., Zeya Hazra, M. and Banerjee, S. 2008. *Rich Lands, Poor People: Is 'Sustainable' Mining Possible?* State of India's Environment and Citizens' Report, 6. New Delhi: Centre for Science and Environment.

Bijoy, C.R. 2012. *Policy Brief on Panchayat Raj (Extension to Scheduled Areas) Act of 1996* [Online: UNDP United Nations Development Programme, New Delhi]. Available at: http://www.undp.org/content/dam/india/docs/UNDP-Policy-Brief-on-PESA.pdf [accessed: 6 March 2013].

Browne, I. and Misra, J. 2003. The Intersection of Gender and Race in the Labour Market. *Annual Journal of Sociology*, 29, 487–513.

Burra, N. 2005. Tribal Women's Struggle for Water in India. *Cultural Survival Quarterly*, 29(4) [Online], Water Rights and Indigenous Peoples. Available at: http://www.culturalsurvival.org/publications/cultural-survival-quarterly/india/tribal-womens-struggle-water-india [accessed: 6 March 2013].

Census of India 2011. *Provisional Population Totals: Orissa Series* 22. New Delhi: Government of India.

Chakrabarty, S. and Bharati, P. 2010. Adult Body Dimension and Determinants of Chronic Energy Deficiency Among Shabar Tribe Living in Urban, Rural and Forest Habitats in Orissa, India. *Annals of Human Biology*, 37, 150–68.

Chakravarty, A. 2012. Conscience of the Constitution and Violence of the Indian State. *Economic & Political Weekly*, XLVII(47–48), 33–8.

Das, V. 2010. Beneath the Gloss and Glitter: A Report from Kashipur. *Economic & Political Weekly*, XLV(44), 17–19.

Das, V. 2011. *Human Rights, Inhuman Wrongs: State of Governance in Tribal Regions*. New Delhi: Sarup Book Publishers.

de Haan, A. and Dubey, A. 2004. *Conceptualising Social Exclusion in the Context of India's Poorest Regions: A Contribution to the Quantitative–Qualitative Debate* [Online] Available at: http://www.utoronto.ca/mcis/q2/papers/II_deHann_Dubey_Qual-Quant-AdH-21m.pdf [accessed: 6 March 2013].

Indian Express 2011. Verdict on Vedanta Expansion, 1 March.

Government of Orissa. 2010. *Orissa Economic Survey*. Bhubaneswar: Government of Orissa.

Jain, S. 1984. Women and People's Ecological Movement: A Case Study of Women's Role in the Chipko Movement in Uttar Pradesh, *Economic & Political Weekly*, XIX(41), 1788–94.

Jayadev, A., Motiram, S. and Vakulabharanam, V. 2007. Patterns of Wealth Disparities in India During the Liberalisation Era. *Economic & Political Weekly*, XLII(3), 3853–63.

Jha, M.K. 2009. Food Security in Perspective: The Significance of Social Action. *Community Development Journal*, 44, 351–66.

Krishna, S. (ed.) 2007. *Women's Livelihood Rights: Recasting Citizenship for Development*. New Delhi: Sage.

Knudsen, S.V. 2006. Intersectionality – A Theoretical Inspiration in the Analysis of Minority Cultures and Identities in Textbooks, in *Caught in the Web or Lost in the Textbook?*, edited by É. Bruillard, B. Aamotsbakken, S.V. Knudsen and M. Horsley. International Conference on Learning and Educational Media [Caen, France, 26–29 October 2005]. Paris: Jouve, 61–76 [Online: International Association for Research on Textbooks and Educational Media (IARTEM)]. Available at: http://www.iartem.no/documents/caught_in_the_web.pdf [accessed: 6 March 2013].

Kumar, K. 2006. *Dispossessed and Displaced: A Brief Paper on Tribal Issues in Orissa*. Bhubaneswar: Vasundhara.

Kumar, K. and Choudhary, P.R. 2005. *A Socio-Economic and Legal Study of Scheduled Tribe's Land in Orissa* [Online]. Available at: http://www.academia.edu/1318494/A_Socio-Economic_and_Legal_Study_of_Scheduled_Tribes_Land_in_Orissa [accessed: 6 March 2013].

Lund, R. and Panda, S.M. 2011. New Activism for Political Recognition: Creation and Expansion of Spaces by Tribal Women in Odisha, India. *Gender, Technology and Development*, 15, 75–99.

Mahmood, S. 2001. Feminist Theory, Embodiment, and the Docile Agent: Some Reflections on the Egyptian Islamic Revival. *Cultural Anthropology*, 16, 202–36.

McCall, L. 2005. The Complexity of Intersectionality. *Signs: Journal of Women in Culture and Society*, 3, 1771–800.

Mishra, B. 2010. Agriculture, Industry and Mining in Orissa in the Post-Liberalisation Era: An Inter-District and Inter-State Panel Analysis. *Economic Political Weekly*, XLV(20), 49–68.

Mishra, R.N. and Maitra, A. 2006. *Industrialization and Protest Movements in Orissa*. [Online]. Available at: http://www.boloji.com/analysis2/0173.htm [accessed: 10 March 2011].

Mohanty, C.T. 2003. 'Under Western Eyes' Revisited: Feminist Solidarity Through Anticapitalist Struggles. *Signs: Journal of Women in Culture and Society*, 28, 499–535.

Mohanty, C.T., Russo, A. and Torres, L. (eds) 1991. *Cartographies of Struggle: Third World Women and the Politics of Feminism*. Bloomington, IN: Indiana University Press.

Mohindra, K.S. and Labonte, R. 2010. Asymmetric Review of Population Health Interventions and Scheduled Tribes in India. *BMC Public Health*, 10, 438–47.

Nayak, N. 2006. Maoists in Orissa Growing Tentacles and a Dormant State, in *Faultlines: Writings on Conflicts and Resolution*, Volume 17, edited by K.P.S. Gill. New Delhi: Institute of Conflict Management, 127–51.

Ota, A.B. and Mohanty, B.N. 2010. *Population Profile of Scheduled Tribes in Orissa*. Bhubaneswar: SCSTRTI.

Padel, F and Das, S. 2010. *Out of this Earth: East India Adivasis and the Aluminium Cartel*. New Delhi: Orient Blackswan.

Padhi, S. and Panigrahi, N. 2011. *Tribal Movements and Livelihoods: Recent Developments in Orissa*. Working Paper 51. New Delhi: Chronic Poverty Centre and Indian Institute of Public Administration.

Panda, S.M. 1996. *Forest Degradation, Changing Livelihoods and Gender Relations: Study of Two Tribal Communities in Orissa, India*. Doctoral dissertation. Bangkok: Asian Institute of Technology.

Pattanayak, S. 2010. *India vs Indians: Revolution Never Ends in Orissa* [Online]. Available at: www.saswat.com/blog/revolution-in-Orissa.html [accessed: 3 February 2011].

Prasad, A. 2011. No More Orissa–Oriya; Its Odisha–Odia Officially: CM Declares State Holiday. *OdishaDiary.tv* [Online, 5 November]. Available at: http://www.orissadiary.com/CurrentNews.asp?id=30113 [accessed: 7 March 2013].

Ramdas, S.R. 2009. Women, Forest Spaces and the Law: Transgressing the Boundaries. *Economic & Political Weekly*, XLIV(44), 65–73.

Roy, A. 2009. *Listening to the Grasshoppers: Field Notes on Democracy*. New Delhi: Penguin.

Sahoo, P.K. 2011. Odisha-Based NGO Takes Initiative to Wipe Out Irregularities in MGNREGA. *OdishaDiary.tv* [Online, 21 May]. Available at: http://orissadiary.com/CurrentNews.asp?id=26785 [accessed: 23 November 2011].

Sarkar, S. and Mehta, B.S. 2010. Income Inequality in India: Pre- and Post-Reform Periods. *Economic & Political Weekly*, XLV(37), 45–55.

SCSTRTI 2004. *Tribes of Orissa*. Bhubaneswar: Scheduled Castes and Scheduled Tribes Research and Training Institute.

Shiva, V. 1988. *Staying Alive: Women, Ecology and Development*. London: Zed Books.

Thakar, S. 2011. The Construction of Political Agency: South Asian Women and Political Activism. *Community Development Journal*, 46. 341–50.

The Forest Rights Act n.d. *The Scheduled Tribes and Other Traditional Forest Dwellers (Recognition of Forest Rights) Act, 2006* [Online: Campaign for Survival and Dignity]. Available at: http://forestrightsact.com/the-act [accessed: 6 March 2013].

The Hindu 2012. Land Acquisition Bill May be Re-introduced [Online, 14 December]. Available at: http://www.thehindu.com/news/national/land-acquisition-bill-may-be-reintroduced/article4196767.ece [accessed: 7 March 2013].

THREAD 2009. *Orissa Nari Samaj*. [Online: Siddharth Village]. Available at: www.siddharthvillage.com/narisamaj.php [accessed: 31 January 2010].

Weisskopf, T.E. 2011. Why Worry About Inequality in the Booming Indian Economy. *Economic & Political Weekly*, XLVI(47), 4–51.

World Bank 2007. *Towards Sustainable Mineral-Intensive Growth in Orissa: Managing Environmental and Social Impacts*. Washington DC: World Bank.

Xaxa, V. 2004. Women and Gender in the Study of Tribes in India. *Indian Journal of Gender Studies*, 11, 345–67.

Chapter 13

The Reemergence of Environmental Causation in Migration Studies and its Relevance for Bangladesh

Haakon Lein

Introduction

In recent decades there has been renewed interest in using environmental factors to explain social phenomena. Such 'neoenvironmental determinist' explanations (Judkins et al. 2008) have been widely disseminated to the public through Jared Diamond's book, *Guns, Germs, and Steel* (1999). In current mainstream social sciences, including political science and economics, natural resources (abundance or scarcity), biogeography, location and climate are commonly presented as causal factors for poverty and global inequality as well as armed conflicts (Sachs 2001, Hibbs and Olsson 2004, Burke et al. 2009).

As part of this general interest in environmental explanations, the concepts of environmentally-induced migration and environmental refugees, first introduced and debated in the 1970s and 1980s, have gained renewed interest – now under the heading climate refugees. In this chapter, I will critically review the role of environmental factors in explaining migration in Bangladesh. The country is used as an example of how environmental change and more recently climate change play a key role in determining migration patterns. This is not surprising given that the country is poor and densely populated, and regularly experiences climatic events such as floods, droughts and storm surges that are assumed to be drivers of (forced) population movements. The present chapter, partly based on research initiated by the Research Group on Forced Migration, led by Ragnhild Lund at the Department of Geography, Norwegian University of Science and Technology, reviews literature on migration in Bangladesh. Contrary to what perhaps might be expected, the literature provides little support for claims that environmental factors have so far played an important role in explaining migration.

The Environment as Driver of Migration

When the concept of environmental refugees was first introduced in the mid-1980s, the focus was on neo-Malthusian arguments that population growth puts pressure on scarce resources, leading to various types of land degradation (e.g. desertification) and, in turn, to poverty and migration. The scale of such phenomena was claimed to be large and growing. One of the key proponents, Norman Myers, argued in the mid-1990s that the number of environmental refugees may be as high as 25 million (Myers 1997), a figure some years later upgraded to possibly 200 million in the not-so-distant future (Myers 2002). Critics, among them Kibreab (1997) and Black (2001), questioned the basic understanding of migration processes underlying the concept and the empirical basis for the numbers presented. Another sceptic, Suhrke (1993), distinguished between the positions of maximalists and minimalists, whereby maximalists see a strong and direct link between environmental degradation, migration and possible conflicts, and minimalists reject the idea of a direct link between environmental change and migration but acknowledge that environmental factors may play a role in migration decisions.

The discussion of environment-driven migration and the concept of environmental refugees have gained renewed focus in the climate change debate, with a rapidly growing body of literature on climate change and migration (e.g. Gleditsch et al. 2007, Perch-Nielsen et al. 2008, Tacoli 2009, Bardsley and Hugo 2010, Hartmann 2010, Piguet 2010, Black et al. 2011). Topics dealt with in this literature include: the concepts of environmental refugees and forced migration; types of drivers and slow versus rapid onset processes; models of explanation; and methodological problems linked to measuring how environmental factors actually influence migration decisions.

The maximalist and minimalist positions are reflected in the current debate on the link between climate change and migration, where there is a huge discrepancy between alarmists positions taken by some academics (Myers 2002) and international organizations (e.g. Christian Aid 2007; Smith and Vivekenada 2007) and the views of sceptics who are critical of the idea that climate change should be seen as a likely driver of future migration (Black et al. 2008, 2011, Tacoli 2009, Hartmann 2010). The underlying premise of the alarmists is that climate change probably will lead to medium- and long-term environmental changes (e.g. drought and sea-level rise) as well as more frequent and devastating natural disasters (e.g. floods and cyclones), and that these will trigger large-scale migration rather than other types of response (e.g. adaptation). Tacoli (2009) argues that this view is based on what Castles (2002: 3) terms a 'common sense' view of migration – 'if water levels rise, or forests disappear, it seems obvious that people will have to move' – rather than an in-depth analysis of the complex relationship between real and perceived environmental changes, human agency and other factors behind migration decisions (Tacoli 2009).

Recent reviews of existing studies on the link between climate change and migration tend to conclude – perhaps not surprisingly – that migration processes are complex and that environmental change is one among many issues accounting for migration (Perch-Nielsen et al. 2008, Bardsley and Hugo 2010). Based on a discussion of three case studies, selected from 23 case studies covered by a large project funded by the European Union (EU), Warner et al. (2010: 707) conclude that environmental degradation currently seems not to be a major cause of migration, but that 'an environmental *signal* [my emphasis] was detected in each area that contributed to migration'. Furthermore, even if authors find limited or ambiguous evidence of environmental factors as a main cause of migration today, the potential future impact of climate change is still perceived as profound:

> It is becoming increasingly clear that environmental pressures resulting from climate change will be so fundamental to societal structures that they will also drive the development of new corridors and new scales of migration ... Environmental change has been, and will increasingly be, an important driver of migration. (Bardsley and Hugo 2010: 241)

Three arguments have been put forward regarding the links between environmental factors and migration in Bangladesh. The first is that poverty forces people to settle in disaster-prone, high-risk areas such as the *chars* (new land formed through accretion) located in the main rivers and in the Bay of Bengal. Spatial marginalization is thus seen as the outcome of economic and social marginalization (Burton et al. 1993, Wisner et al. 2004). Such poverty-induced forced migration to high-risk areas makes the poor vulnerable to disasters and may to some extent explain the high number of deaths caused by tropical cyclones in the Bay of Bengal, as well as the age, gender and class-biased character of such disasters reflected in the fact that the young and old, women and poor households clearly are overrepresented among the deaths in both the cyclone disasters of 1970 (at least 225,000 deaths) and 1991 (between 67,000 and 139,000 deaths) (Sommer and Mosely 1972, Mushtaque et al. 1993, Ikeda 1995).

The second argument is that that disasters force people to move to urban areas, thereby contributing to the rapid growth of urban slums. When crops and land have been destroyed and assets and family members have been lost, people will have few choices other than to become squatters along rural roads and embankments and ultimately to seek a new future in urban slum areas (Wisner et al. 2004).

The third argument is that 'environmental degradation' caused by general population growth (Hazarika 1993) or more specific interventions such as the construction of the Farakka barrage in India (Swain 1996) have led to massive out-migration from Bangladesh into neighbouring countries, first of all to India, and this is causing violent conflict in northwest India (Swain 1996, Reuveny 2008). It has been argued that this type of migration is likely to increase due to climate change and that this may pose a potential threat to regional security, as it is believed that massive inflows of Muslim Bangladeshis into India will inevitably trigger violent conflict between local populations and immigrants (Smith 2007, Warner et al. 2009).

Here, I will mainly limit my discussion to the second and third arguments, focusing on the role of environmental factors as drivers of migration. The first argument that poverty induces settlement in environmentally marginal areas such as the *char* areas is discussed in Lein (2009).

Migration in Bangladesh

In recent decades, Bangladesh has witnessed rapid urban growth, yet still only *c.*37 million out of a total of 146 million people, i.e. approximately one-quarter of the population, live in urban areas. Approximately one third live in the Dhaka Metropolitan Area, which in 2008 had an estimated population of 12.8 million (BBS 2009). In 1951, Dhaka city had a population of 273,459 (Ahmad 1958). The urban population will probably grow to 100 million during the twenty-first century and two thirds of this growth will probably be due to migration from rural areas (Afsar 2003, Streatfield and Karar 2008). Although migration generally takes the form of rural–urban migration, rural–rural migration is still widespread, especially among women who move to live with their husband's family after marriage.

There are relatively few studies on migration in Bangladesh but those that exist portray a fairly consistent picture, showing that migration is driven by economic motives as people move to secure better incomes or jobs (Hug-Hussain 1996, Islam 1996, Afsar 2000, 2003, Lein 2000). Although a large proportion of the migrants end up in slum settlements and in low-paid informal jobs, they perceive this as an improvement compared to their lives in rural areas (Afsar 2002, Alamgir et al. 2009). Afsar (2000) found that the majority of rural–urban migrants were males and were either uneducated agricultural labourers or better educated men. As regards female migrants, the garment sector in the mid-1990s absorbed more than 1.5 million workers (70 per cent of whom were women); of these 1.5 million, more than 90 per cent were migrants (Afsar 2003). A large proportion of migrants to Dhaka came from areas where agricultural productivity was above average, indicating that poor areas may not be the main source of migrants. Districts with good transport links and a long history of out-migration are the main origin of poor migrants (Afsar 2000, 2004).

Apart from migration into India, international migration from Bangladesh today consists primarily of labour migration to the Middle East and Southeast Asia. Approximately five million migrants officially send close to USD 8 billion in remittances (the total is more likely to be USD 10 billion), making labour migration the second largest export industry (Mahmud et al. 2008, Farid et al. 2009, Moses 2009).

Certain areas have a long history of both temporary and permanent internal migration for certain types of jobs and to particular destinations (Rashid 1991). For instance, Gallagher (1992) found that nine out of ten rickshaw-pullers in Dhaka and Chittagong were migrants and that approximately 50 per cent came from Faridpur District. According to Gardner (2009), 95 per cent of 283,000 Bangladeshis living in Great Britain originated from Syleth in northeast Bangladesh. This migration can be traced back to colonial times when Sylethis were recruited as sailors on British ships; this gave them an opportunity to take advantage of a growing demand for labour in Great Britain in the 1950s and 1960s, thus providing a basis for a growing network of newly recruited immigrants. Such patterns reflect the important role that social networks play in migration regarding both the decision to migrate and the choice of destination. These networks may be based on kin groups, lineages and villages, and are important when it comes to finding housing and jobs (Kuhn 2003).

Environmental Change and Migration

It is sometimes assumed that climate change will lead to more frequent and possibly more devastating natural disasters in the form of floods and cyclones accompanied by storm surges in the coastal areas and that this will lead to increased migration to urban areas. Riverbank erosion is also sometimes described as a major driver of migration. However, research has shown that migration tends to be local: when land is lost, people relocate and settle in nearby locations (Indra and Buchignani 1997, Indra 2000, Lein 2009, Kartiki 2011). In one of the few studies focusing specifically on the link between natural disaster and migration, Paul (2005) found no empirical evidence for out-migration following the tornado of 14 April 2004 that affected 38 villages in north-central Bangladesh. Paul's explanation is that relief support had compensated for the losses and in fact the area had experienced in-migration as a result of this.

However, these findings do not rule out that environmental factors may play a role in migration. Surveys of migrants to Dhaka have found that a proportion of respondents claim that disasters are the main reason for migration (Hug-Hussain 1996, Islam 1996). In 1998, I carried out a study in informal settlements in the northern part of Dhaka city, and among 557 households 58 per cent stated lack of land, employment, and income opportunities at their place of origin as the main reason for migrating to Dhaka (Lein 2000). Other reasons for migration were linked to environmental hazards, such as drought (16 per cent), riverbank erosion (10 per cent), cyclones (9 per cent), and floods (2.5 per cent). The limited role of floods as a key driver of migration is somewhat surprising given the prominent role floods have had in the environmental debate in the country.

It is commonly assumed that between 12 and 20 million Bangladeshis have migrated to India since the country gained independence (Afsar 2008). The underlying causes of migration into India are undoubtedly many but can be linked to the exploration of economic opportunities (jobs, trade), religion, trafficking, and legal and illegal cross-border activities (van Schendel 2005, Afsar 2008). However, it has been argued that these primarily reflect migrants being 'pushed' by 'land scarcity' (Homer Dixon 1999: 95) or by 'environmental factors' (Reuveny 2007: 668, 2008).

Such claims are based on the rather uncritical use of two interesting, but very limited case studies (Hazarika 1993, Swain 1996). Both studies refer to the situation in the 1980s and neither explicitly addresses climate change and migration. As a low-lying country, Bangladesh is seen to be particularly vulnerable to sea-level change induced by climate change and this is regarded by some to be a possible future driver of migration, not only within the country but also abroad, primarily to India. According to one report, in 20 or 30 years Bangladesh may:

> ...see mass movement of people from flood-prone areas, possibly to urban centers. The current structures and organizations to help the victims of disasters will not be enough to cope with the increase of migration flows in the future. Given the political instability of the region, population movements associated with climate change could become an issue for regional security. (Warner et al. 2009: 13)

This conclusion is based on the findings of a study of the link between forced migration, environmental degradation and climate change (Poncelet 2009, Poncelet et al. 2010). The empirical material for the Bangladesh case report included interviews with what are termed experts (academics, government employees and NGO staff) as well as 45 migrants and non-migrants in three different parts of the country. The experts argued that most Bangladeshi migration today is internal and predominantly from rural to urban areas. There was also consensus among the experts that 'it would be hazardous to attribute these population movements to the consequences of climate change' (Poncelet 2009: 9). The experts did not consider environmental factors or climate change as influencing the decision to migrate. Yet, contrary to the national experts, the author concludes that people affected by floods, riverbank erosion and other disasters see migration as one of the best strategies for adapting to such disasters (Poncelet 2009).

Estimates of the number of people that may be affected by future sea-level rise vary widely, from a few million to tens of millions.[1] Some figures cited appear to be little more than speculation whereas others, especially the lower estimates, seem to be based on more detailed assessments of sea-level rise and population projection. Nonetheless, the level of uncertainty is high, not only because it is difficult to predict the impact of sea-level rise but also because there is limited understanding of the interaction of sea-level rise with other geophysical processes in the region, such as subsidence, accretion and tectonic changes (Warrick et al. 1996). However, it seems clear that the numbers affected and that potentially will have to migrate due to sea-level rise will be only a modest fraction compared to the growth in the population in the same period: 100 million up to the year 2050 according to Streatfield and Karar (2008).

As pointed out by Black et al. (2008), predictions about future international migration caused by climate change should at best be seen as more or less informed

---

1   Lein, H. 2012. 'A Critical Review of the Bangladesh "Climate Change Migration and Conflict Scenario"'. Trondheim: Department of Geography, Norwegian University of Science and Technology (Mimeo.).

guesses, while predicting where migrants will eventually end up involves even more difficult guesswork. In the case of Bangladesh, Black et al. (2008) argue that, although climate change may lead to increased internal rural–urban migration, international migration from the country will be less clearly impacted by climate change, as this type of migration is associated more with existing migrant networks than environmental change.

Conclusions

This review of literature on migration in Bangladesh does not provide much support for the claim that environmental factors have so far played an important part in migration processes in Bangladesh. While some studies indicate that environmental factors might play a role, migration seems in general to be driven primarily by economic factors and in this process social networks make it possible for people to find jobs and housing. There seems to be a gap in the understanding of migration processes between some members of the climate-change research community on the one hand and the more traditional migration research community on the other. Statements on the possible relationship between climate change and migration are no more than claims about a likely development in the future. Since the future is not known and imagining scenarios or making predictions regarding the future is inherently risky, the idea that climate change may cause massive migration obviously cannot be completely ruled out. However, arguments using 'the future as evidence', i.e. claims that a phenomenon probably will occur and thus is a real phenomenon – which seem to be quite common in some of the climate change literature – are obviously insufficient. Liverman (2009) and Hulme (2008) argue that the current climate change debate may, if not handled carefully, come to echo claims from classical environmental determinism, which saw climate as the key to understanding all types of social phenomena. Rejecting simplistic and largely unfounded claims regarding future causal inks between climate change and migration is one step in the direction away from this determinist trap.

References

Afsar, R. 2000. *Rural-Urban Migration in Bangladesh: Causes, Consequences and Challenges*. Dhaka: University Press.

Afsar, R. 2002. Migration and Rural Livelihoods, in *Hands Not Land: How Livelihoods are Changing in Rural Bangladesh*, edited by K.A. Toufique and C. Turton. Dhaka and London: Bangladesh Institute of Development Studies and DFID, 89–96.

Afsar, R. 2003. Internal Migration and the Development Nexus: The Case of Bangladesh, in *Migration, Development & Pro-Poor Policy Choices in Asia* [Online: Refugee and Migratory Movements Research Unit (RMMRU),

University of Dhaka, and Department for International Development (DFID), London]. Available at: http://www.migrationdrc.org/publications/working_ papers/WP-C2.pdf [accessed: 4 January 2013].

Afsar, R. 2004 Dynamics of Poverty, Development and Population Mobility: The Bangladesh Case. *Asia-Pacific Population Journal*, 19, 69–91.

Afsar, R. 2008. *Population Movement in the Fluid, Fragile and Contentious Borderland Between Bangladesh and India*. Paper Prepared for Presentation at the 20th European Conference on the Modern South Asian Studies, School of Arts, History and Culture, University of Manchester, United Kingdom, July 7–11, 2008 [Online]. Available at: http://archiv.ub.uni-heidelberg.de/ savifadok/volltexte/2008/143/pdf/Afsar_PopulationMovement_2008.pdf [accessed: 4 January 2013].

Ahmad, N. 1958 *An Economic Geography of East Pakistan*. London: Oxford University Press.

Alamgir, M.S., Jabbar, M.A. and Islam, M.S. 2009. Assessing the Livelihood of Slum Dwellers in Dhaka City. *Journal of Bangladesh Agricultural University*, 7, 373–80.

Bardsley, D.K. and Hugo, G.J. 2010. Migration and Climate Change: Examining Thresholds of Change to Guide Effective Adaptation Decision Making. *Population and Environment*, 32, 238–62.

BBS 2009. *Statistical Pocketbook of Bangladesh* [Online: Bangladesh Bureau of Statistics, Dhaka]. Available at: http://203.112.218.65/WebTestApplication/ userfiles/Image/SubjectMatterDataIndex/pk_book_09.pdf [accessed: 3 February 2011].

Black, R. 2001. *Environmental Refugees: Myth or Reality?* Working Paper no. 34, New Issues in Refugee Research. Geneva: United Nations High Commissioner for Refugees.

Black, R., Kniveton, D., Skeldon, R., Coppard, D., Murata, A. and Schmidt-Verkerk, K. 2008. *Demographics and Climate Change: Future Trends and their Policy Implications for Migration*. Working Paper T-27. Brighton: Development Research Centre on Migration, Globalisation and Poverty, University of Sussex.

Black, R., Kniveton, D. and Schnidt-Verkerk, K. 2011. Migration and Climate Change: Towards an Integrated Assessment of Sensitivity. *Environment and Planning A*, 43, 431–50.

Burke, M., Miguel, B.,E., Satyanath, S., Dykema, J.A. and Lobell, D.B. 2009. Warming Increases the Risk of Civil War in Africa. *PNAS: Proceedings of the National Academy of Sciences of the United States of America*, 106, 20670–74.

Burton, I., Kates, R.W. and White, G.F. 1993. *The Environment as Hazard*. New York: Guilford Press.

Castles, S. 2002. *Environmental Change and Forced Migration: Making Sense of the Debate*. PDES Working Papers No. 70. Geneva: United Nations High Commissioner for Refugees.

Christian Aid. 2007. *Human Tide: The Real Migration Tide* [Online: Christian Aid, London]. Available at: http://www.christianaid.org.uk/Images/human-tide.pdf [accessed: 22 February 2011].

Diamond, J. 1999. *Guns, Germs, and Steel: The Fates of Human Societies*. New York: Norton.

Farid, K.S., Mozumdar, L., Kabir, M.S. and Hossaink, B. 2009. Trends in International Migration and Remittance Flows: Case of Bangladesh. *Journal of the Bangladesh Agricultural University*, 7, 387–94.

Gallagher, R. 1992. *The Rickshaws of Bangladesh*. Dhaka: University Press.

Gardner, K. 2009. Lives in Motion: The Life-Course, Movement and Migration in Bangladesh. *Journal of South Asian Development*, 4, 229–51.

Gleditsch, N.P., Nordås, R. and Salehyan, I. 2007. *Climate Change and Conflict: The Migration Link*. Coping with Crisis Working Paper Series. New York: International Peace Academy.

Hartmann, B. 2010. Rethinking Climate Refugees and Climate Conflict: Rhetoric, Reality and the Politics of Policy Discourse. *Journal of International Development*, 22, 233–46.

Hazarika, S. 1993. Bangladesh and Assam: Land Pressures, Migration and Ethnic Conflict. *Occasional Paper Series of the Project on Environmental Change and Acute Conflict*, 3(March), 45–65.

Hibbs, D. and Olsson, O. 2004. Biogeography and Why Some Countries are Rich and Others Poor. *PNAS: Proceedings of the National Academy of Sciences of the United States of America*, 101, 3715–20.

Homer-Dixon, T.F. 1999. *Environment, Scarcity, and Violence*. Princeton, NJ: Princeton University Press.

Hug-Hussain, S. 1996. *Female Migrants' Adaptation to Dhaka: A Case of the Processes of Urban Socio-Economic Change*. Dhaka: Urban Studies Programme, Department of Geography, University of Dhaka.

Hulme, M. 2008. The Conquering of Climate: Discourses of Fear and their Dissolution. *Geographical Journal*, 174, 5–16.

Ikeda, K. 1995. Gender Differences in Human Loss and Vulnerability in Natural Disasters: A Case Study from Bangladesh. *Indian Journal of Gender Studies*, 2, 171–93.

Indra, D. 2000. Not Just Displaced and Poor: How Environmentally Forced Migrants in Rural Bangladesh Recreate Space and Place under Trying Conditions, in *Rethinking Refuge and Displacement: Selected Papers of Refugees and Immigrants*, Volume VIII, edited by E.M. Gozdziak and D. Shandy. Washington, DC: American Anthropological Association, 163–91.

Indra, D.M. and Buchignani, N. 1997. Rural Landlessness, Extended Entitlements and Inter-Household Relations in South Asia: A Bangladesh Case. *Journal of Peasant Studies*, 24, 25–64.

Islam, N. 1996. *Dhaka: From City to Megacity: Perspectives on People, Places, Planning and Development Issues*. Dhaka: Urban Studies Programme, Department of Geography, University of Dhaka.

Judkins, G., Smith, M. and Keys, E. 2008. Determinism Within Human–Environment Research and the Rediscovery of Environmental Causation. *Geographical Journal*, 174, 17–29.

Kartiki, K. 2011. Climate Change and Migration: A Case Study From Rural Bangladesh. *Gender & Development*, 19, 23–38.

Kibreab, G. 1997. Environmental Causes and Impact of Refugee Movements: A Critique of the Debate. *Disasters*, 21, 20–38.

Kuhn, R. 2003. Identities in Motion: Social Exchange Networks and Rural–Urban Migration in Bangladesh. *Contributions to Indian Sociology*, 37, 311–37.

Lein, H. 2000 Hazards and 'Forced' Migration in Bangladesh. *Norsk Geografisk Tidsskrift–Norwegian Journal of Geography*, 52, 122–7.

Lein, H. 2009. The Poorest and Most Vulnerable? On Hazards, Livelihood and Labelling of Riverine Communities in Bangladesh. *Singapore Journal of Tropical Geography*, 30, 98–113.

Liverman D. 2009. Conventions of Climate Change: Constructions of Danger and the Dispossession of the Atmosphere. *Journal of Historical Geography*, 35, 279–96.

Mahmud, W., Ahmed, S. and Mahajan, S. 2008. *Economic Reforms, Growth, and Governance: The Political Economy Aspects of Bangladesh's Development Surprise*. Commission on Growth and Development Working Paper No 22. Washington: International Bank for Reconstruction and Development/ World Bank.

Moses, J. W. 2009. Leaving Poverty Behind: A Radical Proposal for Developing Bangladesh Through Emigration. *Development Policy Review*, 27, 457–79.

Mushtaque, A., Chowdhury, R., Bhuyia, A., Yusuf, A. and Sen, R. 1993. The Bangladesh Cyclone of 1991: Why So Many People Died. *Disasters*, 7, 291–303.

Myers, N. 1997. Environmental Refugees. *Population and Environment*, 19, 167–82.

Myers, N. 2002. Environmental Refugees: A Growing Phenomenon of the 21st century. *Philosophical Transactions of the Royal Society of London, Series B Biological Sciences*, 357, 609–15.

Paul, B.K. 2005. Evidence Against Disaster-Induced Migration: The 2004 Tornado in North Central Bangladesh. *Disasters*, 29, 370–85.

Perch-Nielsen, S.L., Bättig, M.B. and Imboden, D. 2008. Exploring the Link Between Climate Change and Migration. *Climatic Change*, 91, 375–93.

Piguet, E. 2010. Linking Climate Change, Environmental Degradation, and Migration: A Methodological Overview. *Wiley Interdisciplinary Reviews: Climate Change*, 1, 517–24.

Poncelet, A. 2009. *Bangladesh Case Study Report: 'The Land of Mad Rivers'* [Online: EACH-FOR Environmental Change and Forced Migration Scenarios]. Available at: http://www.each-for.eu/documents/CSR_Bangladesh_090126.pdf [accessed: 22 February 2011].

Poncelet, A., Gemenne, F., Martiniello, M. and Bousetta, H. 2010. A Country Made for Disasters: Environmental Vulnerability and Forced Migration in Bangladesh, in *Environment, Forced Migration and Social Vulnerability*, edited by T. Afifi and J. Jäger. Berlin and Heidelberg: Springer, 211–22.

Rashid, H. E. 1991. *Geography of Bangladesh*. Dhaka: University Press Limited.

Reuveny, R. 2007. Climate Change-Induced Migration and Violent Conflict. *Political Geography*, 26, 656–73.

Reuveny, R. 2008. Ecomigration and Violent Conflict: Case Studies and Public Policy Responses. *Human Ecology*, 36, 1–13.

Sachs, J. 2001. *Tropical Underdevelopment*. NBER working paper 8119. Cambridge MA: National Bureau of Economic Research.

Smith, D., and Vivekenada, J. 2007. *A Climate of Conflict: The Links between Climate Change, Peace and War*. London: International Alert.

Smith, P.J. 2007. Climate Change, Weak States and the 'War on Terrorism' in South and Southeast Asia. *Contemporary Southeast Asia*, 29, 264–85.

Sommer, A. and Mosely, W.H. 1972. East Bengal Cyclone of November 1970: Epidemiological Approach to Disaster Assessment. *Lancet*, 299(7759), 1030–36.

Streatfield, P.K. and Karar, Z.A. 2008. Population Challenges for Bangladesh in the Coming Decade. *Journal of Health, Population and Nutrition*, 26, 261–72.

Suhrke, A. 1993. Pressure Points: Environmental Degradation, Migration and Conflict. *Occasional Paper Series of the Project on Environmental Change and Acute Conflict*, 3(March), 3–43.

Swain, A. 1996. Displacing the Conflict: Environmental Destruction in Bangladesh and Ethnic Conflict in India. *Journal of Peace Research*, 33, 189–204.

Tacoli, C. 2009. Crisis or Adaption? Migration and Climate Change in a Context of High Mobility. *Environment and Urbanization*, 21, 513–25.

van Schendel, W. 2005. *The Bengal Borderland*. London: Anthem Press.

Warner, K., Ehrhart, C., de Sherbinin, A., Adamo, S.B. and Onn, T.C. 2009. *In Search of Shelter: Mapping the Effects of Climate Change on Human Migration and Displacement*. A Policy Paper Prepared for the 2009 Climate Negotiations. Bonn: United Nations University, CARE, and CIESIN-Columbia University and in Close Collaboration with the European Commission 'Environmental Change and Forced Migration Scenarios Project', the UNHCR, and the World Bank.

Warner, K., Hamza, M., Oliver-Smith, A., Renaud, F. and Julca, A. 2010. Climate Change, Environmental Degradation and Migration. *Natural Hazards*, 55, 689–715

Warrick, R.A., Azizul Hoq Bhuiya, A.K., Mitchell, W.M., Murty, T.S. and Rasheed, K.B.S. 1996. Sea-Level Changes in the Bay of Bengal, in *The Implications of Climate and Sea-Level Change for Bangladesh*, edited by R.A. Warrick and Q.K. Ahmad. Dordrecht/Boston/London: Kluwer Academic Publishers, 97–142.

Wisner, B., Blaikie, P., Cannon, T. and Davis, I. 2004. *At Risk: Natural Hazards, People's Vulnerability and Disasters*, 2nd edition. London: Routledge.

Chapter 14

Discourses that Hide: Gender, Migration and Security in Climate Change

Bernadette P. Resurrección

Introduction

In the global debate on climate change, many voices deliberate on and hammer out agreements on emission targets, adaptation programmes and financing as well as carbon trade scenarios. Recent discussions have increasingly raised awareness of issues of 'climate justice' (GenderCC 2010) and there is now greater propensity to evaluate the uneven conditions of social systems that exacerbate the vulnerability and inhibit adaptive responses of women and the poor. The discourse on climate justice is being played out concurrently with separate discourses on the security risks and other implications of climate-induced population migration. Two disparate sets of anxious actors have taken up these discourses: civil society and feminist groups concerned with the paucity of climate justice consciousness in climate change agendas on one hand; and international security institutions worried by threats of climate-induced migrations on the other. This indicates that the themes of migration, gender and adaptation have become more crucial today as more and more planners and scholars are collectively convinced of the need to ensure human security in the face of threats and dislocations caused by climate change, and the need to bring these potential threats to the negotiating table.

This chapter aims to (i) review international agreements, advocacy literature and policy statements in order to understand the conceptual underpinnings of discourses on gender, migration and climate change; and (ii) discuss possible pathways of understanding their linkages, informed by earlier more grounded work on gender, migration and environment.

The 'Women and Environment' Lobby in International Agreements

The United Nations Conference on Environment and Development (UNCED), or Earth Summit, in Rio de Janeiro in 1992 offers convincing evidence of strong feminist presence in international agreements on environmental degradation and efforts to reduce and mitigate its effects on developing regions. Four international agreements grew out of the Earth Summit: Agenda 21 (1992) (UN 2009), the United Nations (UN) Convention on Biological Diversity (1993) (CBD n.d.), the

UN Convention to Combat Desertification (1994) (UNCCD 2012) and the UN Framework Convention on Climate Change (UNFCCC) (1994) (UN 1992). The UNFCCC has begun an international process of climate change negotiations and committed parties to a universal objective to reduce emissions with a benchmark of 1990 emission levels. However, only Agenda 21 and the Convention on Biological Diversity contain explicit, albeit few, clauses that recognize the gender-specific effects of environmental change and ways to reduce these. The Beijing Platform for Action 1995 from the UN World Summit on Women contained a separate section on 'Women and Environment' (Section K) (UN Women n.d.). Subsequent UNFCCC agreements and treaties did not articulate any concern for gender issues, except for the need to include gender experts in the National Adaptation Programmes for Action (NAPAs) among Annex I countries.[1] Efforts to bring in a gender or feminist agenda mostly fell by the wayside in meetings of the Conference of the Parties (COP). The first stirrings of a gender coalition were felt only during COP11 in Montreal in 2005.

The thin presence of gender advocacy groups was in large part due to the global and transboundary nature of the problems identified by the climate change actors. These require international and multilevel approaches, unlike in the post-UNCED period when more community-based and localized responses to environmental degradation were proposed (Leach 2007). Skutsch (2002: 31) confirms that there was felt a need to coalesce around universal issues 'and not divert attention to gender aspects', given limited resources and the uncooperative behaviour of the USA during the signing of the Kyoto Protocol to the UNFCCC. Villagrasa (2002), however, notes that women were active in the negotiations for the signing of the Kyoto Protocol, but it was unclear whether there was a clear feminist agenda during the negotiations or side events thereof. Delegates celebrated adoption of the Kyoto Protocol in 1997 (UN 1998).

The Intergovernmental Panel on Climate Change (IPCC), founded in 1988 under the UN Environment Programme (UNEP) and the World Meteorological Organization (WMO), became the scientific bedrock of the UNFCCC but for long did not discuss the gender dimensions of climate change; for decades, it devoted its discussions to the technical aspects of climate changes such as mitigation measures and scales of impact through modelling approaches (Terry 2009). A recent scoping study on climate change adaptation confirms this: 'Adaptation is understood as primarily a technical means with which to reduce and minimize the impact of climate change rather than as a complex set of responses to existing climatic and non-climatic factors that contribute to people's vulnerability' (Resurrección et al. 2008: 19). However, in its Fourth Assessment Report, Chapter

---

1   Annex 1 countries are the countries included in Annex I (as amended in 1998) to the UNFCCC, including all the developed countries in the Organization of Economic Cooperation and Development (OECD) and economies in transition. These countries committed themselves to reduce by 2000 their greenhouse gas emissions to 1990 levels (UNEP/GRID-Arendal 2011).

17 on Adaptation, the IPCC has discussed gender as a differentiating social category (Adger et al. 2007). It was only in 2007 at COP13 in Bali, Indonesia, that global network organizations more visibly emerged, such as the Women's Environment & Development Organization (WEDO), ENERGIA (International Network for Gender and Sustainable Energy) and gender advocates within the International Union for the Conservation of Nature (IUCN) and the Food and Agriculture Organization (FAO). The network GenderCC: Women for Climate Justice was formally constituted at Bali and put forward a definitive women's and gender agenda to UNFCCC negotiations and meetings (GenderCC 2010). The emerging discourse of climate justice emerged from these new formations, drawing from earlier concerns on the critical gaps in energy consumption patterns between North and South, responsibility for payments for adaptation programmes in view of earlier huge investments into mitigation efforts, risky trade-offs between new initiatives for carbon sequestration such as Reduced Emissions from Deforestation and Forest Degradation (REDD), and sustaining local livelihoods of communities in the face of climate change.

The Gender and Climate Justice Discourse

Early in the climate debate, O'Riordan and Jordan (1999) posited that climate change is a context through which institutions employ 'social devices' such as creating and interpreting scientific knowledge and selecting politically tolerable adaptation strategies. The discourse on gender and climate justice in these initial international events has revolved around a feminine subject – poor rural women of the Global South, who have been adversely affected by climate change. A running logic permeates the discussions: climate change is most adversely felt by vulnerable people in climate hotspots of the Global South, chiefly among women, who constitute the largest percentage of the world's poorest. An example of this discourse is from WEDO:

> The brunt … will be borne by poor women and their communities who are most dependent on the land and natural resources for their food, livelihood, fuel and medicine yet less equipped to cope with natural disasters and weather variations. Women are particularly affected because of socially ascribed roles resulting from entrenched feudal-patriarchal discrimination on them. Rural women also take a heavy toll being the ones engaging in various remedies to make ends meet. (WEDO 2008: 5)

Awareness raising, marshalling evidence through a collection of case studies on climate change impacts on women and capacity-building are some of the activities that build around this logic. This same thinking argues that women are powerful agents of change and that their full participation is critical in adaptation and mitigation policies and programmes, and hence it is important that women and gender experts participate in all decisions related to climate change (GenderCC 2007).

The discourse of women as chief victim and caretaker in climate change debates and programmes resonates with women-environment-and-development (WED) thinking in the global discussions on environmental degradation in the 1990s. WED was a corrective to earlier gender-blindness in these discussions as it emphasized relational perspectives between women and men, where experiences of the environment are differentiated by gender through materially distinct daily work activities and responsibilities of women and men. As a result, it was assumed that women and men hold gender-specific interests in natural resource management through distinctive roles, responsibilities and knowledge (Elmhirst and Resurrección 2008). Women were also recognized as a natural constituency for environmental 'care', especially since it was fundamentally assumed that their livelihoods were disrupted by environmental stresses.

Strands from the WED genre celebrate women's agency in adapting to climate change. The following are striking examples:

- *Adaptation strategies*: Scholars report that women have been capable of mobilizing the community in the different phases of the disaster risk cycle (Guha-Sapir 1997, Enarson 2001, Yonder et al. 2005), and thus show visible signs of adapting to climate changes in the long run. Speranza et al. (2006: 121) report that, among agropastoralists in Kenya and Tanzania, rural actors, 'especially women, organize themselves in Self Help Financial Groups (SHFGs) to increase their financial capacity' in order to find alternatives for adjusting to climate change and climate variability impacts on their household and livelihoods. Their activities included intercropping, planting crops to coincide with the rains, or possibly even forfeiting planting for the season if they think that planting will anyway result in crop loss. In Zaheerabad, Dalit women, who form the lowest rung of India's stratified society, demonstrate adaptation to climate change through the formation of local self-help groups that convene regularly. They follow a system of interplanting crops that do not need extra water, chemical inputs or pesticides for production. The women grow as many as 19 types of indigenous crops to an acre on arid, degraded lands (Acharya 2009).
- *Knowledge*: Women are thought to have greater clarity about risk surrounding their environment. Their high adaptive knowledge plunges them into action when a community is at risk (Enarson and Fordham 2001). Women develop broad knowledge and experiences regarding their environment (Ariyabandu 2004), which they evaluate constantly and change with changing environmental and social conditions, a result of the responsibilities that they assume within their families and in their communities. Around 5,000 women spread across 75 villages in the arid, interior parts of southern India are now practising chemical-free, non-

irrigated, organic agriculture as one method combating global warming (Acharya 2009).

The danger with such 'women only' assertions when translated into policy is that they 'naturalize' and reinforce inequitable gender divisions of labour, thus inadvertently increasing women's workloads in programmes that had aimed to empower them. In short, they add 'environment' and 'climate adaptation/ mitigation' to women's already long list of caring roles.

In the past, scholars expressed their disquiet with WED for a number of reasons. First, WED was seen as holding essentialist views and simplifications that posit the idea of women as being a natural constituency for environmental projects. Second, WED tended to add 'environment' to women's long list of caring roles (Leach 1992, Jackson 1993). Third, WED was seen as making universalist assumptions on women's environmental roles that do not match ground realities and their intersecting class, ethnic or age-related subjectivities (Rao 1991). Fourth, there was a marked absence of men in the analysis, which can raise issues of power more clearly or draw attention to other subjects of vulnerability. Fifth, special emphasis was placed on women's knowledge of the environment without investigating whether this emanates from a position of subordinate obligation and power configurations (Jewitt 2002).

Despite criticism levelled at the theoretical premises, simplifications and policy applications of WED in the 1990s, the women–environment linkage has been reinscribed into the contemporary climate change debates. However, a more critical look at these debates reveals that summoning simplifications is sometimes useful for political projects such as feminism, to carve the space it sorely needs in a discursive arena that thrives on the homogenization of its subjects and technical fixes to ameliorate damage from climate stresses.

While 'women' as the sole subject in the growing gender and climate change literature still persists, other studies point to the complex influences of social and cultural norms that dynamically shape the gender divisions of labour, labour mobility and decision-making patterns in households and communities, and which may create situations where men may also suffer from gender-specific vulnerability due to their relatively limited access to resources and their resulting poverty (Lambrou and Piana 2006, BRIDGE 2008, Terry 2009). This stream calls for a more critical and nuanced understanding of the inequalities between women and men, and the ways that climate change could exacerbate the effects of these inequalities (BRIDGE 2008). Demetriades and Esplen (2009) argue for more context-specific research, drawing on local realities and adaptation strategies and a plea for understanding the complex relational nature of gendered power. Studies by Norm (2009) and Sreng (2010) on Cambodia demonstrate this complexity.

Subsistence farmers are experiencing longer dry spells in Cambodia. Norm's (2009) study, focusing on Battambang Province, reveals that the province's average rainfall has been irregular, based on average yearly rainfall data from 1982 to 2008. The last heavy rainfall was in 1999 (1,500 mm). Before and after then, rainfall fell to low levels of 1,000 to 1,200 mm, with the lowest level recorded in 2004. The average maximum temperature in the same period has been on the rise, from a benchmark of 32°C in 1982 to an average of 34–35°C in recent years, especially in 2004 and 2006.

Farmers in Kors Krolar District, for instance, said that they were experiencing hotter days during both the dry and rainy seasons. The rising temperature and decreased rainfall had a tremendous impact on rain-fed rice cultivation in the area. Farmers no longer transplanted seedlings from seedbeds as they did in the past when there was more regular rainfall. Instead, they turned to sowing seeds and harvesting them directly from the paddy fields in order to maximize the short period of rainfall that occurred irregularly between May and October. In former times, they used to cultivate a second crop during the rainy season to ensure adequate rice supply as well as a slight surplus to sell in markets. This was no longer possible. To replace this activity and to redress shortfalls in their own households' rice supply, men more frequently ventured into the forests to cut trees, collect fuel wood for charcoal production, cut bamboo and collect non-timber forest products. Men largely bore the heavy responsibility of earning money to support their households, taking other jobs such as driving motorcycle taxis and working as maize harvesters in neighbouring farms. Some temporarily migrated to Thailand to work as construction workers. Women also worked as wage labourers, performing work such as clearing bush in the plantations of other villagers and planting maize for other landowners in the commune. In addition, they collected forest products such as vegetables and wild mushrooms as well as digging up wild potatoes for household consumption. They also raised livestock at home, made rice wine and assisted their husbands in charcoal production.

Norm's study also indicated that some wives urged their husbands to log and sell trees. Husbands were reluctant but realized there was little choice. Khmer men, who are recognized as heads of their households, usually do not undertake trading activities. Nhe Houy, aged 36, who had five children ranging from three to 13 years old, recalled the tough times:

> When my family had money shortage and no rice to eat and my husband did not go to work anywhere, I always pushed him to go and cut wood in the forest and then sell them to wood merchants. After I talked to him, he went to cut wood and sold them for about 30,000R to 40,000R (= USD7.50 to USD10) each time. With this money I bought rice for the children. (Norm 2009: 58)

Sons were also urged to go to the forest to cut trees for the timber market, and many of them dropped out of school in order to earn money to help their families due to crop failures on their parents' farms. The drought was so severe in 2004 that the Battambang Provincial Rural Development Department was compelled to distribute water to affected residents. An officer of the department recalled that conflicts arose due in large part to the scarcity of water itself, but also because of inequitable distribution. Those who lived near water distribution stations received larger amounts. Also, those who knew and were close to the provincial department's water distributors were able to access information on distribution schedules before others and thus received water earlier and in larger amounts than the rest. Those who lived farther away and had no such relations received very little or no water at all.

The survey by Sreng (2010) in Ratanakiri Province reports that floods inundated villages both due to unusually heavy rainfall and allegedly unexpected water releases from the hydropower dam at the Lower Sesan River in September 2009. Among the respondents, women were chiefly responsible for 60 per cent of home-based businesses such as selling fruit, homemade cakes, noodle dishes, fried bananas and other cooked food. These women also collected non-timber forest products, largely wild vegetables. When the floods came, more male residents took on collection of non-timber forest products. After the flood, only 10 per cent of women respondents resumed collecting non-timber forest resources, whereas 18 per cent of male respondents admitted that they now pursued this more intensively, as it included logging activities. Women obtained loans from relatives and moneylenders to tide them over the crisis period of food scarcity and disruptions in farming and fishing livelihoods in which both women and men were formerly involved.

The Cambodian studies discuss climate change impacts and how people respond and adapt. They indicate that men, as well as women, are affected by longer dry spells and drought. They dispel earlier WED 'women only' views that women are often the hardest hit and are thereby the main actors and stakeholders in effective climate change adaptation. These cases demonstrate that, in adaptation, gender roles are far from fixed; they are changing, and are often contingent on changing social, political and even climate conditions. Women and men negotiate their roles, which may be contested at times, as seen in the reluctance of the men in one of the studies to take up more active marketing activities in the wake of floods.

The rapidly growing literature on gender and climate change remains largely silent on migration issues. It appears that scholars who work on gender and environment issues do not work with gender and migration issues. This is a reflection of how these are generally regarded as separate fields of study. Hugo (2008) confirms that migration and environmental scholars tend to work separately, resulting in the lack of a migration–environment–development interface. Gender-blind research neglects the fundamental ways in which climate-shaped migratory experiences and impacts will differ for women and men (Hunter and David 2009). There is, however, much to learn from the literature on gender, disaster and hazards, where displacement and resettlement figure as responses to hazards and extreme events.

Adaptation and Vulnerability

Cleaver (2000) earlier cautioned against essentialist assumptions about men's and women's roles in natural resource management. While culturally defined gender roles in response to climate stresses exist, they may be more flexible than at first appears and subject to negotiation and change that go beyond fixed definitions of 'women' and 'men'.

Nightingale (2009) suggests that climate adaptation, as a concept drawn from the ecological sciences, is fundamentally an aggregated concept referring to the ability of human societies and ecological systems to cope with climate variation. This ability is premised on the notion of the 'adaptive capacity' of human and ecological systems', whereby people's adaptive capacities are determined by their socioeconomic characteristics. For instance, the IPCC (Watson et al. 1998) states that the determinants of adaptive capacity are directly correlated with measures of economic development (GDP per capita). Developing countries are recognized to be more vulnerable to climate change because of their 'lack of institutional capacity' among other things (this is usually interpreted as a lack of capacity of government) (Watson et al. 1998). This reasoning connects well with the logic that women from developing and marginal groups and regions are the hardest hit by climate stresses, and that in their hands rests the challenge of adaptation and mitigating the effects of climate change. Women are hence seen as possessing the skills for adaptation and as having enormous stakes in ensuring the survival of their livelihoods and households such as in studies highlighting the feminization of agriculture where men out-migrate, leaving the care for children and elderly to older women (Resurrección et al. 2008). The gender and disaster literature has identified several vulnerability characteristics of women (Enarson 1998, Bradshaw 2004), sensitizing disaster risk managers to mitigate these characteristics.

Nightingale departs from a focus on individual characteristics, arguing that attention should instead shift to the kinds of climate-related hardships that will result for specific classes and ethnic groups of women and men due to their different economic and political positions and uneven power relations in society. She remarked, 'the biggest impact of climate change will be on differentiation within human societies, closely linked to resource availability. This would mean increases in inequality based on gender, class, caste, geography and ethnicity, which are some of the key axes of difference by which resources are currently distributed' (Nightingale 2009: 85). Farmers' adaptive capacity is more than just their ability and knowledge to cultivate and select crops that are drought-resistant, as they are used to coping with climate variations. This, says Nightingale, is not the crucial element to consider. Instead, farmers' adaptive capacity will depend largely on whether women and men are equally able to gain access to and control over household and community decision-making processes in managing threatened or scarce resources as a result of climate stresses. A preoccupation with power and differentiation appears to be the blind spot in much of the literature on climate adaptation but has nevertheless had strong resonance in earlier conceptualizations of vulnerability in the disaster literature.

Similarly preoccupied with people's characteristics, Wisner et al. (2007: 4, 11) view vulnerability as 'the characteristics of a person or group and their situation influencing their capacity to anticipate, cope with, resist and recover from the impact of natural hazard'. This definition considers people's characteristics as central to the shaping of their vulnerability. In contrast, others view vulnerability as a process rather than a set of sometimes assumed fixed characteristics. For instance, in the hazards literature, Blaikie et al. (1994) state that vulnerability is a key concept in predicting and understanding differentiated impacts of disasters on groups in a society, as it accounts for people differences among people and affirms that people's circumstances change and can be changed by a disaster. Social and gender processes generate unequal exposure to risk, rendering some people more prone to disaster than others. These inequalities are a function of power relations that exist in society, making an individual, household or community vulnerable to disasters (Helmer and Hilhorst 2006). Vulnerability is therefore a dynamic condition shaped by existing and emerging inequities in resource distribution and access, the control individuals can exert over choices and opportunities, and historical patterns of social domination and marginalization (Eakin and Luers 2006), and not solely a set of properties that individuals or groups possess.

In addition to understanding vulnerability as process-oriented, Nightingale (2009) shifts and complicates the conceptualization of gender from a set of fixed binary roles assigned to women and men to viewing resource management and climate adaptation as processes where gender and social inequalities are contested, changed and reinforced. Through these processes, the social meanings of what it means to be a 'man', 'woman', 'ethnic group member' and other social categories of difference are played out, and power is produced and performed. Scott (1988) presages gender-essentializing tendencies when she remarks that invoking essentialized social difference is an act imbued with power. Cornwall (2007) laments the 'gross essentialism' that has stalked the gender and development industry for decades and suggests that it may be more instructive to focus on and transform social practices that constitute gender inequality rather than assume fixed, assigned and perpetually oppositional characteristics in women and men. These contingent and unpredictable gender and social dynamics and processes within society–nature interactions are often lost in the climate and gender discussions, which tend to oversimplify human behaviour.

Adaptation and vulnerability are closely interlocking, where vulnerabilities stem from social and gender inequalities that materialize when people attempt to adapt to a changing climate through various immediate and long-term strategies. People – including women – are not essentially vulnerable and cannot be attributed distinct or fixed properties of vulnerability; instead, they become vulnerable as they adapt to changing conditions because, in doing so, they become subject to social biases and discriminatory institutional practices that render them less able to adapt adequately or fully in concrete ways. These practices are the elements worth mitigating, rather than creating programmes and advocacies focusing on women only, tapping an imagined special agency, and thus passing on to them the additional burden of adapting to changed conditions resulting from climate change. Planned programmes should enable women and men to respond adequately to the gradual and short-term effects of climate change, but in ways that do not increase inequalities in their workloads, stoke discriminatory attitudes, and/or unevenly distribute risks and costs.

Other forms of politically expedient simplification define the climate change agenda, posing even more grim and dramatic scenarios that stimulate perceptions of immediate and long-term security responses and action. Migration is one of them.

Climate Change and Migration: The Security Debate

In recent times, the notion of migration as an adaptation to the deleterious effects of climate change has pervaded global policy discussions and research. Many of the pronouncements predict massive population movements and shifts in human demographic configurations with severe and frightening implications. A few of these noteworthy pronouncements follow:

- When global warming takes hold, there could be as many as 200 million people overtaken by disruptions of monsoon systems and other rainfall regimes, by droughts of unprecedented severity and duration, and by sea-level rise and coastal flooding (Myers 2005: 1).
- Climate change will lead to hundreds of millions more people without sufficient water or food to survive or threatened by dangerous floods and increased disease (Stern 2007: 77).
- Movement on this scale has the potential to destabilize whole regions where increasingly desperate populations compete for dwindling food and water ... Let Darfur stand as the starkest of warnings about what the future could bring (Christian Aid 2007: 2).
- The United States and Australia are likely to build defensive fortresses around their countries because they have the resources and reserves to achieve self-sufficiency ... Borders will be strengthened around the country to hold back unwanted starving immigrants from the Caribbean islands (an especially severe problem), Mexico, and South America (Schwartz and Randall 2003: 18).
- Climate-related disruptions of human populations and consequent migrations can be expected over coming decades. Such climate-induced movements can have effects in source areas, along migration routes and in the receiving areas, often well beyond national borders. Many of the consequent stresses, risks and opportunities have implications for the security community. The incorporation of climate change vulnerability assessments into existing risk assessment procedures offers an opportunity

for the Service to enhance its ability to anticipate and prepare for climate-related movements of people (McLeman and Smit 2004: 14).

- In assessing the situation speakers underlined the alarming projections and stressed that climate change is expected to bring about major global environmental change, which may have extensive humanitarian and human mobility consequences. According to some estimates:
  - By 2050, between 25 million to 1 billion people may migrate or be displaced due to environmental degradation and climate change;
  - areas that may potentially be flooded might be 1.3 million square kilometres;
  - nine out 10 extreme environmental events are argued to be related to climate change;
  - developing countries might suffer 98 per cent of the casualties resulting from natural disasters;
  - within these countries, the most vulnerable groups of population, especially women, are likely to be the most affected;
  - South and East Asia, Africa and small island states will be most severely affected;
  - climate change in combination with the current demographic trends will intensify the already existing migration pressures;
  - large scale migration due to climate change and environmental degradation will have adverse effects on the environment in areas/countries of destination and will subsequently increase the potential for conflict in these areas/countries. It can be expected to provoke resource scarcity in some areas; for example through disruption of production cycles as well as water scarcity (IOM 2008: 7–8).
- It is clear that climate change will 'worsen in the near future' and force us to rethink our policy, said Rudy de Leon, Senior Vice President for National Security and International Policy ... Climate change is an 'accelerant' to conflict in already volatile and unstable regions, and tackling the issue will mean reconsidering the evolving 21st century connections among diplomacy, development, security, climate change, and migration. 'Our government needs to use its resources to get out ahead of crises such as those exacerbated by climate change. It also needs to thoroughly rethink the way development, diplomacy, and defense are interconnected,' de Leon said (Center for American Progress 2010: 1).

In March 2008, High Representative Javier Solana and the European Commission presented a Joint Report on *Climate Change and International Security* (Council of the European Union 2008). The report summarizes the main threats stemming from climate change to the security of the European Union (EU) and its partners abroad. Climate change was viewed as a 'threat multiplier', exacerbating existing tensions and potentially creating new ones over time. The following main threats from climate change were identified:

- Conflict over depleting resources could ensue, particularly in relation to water and food, and where access to resources is politicized.
- Economic damage and risks to coastal cities and critical infrastructure could arise as a result of sea level rise and an increase in the strength and frequency of extreme weather events.
- Loss of territory and border disputes may result from receding coastlines and submergence of large areas and even entire countries such as small island states.
- Environmentally-induced migration may be amplified by the impacts of climate change.
- Situations of fragility and radicalization may become exacerbated by climate change, resulting in increased instability in weak or failing states and potentially destabilizing countries and even entire regions.

Such pronouncements are, however, contested. To start with, the IPCC has recently altered its earlier 1990 position on climate-induced migration and 'climate refugees' (Raleigh et al. 2008). The IPCC's Fourth Assessment Report (Parry et al. 2007) describes the estimates of migrants as 'at best guesswork' because of a host of intervening factors that influence climate change impacts and migration patterns, thus suggesting the need for caution (Black et al. 2008). Since 1990, there have been significant changes in the IPCC's position as it has increasingly recognized that multiple and complex interactions mediate migration decision-making. Subsequent reports have adopted more nuanced depictions of migration, primarily by redirecting the focus in terms of 'human vulnerability' (Raleigh et al. 2008). Positioning climate-driven migration as a trigger of security risks has therefore come under severe challenge.

Alternative views hint at a more complex configuration of climate change responses, of which migration is only one of many options. Establishing a clear linear and causative relationship between climate change and migratory movements is therefore fraught with difficulties since the evidence from past studies is inconclusive and migration results from a confluence of factors and conditions (Black et al. 2008). There are inherent difficulties in predicting with any precision the impact of climate change on population movements. This is partly because of the relatively high level of uncertainty and unpredictability about the specific effects of climate change and partly because of the lack of comprehensive data on migration flows, especially movements within national boundaries. This applies particularly for low-income countries that are likely to be most affected by climate change (Tacoli 2009).

At a conference on climate change and migration convened by the Human Security Network and the International Organization for Migration (IOM) in Geneva in February 2008, participants recommended a clear definition of 'environmental refugee'. This would require revisions in international treaties and their legal instruments (IOM 2008: 18). Problems arise, a number of scholars say, due to crude population estimates. Further, past empirical studies do not demonstrate massive population movements to industrialized countries where it is expected and feared that climate 'refugees' will seek asylum. Renaud et al. (2007) point out that the critics argue that the terminology of 'climate refugees' is misleading and too narrow since it focuses on only one of many potential 'push factors'. This said, environmental and climate factors can be exacerbated by social, economic and political factors and thus can potentially become a major push factor.

Climate-induced migration also conjures scenarios of conflict and security risk, and hence the debate has become a serious national concern for some countries. A number of scholars contend, however, that recent pronouncements on climate-induced migration and their potential for conflicts are still inconclusive since they have found conflicting evidence linking environmental pressures to conflict situations. For the most part, scholars have relied on anecdotal case studies which often suffer from omitted variables and overdetermination (Black et al. 2008, Raleigh et al. 2008).[2] A critic in the Global North, Betsy Hartmann (2009: 8), reports:

> In September the CIA launched a new Center on Climate Change and National Security, reflecting growing concern in U.S. and European security circles that climate change could trigger violent conflict over scarce environmental resources in the global South, mass migrations of poor, unruly 'climate refugees' towards Western borders, and even wars between states. This linkage between climate and security threatens to militarize climate policy and subvert humanitarian and development aid.

Hartmann (2007: 3) also notes:

> The images and narratives in the articles and reports ... have an all too familiar ring, drawing on neo-Malthusian environmental security discourses of the 1980s and 1990s that blamed intrastate conflict in the Global South on environmental degradation, resource scarcity and migration. Then, as now, this line of reasoning not only naturalizes profoundly political conflicts, but casts poor people as victims-turned-villains, a dark, uncontrollable force whose movement ultimately threatens our borders and way of life.

A recent report from the UN Refugee Agency (UNCHR) further warns:

> the very concept of climate or environmental refugees, because of its connotations of urgency and unavoidability, is to be handled with care. It actually evokes fantasies of uncontrollable waves of migration that run the

---

2   Also: Salehyan, I. 'Refugees, Climate Change, and Instability.' Paper presented at the International Workshop on Human Security and Climate Change, Asker, Norway, 21–23 June, 2005.

> risk of stoking xenophobic reactions or serving as justification for generalized
> policies of restriction for migrants. (Piguet 2008: 8)

This turn in the climate change debate reminds us of the critical deconstructionist framework established by Ferguson (1994) and others (Hajer 1995, Scott 1998), who posit the workings of the techno-managerial discourse as an instrument of cognitive control and social regulation while concealing real political relations. The securitization of climate-induced migration becomes a 'power–knowledge' regime that produces simplifications deflecting attention from more complex configurations of causes and effects of migration by vulnerable people in vulnerable places affected by different forms and degrees of climate change (Lewis and Mosse 2006).

Climate-induced migration is a complex subject, often viewed through the prism of forced displacement from extreme climatic events and overlooking a more 'solution-oriented' role that migration can play in adapting to a new climate reality. Additionally, these discourses on conflict, resource scarcity and mass migration triggered by extreme climate events tend to represent affected populations in largely aggregate terms. They are often blind to socially differentiated outcomes of climate change and processes of adaptation whereby women and men and different ethnic groups and classes respond differently to the effects of a changing climate on their lives and livelihoods. The gender and migration literature is replete with cases where women opt to stay behind while others in their households migrate, thus feminizing agriculture under more difficult economic and climatic conditions. Should they migrate and use unsafe and irregular channels to enable their movement, this may subject them to exploitative and coercive forms of labour, including sex work. Thus, current discourses that raise the alarm for mass migration due to climate change sidestep contemporary migration studies, which show migration as a complex and differentiated process, fraught with multiple drivers and outcomes.

Concluding Remarks: Going Beyond Simplifications

This chapter argues that the 'women only' approach in WED and the securitization of climate-induced migration serve as smokescreens to more complex issues that can illumine our understanding of people's responses to climate change effects and the implications for human security. A number of points can be made about these two discourses:

1. They are simplifications that hide more than reveal the complex workings of climate change on people's decisions, welfare and livelihoods. Migration is only a part of a spectrum of possible responses to environmental stresses (Warner 2010). Multifaceted nature and causes of migration are often hidden.

A recent case study (Dun 2009) was conducted in the Mekong Delta, Vietnam, on the linkages between flooding and population movement. This research found that people are used to living with floods, and people, usually women, undertake seasonal labour migration and movement towards urban areas, while the men stay behind. In the face of increasing environmental stress and wider market-oriented socioeconomic changes, people – women and men – in the Mekong Delta adapt in various ways. One type of coping mechanism may be migration (mainly seasonal or internal migration). In recent years, successive flooding led to the destruction of crops and on more than one occasion drove poor people solely dependent on agriculture to migrate in search of alternative livelihoods in Ho Chi Minh City or across the border in Phnom Penh, Cambodia. There were some cases of families 'selling' their young daughters for commercial sex work in Cambodia to generate income. This case study highlights environmental factors that together with other factors can lead to migration. However, the case study does not show that the environment is a direct cause of migration. Rather, environmental change (increased flooding) is shown to be a trigger for independent migration decisions when livelihoods are negatively affected, supporting the argument that migration is essentially multifaceted.

2. The discourses blur the merits of voluminous scholarship that addresses the complex and converging dynamics behind gender and migration.

The literature on migration identifies three categories of factors driving migration: (i) factors related to country or place of origin, e.g. political instability, lack of economic opportunities or lack of access to resources; (ii) factors associated with destination countries or places, e.g. labour market demand and opportunities, higher wages or political stability and order, and (iii) intervening factors that facilitate or restrict migration, such as ease and affordability of transportation, social networks, state immigration and emigration policies, economic ties and cultural exchanges. Only some of these factors are sensitive to climate change (Black et al. 2008), for instance lack of economic opportunities in natural resource-based livelihoods; subsistence agriculture and fisheries are most directly affected, while specific types of agriculture are especially affected in coastal wetlands and floodplains.

Multiple responses to environmental stressors include intensification of production, diversification of livelihoods and variable types of migration (short-term, circular, permanent, cyclical, irregular, forced, rural–rural, rural–urban, crossborder, transnational etc.). Further, groups are diversely affected according to gender, class, caste, ethnicity and age – not all will move, but some will stay behind.

3. The discourse may prompt action, which in the end may turn out to be counterproductive. First, the women-only approach to climate change adaptation targets women as a distinct constituency for climate adaptation programmes while 'letting men off the hook,' hence burdening women further without understanding and responding to practices that disadvantage

them socially. Second, fear of climate-induced migration leads to the deployment and strengthening of security apparatuses that aim to thwart impending and massive population movements through more stringent immigration policies and measures.

4. The literature tends to deflect attention from further understanding the nature of vulnerability and the workings of power and governance at various scales and in various arenas. This constrains holistic adaptation measures and mitigation efforts that are empowering and equitable among environmentally and economically vulnerable people and groups located in vulnerable places.

    Climate-induced migration is seen as displacement of people rather than as part of a number of adaptation options and strategies of people. Women-only analyses, in which women are depicted as victims of climate change and as agents of mitigation and adaptation, assign women heavier responsibilities without addressing the imbalances of power in their lives.

5. The discourses indicate the paucity of interdisciplinary teams on cross-disciplinary projects. Very little work has been done on migration, climate and gender. There has been very little cross-fertilization between gender and climate change on one hand and gender and migration on the other.

    In migration and gender studies, drivers of migration are usually related to political economy and social factors. Only a few studies draw linkages between migration, gender and environmental issues. The growing number of studies on gender and climate change relate minimally with migration issues (Ge et al. 2011). In the absence of serious interdisciplinary research, there are many unsubstantiated claims by interest groups and the media about the scale of the impacts of environmental change on population dynamics and outcomes (Hugo 2008).

6. The discourses beg explanations as to why they have become hegemonic and seductive. There is need for continued investigation into power–knowledge regimes, such as the securitization of climate-induced migration and essentialized feminist politics in environmental discourses, which departs from associating them with some inner logic or 'order'. Instead, social scientists are increasingly concerned with the primacy of contingent practice in these arenas, setting aside notions of rationality in such regimes, in order to uncover their inner workings and how they are much more negotiated than given. As Li (1999) points out, 'claims to order are always fragile, contested, built on compromise; hegemony is not imposed but has to be worked out.'

References

Acharya, K. 2009. Women Farmers Ready to Beat Climate Change. *IPS News* [Online, 17 March]. Available at: http://ipsnews.net/news.asp?idnews=46131 [accessed: 7 June 2010].

Adger, W.N., Agrawala, S., Mirza, M.M.Q., Conde, C., O'Brien, K., Pulhin, J., Pulwarty, R., Smit, B. and Takahashi, K. 2007. Assessment of Adaptation Practices, Options, Constraints and Capacity, in *Climate Change 2007: Impacts, Adaptation and Vulnerability: Contribution of Working Group II to the Fourth Assessment Report of the Intergovernmental Panel on Climate Change*, edited by M.L. Parry, O.F. Canziani, J.P. Palutikof, P.J. van der Linden and C.E. Hanson. Cambridge: Cambridge University Press, 717–43.

Ariyabandu, M.M. 2004. *Women: The Risk Managers in Natural Disasters* [Online: Presentation prepared for the workshop, 'Gender Equality and Disaster Risk Reduction', East West Centre, Hawaii 10–12 August 2004]. Available at: http://www.ssri.hawaii.edu/research/GDWwebsite/pdf/Ariyabandu.pdf [accessed: 27 November 2012].

Blaikie, P., Cannon, T., Davis, I. and Wisner, B. 1994. *At Risk: Natural Hazards, People's Vulnerability and Disasters*. London and New York: Routledge.

Black, R., Kniveton, D., Skeldon, R., Coppard, D., Murata, A. and Schmidt-Verkerk, K. 2008. *Demographics and Climate Change: Future Trends and their Policy Implications for Migration.* Working Paper T-27. Sussex: Development Research Centre on Migration, Globalisation and Poverty.

Bradshaw, S. 2004. *Socio-Economic Impacts of Natural Disasters: A Gender Analysis*. Santiago: Women's Unit, Economic Commission for Latin America and the Caribbean (ECLAC).

BRIDGE 2008. *Gender and Climate Change: Mapping the Linkages: A Scoping Study on Knowledge and Gaps*. Sussex: Institute of Development Studies, University of Sussex.

CBD n.d. *Text of the CBD* [Online: Convention on Biological Diversity]. Available at: http://www.cbd.int/convention/text/ [accessed: 21 November 2012].

Center for American Progress 2010. *The Global Implications of Climate Migration* [Online: Center for American Progress]. Available at: http://www.americanprogress.org/events/2010/03/01/16915/the-global-implications-of-climate-migration/ [accessed: 27 November 2012].

Christian Aid 2007. *Human Tide: The Real Migration Crisis* [Online: Christian Aid]. Available at: http://www.christian-aid.org.uk [accessed: 27 June 2010).

Cleaver, F. 2000. Analysing Gender Roles in Community Natural Resource Management Negotiation, Life Courses and Social Inclusion. *IDS Bulletin*, 31(2), 60–67.

Cornwall, A. 2007. Revisiting the 'Gender Agenda'. *IDS Bulletin*, 38(2), 69–78.

Council of the European Union 2008. *Climate Change and International Security: Paper from the High Representative and the European Commission to the European Council* [Online: Council of the European Union]. Available at: http://www.envirosecurity.org/activities/diplomacy/gfsp/documents/Solana_security_report.pdf [accessed: 27 November 2012].

Demetriades, J. and Esplen, E. 2009. The Gender Dimensions of Poverty and Climate Change Adaptation. *IDS Bulletin*, 39(4), 24–31.

Dun, O. 2009. *Viet Nam Case Study Report: Linkages between Flooding, Migration and Resettlement* [Online: EACH-FOR Environmental Change and Forced Migration Scenarios]. Available at: http://www.each-for.eu/documents/ CSR_Vietnam_090212.pdf [accessed: 27 November 2012].

Eakin, H. and Luers, A.L. 2006. Assessing the Vulnerability of Social–Environmental Systems. *Annual Review of Environmental Resources*, 31, 365–94.

Elmhirst, R. and Resurrección, B.P. 2008. Gender, Environment and Natural Resource Management: New Dimensions, New Debates, in *Gender and Natural Resource Management: Livelihoods, Mobility and Interventions*, edited by B.P. Resurrección and R. Elmhirst. London and Ottawa: Earthscan and International Development Research Centre, 3–32.

Enarson, E. 1998. Through Women's Eyes: A Gendered Research Agenda for Social Science. *Journal of Disaster Studies, Policy and Management*, 22, 157–73.

Enarson, E. 2001. What Women Do: Gendered Labor in the Red River Valley Flood. *Environmental Hazards*, 3, 1–18.

Enarson, E. and Fordham, M. 2001. From Women's Needs to Women's Rights in Disasters. *Environmental Hazards*, 3, 133–6.

Ferguson, J. 1994. *The Anti-Politics Machine: 'Development', Depoliticisation and Bureaucratic Power in Lesotho*. Minneapolis: University of Minnesota Press.

Ge, J., Resurrección, B.P. and Elmhirst, R. 2011. Return Migration and the Reiteration of Gender Norms in Water Management Politics: Insights from a Chinese Village. *Geoforum*, 42, 133–42.

GenderCC 2007. *Recommendations of Women Ministers and Leaders for the Environment, Bali, Indonesia, 11 December 2007* [Online: GenderCC, Women for Climate Justice]. Available at: http://www.gendercc.net/fileadmin/inhalte/ Dokumente/UNFCCC_conferences/Women_ministers_Bali-Declaration_ COP13.pdf [acccessed 6 June 2010].

GenderCC 2010. *GenderCC – Women for Climate Justice* [Online: GenderCC]. Available at: http://www.gendercc.net/about-gendercc.html [accessed: 21 November 2012].

Guha-Sapir, D. 1997. Women in the Front Line: Women in Developing Countries Should Be Key Players in Disaster Preparedness and Relief. *Le Courrier de l'UNESCO*, 50(10), 27–29.

Hajer, M. 1997. *The Politics of Environmental Discourse: Ecological Modernization and the Policy Process*. Oxford: Oxford University Press.

Hartmann, B. 2007. *Climate Refugees and Climate Conflict: Who's Taking the Heat for Global Warming? Paper Delivered at the Panel on Climate Change, 4S Annual Conference, Montreal, Quebec, October 11, 2007* [Online]. Available at: http://www.radixonline.org/resources/Betsy%20Hartmann%20 Climate%20Refugees%20and%20Climate%20Conflict%202007.doc [accessed: 27 November 2012].

Hartmann, B. 2009. Don't Beat the Climate War Drums. *Climate Chronicle* [Online, 9 December]. Available at: http://www.tni.org/sites/www.tni.org/files/download/ClimateChronicle_issue2.pdf [accessed: 29 June 2010].

Helmer, M. and Hilhorst, D. 2006. Natural Disasters and Climate Change. *Disasters*, 30, 1–4.

Hugo, G. 2008. *Migration, Development and Environment* [Online: Paper submitted to the PERN Cyberseminar 'Environmentally Induced Population Displacements', 18–19 August]. Available at: http://www.populationenvironmentresearch.org/papers/hugo_statement.pdf [accessed: 27 November 2012].

Hunter, L.M. and David, E. 2009. *Climate Change and Migration: Considering the Gender Dimensions* [Online: IBS Population Program, Institute of Behavioral Science, University of Colorado at Boulder]. Available at: http://www.colorado.edu/ibs/pubs/pop/pop2009-0013.pdf. [accessed: 29 June 2010].

IOM 2008. *Climate Change, Environmental Degradation and Migration: Addressing Vulnerabilities and Harnessing Opportunities: Geneva, 19 February 2008: Report of the Conference* [Online: Permanent Mission of Greece, Geneva, and International Organization for Migration (IOM)]. Available at: http://www.reseau-terra.eu/IMG/pdf/Rapport-IOM-2008.pdf [accessed: 27 November 2012].

Jackson, C. 1993. Doing What Comes Naturally? Women and Environment in Development. *World Development*, 21, 1947–63.

Jewitt, S. 2002. *Environment, Knowledge and Gender: Local Development in India's Jharkhand.* Aldershot: Ashgate.

Lambrou, Y. and Piana, G. 2006. *Gender: The Missing Component of the Response to Climate Change.* Rome: Food and Agriculture Organisation of the United Nations (FAO).

Leach, M. 1992. Gender and the Environment: Traps and Opportunities. *Development in Practice*, 2(1), 12–22.

Leach, M. 2007. Earth Mother Myths and Other Eco-Feminist Fables: How a Strategic Notion Rose and Fell. *Development and Change*, 38, 67–85.

Lewis, D. and Mosse, D. 2006. Encountering Order and Disjuncture: Contemporary Anthropological Perspectives on the Organization of Development. *Oxford Development Studies*, 34, 1–13.

Li, T.M. 1999. Compromising Power: Development, Culture and Rule in Indonesia. *Cultural Anthropology*, 14, 295–322.

McLeman, R. and Smit, B. 2004. *Commentary No. 86: Climate Change, Migration and Security* [Online: Canadian Security Intelligence Service]. Available at: http://www.csis-scrs.gc.ca/pblctns/cmmntr/cm86-eng.asp [accessed: 29 June 2010].

Myers, N. 2005. *Environmental Refugees: An Emergent Security Issue* [Online]. Available at: http://www.osce.org/eea/14851 [accessed: 27 November 2012].

Nightingale, A. 2009. Warming Up the Climate Change Debate: A Challenge to Policy Based Adaptation. *Journal of Forest and Livelihood*, 8, 84–9.

Norm, S. 2009. *Gender & Climate Change Adaptation: Changing Rural Farmers'* *Livelihood Patterns in Maung Russey District, Battambang Province, Cambodia.* MSc thesis. Bangkok: Gender & Development Studies, Asian Institute of Technology.

O'Riordan, T. and Jordan, A. 1999. Institutions, Climate Change and Cultural Theory: Towards a Common Analytical Framework. *Global Environmental Change*, 9, 81–93.

Parry, M.L., Canziani, O.F., Palutikof, J.P., van der Linden, P.J. and Hanson, C.E. (eds) 2007. *Climate Change 2007: Impacts, Adaptation and Vulnerability: Contribution of Working Group II to the Fourth Assessment Report of the Intergovernmental Panel on Climate Change.* Cambridge: Cambridge University Press.

Piguet, E. 2008. *Climate Change and Forced Migration.* New Issues in Refugee Research, Research Paper No. 153. Geneva: United Nations High Commissioner for Refugees.

Raleigh, C., Jordan, L. and Salehyan, I. 2008. *Assessing the Impact of Climate Change on Migration and Conflict.* Washington: Social Development Department, World Bank.

Rao, B. 1991. *Dominant Constructions of Women and Nature in the Social Science Literature.* CES/CNS Pamphlet 2. Santa Cruz: University of California.

Renaud, F., Bogardi, J.J., Dun, O. and Warner, K. 2007. *Control, Adapt or Flee: How to Face Environmental Migration.* Geneva: International Organization for Migration (IOM).

Resurrección, B.P., Sajor, E., and Fajber, E. 2008. *Climate Adaptation in Asia: Knowledge Gaps and Research Issues in Southeast Asia.* Colorado: Institute for Social and Environmental Transitions (ISET).

Schwartz, P. and Randall, D. 2003. *An Abrupt Climate Change Scenario and its Implications for United States National Security* [Online]. Available at: http://www.gbn.com/articles/pdfs/Abrupt%20Climate%20Change%20February%20 2004.pdf [accessed: 27 November 2012].

Scott, J.W. 1988. *Gender and the Politics of History.* New York: Columbia University Press.

Scott, J. 1998. *Seeing Like a State: How Certain Schemes to Improve the Human Condition Have Failed.* New Haven: Yale University Press.

Skutsch, M. 2002. Protocols, Treaties, and Action: The 'Climate Change Process' Viewed Through Gender Spectacles. *Gender and Development*, 10(2), 30–39.

Speranza, I.C., Ayiemba, E., Mbeyale, G., Ludi. E., Ong'any, P. and Mwamfupe, D. 2006. Strengthening Policies and Institutions to Support Adaptation to Climate Variability and Change in the Drylands of East Africa, in *Global Change and Sustainable Development: A Synthesis of Regional Experiences from Research Partnerships*, edited by H. Hurni. Perspectives of the Swiss National Centre of Competence in Research (NCCR) North-South, 5. Bern: Geographica Bernensia, 107–30.

Sreng, S. 2010. *ICTs in Flood Vulnerability Mitigation and its Impacts on Rural Livelihoods: Gender and Ethnicity Dimensions in Ratanakiri Province, Cambodia.* MSc thesis. Bangkok: Gender & Development Studies, Asian Institute of Technology.

Stern, N. 2007. *The Economics of Climate Change: The Stern Review.* Cambridge: Cambridge University Press.

Tacoli, C. 2009. Crisis or Adaptation? Migration and Climate Change in a Context of High Mobility, in *Population Dynamics and Climate Change*, edited by J.M. Guzman, G. Martine, G. McGranahan, D. Schensul and C. Tacoli. London: International Institute for Environment and Development and UNFPA, 104–18.

Terry, G. 2009. No Climate Justice Without Gender Justice: An Overview of the Issues. *Gender and Development*, 17, 5–18.

UN 1992. *United Nations Framework Convention on Climatic Change* [Online: United Nations]. Available at: http://www.unfccc.int/resource/docs/convkp/conveng.pdf [accessed: 21 November 2012].

UN 1998. *Kyoto Protocol to the United Nations Framework Convention on Climatic Change* [Online: United Nations]. Available at: http://www.unfccc.int/resource/docs/convkp/kpeng.pdf [accessed: 21 November 2012].

UN 2009. *Agenda 21* [Online: UN Department of Economic and Social Affairs, D ivision for Sustainable Development]. Available at: http://www.un.org/esa/dsd/agenda21/index.shtml [accessed: 21 November 2012].

UNCCD 2012. *Text of the Convention Including All Annexes* [Online: UNCCD United Nations Convention to Combat Desertification]. Available at: http://www.unccd.int/en/about-the-convention/Pages/Text-overview.aspx [accessed: 21 November 2012].

UNEP/GRID-Arendal 2011. *IPCC Third Assessment Report – Climate Change 2001: Working Group III: Mitigation* [Online: UNEP/GRID-Arendal]. Available at: http://www.grida.no/publications/other/ipcc_tar/?src=/climate/ipcc_tar/wg3/454.htm [accessed: 16 February 2011].

UN Women n.d. *The United Nations Fourth World Conference on Women, Beijing, China, September 1995, Action for Equity, Development and Peace: Platform for Action* [Online: UN Women, United Nations Entity for Gender Equality and the Empowerment of Women]. Available at: http://www.un.org/womenwatch/daw/beijing/platform/index.html [accessed: 21 November 2012].

Villagrasa, D. 2002. Kyoto Protocol Negotiations: Reflections on the Role of Women. *Gender and Development*, 10(2), 40–44.

Warner, K. 2010. Global Environmental Change and Migration: Governance Challenges. *Global Environmental Change*, 20, 402–13.

Watson, R.T., Zinyowera, M.C. and Moss, R.H. (eds) 1998. *The Regional Impacts of Climate Change: An Assessment of Vulnerability.* A Special Report of IPCC Working Group II. Published for the Intergovernmental Panel on Climate Change. Cambridge: Cambridge University Press.

WEDO 2008. *Gender and Climate Change Finance: A Case Study from the Philippines.* New York: WEDO Women's Environment & Development Organization.

Wisner, B., Fordham, M., Kelman, I., Johnston, B.R., Simon, D., Lavell, A., Brauch, H.G., Spring, U.O., Wiches-Chaux, G., Moench, M. and Weiner, D. 2007. *Climate Change and Human Security* [Online]. Available at: http://www.radixonline.org/cchs.html [accessed: 6 June 2010].

Yonder, A., Akcar, S. and Gopalan, P. 2005. *Women's Participation in Disaster Relief and Recovery.* Seeds No. 22. New York: The Population Council.

PART IV
On the Margins: Conflict, Migration and Development

PART IV

Chapter 15

Impacts of Internal Displacement on Women's Agency in Two Resettlement Contexts in Sri Lanka

Fazeeha Azmi

Introduction

This chapter investigates the agency of internally displaced women in establishing and reestablishing their livelihoods and highlights the importance of the transformative nature of agency. The chapter analyses representations, realities and livelihoods pursued by internally displaced women in two different internal displacement contexts in Sri Lanka. It challenges the general assumption about women's passivity in an internal displacement context (Rajagopalan 2010). The study was undertaken in System H of the Accelerated Mahaweli Development Project, focusing on development-induced displacement, and in Vavuniya, focusing on displacement induced by war and natural disaster. Internal displacement generally has serious consequences for those affected. The material losses created by internal displacement are one aspect. However, when the repercussions of internal displacement are examined from a gender perspective, the depth and magnitude of displacement particularly affect women, who may lose their individual and collective social identities. Sometimes these women are assisted and sometimes hindered by existing structures and processes in the new places to which they are displaced. This present study shows through collected narratives how internally displaced women use their agency in establishing or reestablishing their livelihoods in new places and emphasizes the transformative nature of agency.

In recent years geographers have been active in research into questions of internal displacement. Substantial scholarly work has been undertaken on several internal displacement issues in Sri Lanka, including relief, repatriation, resettlement, aid and development, humanitarianism, refugee integration and transnationalism. Studies have focused on internally displaced persons (IDPs) as both victims and agents (Lund 2000, Shanmugaratnam 2001, Zackariya and Shanmugaratnam 2002, Shanmugaratnam et al. 2003, Brun 2008, Walker 2010). My study investigates how women in different internal displacement contexts negotiate gender and agency for the betterment of their families. By making visible their agency in managing their lives and livelihoods in new places, I treat their experiences as a resource and a space within which the role of gender is manifested in the postconflict reconstruction of the country.

Internal Displacement in Sri Lanka

Sri Lanka provides a case study for different types of displacement. Internal displacement is not a recent phenomenon in Sri Lanka. There are three main categories: development-induced, conflict-induced and disaster-induced internal displacement. The country has experienced all these forms of displacement during the last fifty years. Although internal displacement in Sri Lanka has a long history, starting with colonization projects in the dry zone, it has received most attention since the late 1970s. IDPs have become a recognized community in Sri Lanka since then.

Displacement of communities due to large-scale development projects has been a strategy adopted by many countries in South Asia. Sri Lanka is no exception. During the 1930s a policy decision was made to develop the impoverished dry zone of the country through state-aided colonization schemes. Since then both small-scale and large-scale colonization schemes have been implemented in the dry zone as well as elsewhere in the country, resulting in the resettlement of several thousand families. To encourage the settlers the state provided necessary infrastructure (de Zoysa 1995, Peiris 1996, Scudder 2005).

Until the implementation of the Mahaweli Development Project in 1969, followed by the Accelerated Mahaweli Development Project (AMDP) in 1977, there were no major forms of development-induced displacement in the country that created a category of settlers that could be called a 'forcibly evicted group'. The majority of them were from the cultivator caste, which ranks highest in the Sinhalese caste system. In general these evacuees were eager to own a plot of land, as land had determined the social hierarchy in their previous villages. When they decided to migrate to the 'unknown' land, most of them had ambitions of becoming owners of large tracts of land. Although these settlers arrived with compensation money for their former property, it took them a considerable time to adjust to their new environment in the settlements. Many researchers have highlighted socioeconomic issues existing in Mahaweli settlements (Lund 1981, Siriwardane 1981, Wanigaratne 1987, Muller and Hettige 1995) and paid attention to different ways people overcome difficulties (Raby 1995, Sørensen 1996, Azmi 2007).

The history of conflict-related internal displacement goes back to the bloody communal riots of 1983. After this, as a result of terrorist and counter-terrorist activities, especially in the northern and eastern parts of the country, thousands of people regardless of ethnicity were killed or had to flee their native places and settle elsewhere within the district or outside the district, wherever they felt secure. Since the 1980s, conflict-related displacement has characterized the internal displacement pattern of the country. The 30-year-old war between the Government of Sri Lanka (GOSL) and the Liberation Tigers of Tamil Eelam

(LTTE) has resulted in massive suffering for the whole country, as the impact of the conflict was not limited to the north and east.

According to the International Displacement Monitoring Centre, over 115,000 people were estimated to be internally displaced in Sri Lanka as of October 2012 (IDMC 2012). These people were mainly from Jaffna, Mannar, Kilinochchi, Trincomalee, Mullaitivu, Batticaloa and Ampara Districts. Due to the ethnic riots in 1983, Tamils living in Sinhala-dominated areas were displaced internally. In 1990, almost all of the Muslims in the North were forced to leave by the LTTE. About 80,000 Muslims have been living in displacement for more than 20 years. Muslims who were forced to move from Jaffna were given only two hours to leave and in other areas 48 hours. The displaced Muslims made their way towards government-controlled areas in Vavuniya and Anuradhapura, as well as to Puttalam District on the northwestern coast (Brun 2008). Over the course of these two decades, they have continued to live in open camps in Puttalam, which have not been further affected by the conflict. With the defeat of the LTTE in May 2009, there is now hope within this community to be able to assert their right of return.

The worst internal displacement due to war came on 17 May 2009, when the GOSL declared victory over the LTTE. Over 280,000 people fled from LTTE-controlled areas and sought refuge in Vavuniya. Within a few days the country had the highest number of IDPs on record. They are now gradually being resettled. The government has to restore their livelihoods, reconstruct the infrastructure and create normalcy (CEPA 2009). The postwar situation has given the Sri Lankan government an opportunity to move into a new era of reconstruction and reconciliation. It is implementing two major projects, termed 'Eastern Awakening' and 'Northern Spring' in the war-affected Eastern and Northern Provinces.

Due to these different types of internal displacement, a wide variety of social problems has arisen. When displacement occurs, far more damage results than simply the loss of physical property in situ. The displaced community suffers not only physically but also economically, socially and psychosocially. Many people lost their houses and other movable and immovable property which they had built up and secured with their earnings over the past decades, and many also lost their traditional livelihoods.

Displacement breaks up entire communities and families, making it more difficult for them to cope with the uncertainty of resettlement. Risks are usually higher for vulnerable groups, such as children, women, the elderly, ethnic minorities and indigenous people. They are forced to experience new living environments and sometimes are forced to compromise their needs. People's lives, social networks and social capital are shattered. Unfamiliar living environments affect the gender roles of women and men. They are forced into new social and cultural worlds. They have to live under trauma and overwhelming uncertainty and have to earn their livelihoods in the new place as best they can.

Conceptualizing Transformatory Agency

Gender, seen as the socially constructed roles of men and women, differs from society to society (Moser 1993). This has deep-seated implications in the context of forced internal displacement. The experience of being displaced affects men and women in different ways when they are forced to live in new environments (El-Bushra 2000, Manchanda 2004). When men and women are forced to live in alien cultures, their lives change dramatically. They often lose their negotiated positions and generally revert to a less equitable social status. Gender and forced migration constitute a recent paradigm, which looks at the disadvantages and vulnerabilities of displaced girls and women as well as their opportunities for addressing gender inequalities in forced migration contexts. This concept focuses attention on situation-specific interventions and an understanding of the shifting social worlds of the displaced (El-Bushra 2000).

The research literature exploring the experiences and practices of displaced populations has long been informed by the actor-oriented perspective, which pays attention to the agency of the displaced. The influential work on actor-oriented approaches comes from Long (1984, 1997, 2001, Long and Long 1992), who placed high priority on actors through recognizing the central role played by human actions and consciousness through the concept of agency. Long (1997, 2001) asserts that agency is the capacity of an individual actor to process social experience and devise ways of coping with life even under the most extreme situations. Long further emphasizes the importance of acknowledging the potential strength of people and how they actively make choices in different contexts.

Inherent in the actor-oriented perspective is an understanding of the individual as an active subject, with the capacity to process social experience and invent new ways of coping with life, even under extreme coercion. According to Long and Long (1992: 23, citing Giddens 1984), the notion of agency presupposes that actors are 'knowledgeable' and 'capable', and 'attempt to solve problems, learn how to intervene in the flow of social events around them and monitor continuously their own actions, observing how others react to their behaviour and taking note of the various contingent circumstances'. Agency concerns an individual's own capacities and competencies (Kabeer 1999). It comprises the ability to make important decisions. It is through agency that people navigate different contexts in order to fulfill their economic, social and cultural needs in the present and for the future.

Kabeer (2003) distinguishes between the 'effectiveness of agency' and the 'transformative nature of agency'. The former reflects the ability of women to carry out given roles and responsibilities, while the latter refers to their ability to question or challenge various structures. Transformatory agency refers to long-term objectives. In patriarchal societies, this may bring about changes in social roles and relationships in the long run (Kabeer 1994, 1999). This applies especially in the context of resettlement and reconstruction where women find themselves in very different social worlds from their former home environment. Although the agency of victims has received attention in Sri Lanka, the transformative nature of agency has in general received less attention.

Methodology

The present chapter is based on four qualitative interviews with Sri Lankan internally displaced women, conducted between March and May 2010 in two different internal displacement contexts. In the latter part of March 2010, I went back to System H of the Accelerated Mahaweli Development Project, where I had undertaken earlier research (Azmi 2008), in order to interview two women from a development-induced displacement background. In early May, with the help of one of my postgraduate students, I interviewed two women from Vavuniya, who were displaced due to war. This study used qualitative methods such as life histories, observations and informal discussions. This allowed me to learn from and validate the knowledge of internally displaced women. The objective was to gain in-depth information from a small number of participants (Mason 1996). When discussing the characteristics of life histories, Brettell (2002: 439) argues that the 'goal is not to find a typical or representative individual but to assess how individuals interpret and understand their own lived experiences.'

In combination with the narratives obtained from the women, I also collected secondary data concerning the social position of the interviewees. The issues of gender, internal displacement, vulnerabilities and capabilities that I discuss below are based on the knowledge that I was able to gain from the women that I interviewed in two different displacement contexts. I demonstrate the importance of recognizing the transformative nature of agency, and recommend that it should be fully incorporated into the ongoing resettlement programmes. I emphasize that, unless development planners and policy-makers fully take account of the theories and terminologies of postwar reconstruction in practice in different contexts, the impacts of their projects will be marginal.

Gendered Actors in Displacement

Through the narratives of internally displaced women, the following section seeks to highlight their agency. Table 15.1 gives some basic information on the interviewed internally displaced women.

Table 15.1   Internally displaced women interviewed March–May 2010

Name	Age	Marital status	Reason for displacement	Place of origin	Place where they live now	No. of children	Present employment
Arulmani	50	Widow	War, Tsunami	Kilinochi	Vavuniya	4	Wrapping popsicles
Thangamma	41	Married	War	Kilinochi	Vavuniya	6	Coolie work
Dayawathie	57	Widow	Development project	Kothmale	Maliyadevapura	4	Fish trader
Podimenike (Menike teacher)	65	Widow	Development project	Kothmale	Maliyadevapura	3	Retired teacher/ farmer

Note: Names are fictitious.
Source: Fieldwork May 2010.

Recent War IDPs

Arulmani was born in Kilinochchi. She has three daughters and a son. Her husband was a mason. He was killed in cross-fire at Puthumathalan while trying to escape from Kilinochchi. Her family had been displaced three times since 1988. Arulmani told her experience with tears:

> We left our village on the 9 May 2008. We came to Dharmapuram, Visuvamadu, Puthukudiyrippu, Maathalan and Pokkana. We arrived at Vavuniya Arunachalam camp on the 20 April 2009. My husband was killed on our way to Vavuniya. I could not bury him. I saw many dead bodies and wounded people on the way. But we could not do anything. Everybody was worried about their life. I am still worried that I could not give him a decent funeral. Life in the camp was extremely difficult. No water, not enough food, presence of security forces, no education for children. I decided to live outside the camp with a relative and build a small hut on her land. My children are going to school. They could not go to school continuously due to war. Their education was disturbed during the war. My second daughter lost her eye sight in a bomb blast. The third one is working in a local tailor shop after school. I am wrapping popsicles in an ice factory. When I first went to work I really felt uncomfortable. If my husband was alive he would not have allowed me to do this type of work. But I cannot see my children starving. We receive a very poor income. That is hardly enough to run a family. As we have our land in our village we can go back and start agriculture again. While we stay in Vavuniya, we have to earn money to cover the daily expenses. Going out for work is not safe here. Many women are facing problems that they can't talk openly. What to do? Although I am a widow, I wear *thali*. If people ask about my husband I am telling that he is in a detention camp.

As a result of the brutal last days of fighting between the GOSL and LTTE in May 2009, Menik Farm IDP camp in the northern district of Vavuniya became the largest IDP site in the world. At that time there were 16 camps in Vavuniya alone, holding nearly 300,000 IDPs. Conditions in the camps were not favourable for the IDPs, who had endured war zone conditions for several months. After a few months, insecurity and worsening health conditions forced many IDPs to take shelter with their relatives outside the camp, moving to adjoining or faraway places to live with their relatives, as in the case of Arulmani. She was deprived of compensation and the monthly benefits that are paid by the government to the affected families, as she is living outside the camp. Although she felt very uncomfortable working in a factory, she hardly had any other choice.

Arulmani decided to wear the *thali* (a type of neckless that is a symbol reserved for married women) in order to cover her widow identity. It made it easier for her to find a job and work in alien environments. Rajasingham (2007), in her study of Sidamparapuram refugee camp outside Vavuniya, demonstrated how displaced young Hindu widows working in Vavuniya displayed their sense of independence and challenged the conventional negative Hindu construction of widowhood by wearing the red *pottu*, a spot on the forehead that is also a symbol reserved for married women. While it challenges the conventional negative construction of widowhood, wearing *thali* and *pottu* is used by recently widowed IDPs as a strategy to seek employment and to avoid sexual harassment. Widows also attend weddings, auspicious religious ceremonies, dress in bright coloured *sarees,* engage in small-scale businesses and deal with male officials and businessmen on equal terms. Arulmani has already started to work and save some money. She is not just getting on with her daily income needs, but has devised her own strategy to conceal her identity. She thinks that if she can go back she can undertake agriculture on her land, which would be an example of challenging existing social roles and responsibilities. Further, Arulmani's survival in the new environment and her future plans are signs of the transformative nature of agency discussed by Kabeer (2003).

Thangamma is one of many women forced to leave their homes by the conflict. With a wrinkled face, she described her experience during the final stages of the war. She started her story with a romanticized past:

> When we were in Kilinochchi, we had income from agriculture; we had everything from our paddy fields and gardens. We were happy there. Now I can't explain all the good things we had in our village. Now I have forgotten what happiness was. I had three sons and two daughters. We kept our sons at home as we feared abduction [by the LTTE]. But two of my sons were taken by the *iyakkam* [LTTE] when they were coming from the *kovil* [Hindu temple]. At that time we did not have control over such abductions. We do not know the whereabouts of them. I am really worried … but that is the way many parents with young children lived there. Suddenly, things fell apart when the war started. The circumstances became unbearable in our village. Our house was destroyed

at the beginning of the war, burnt to the ground. We left everything behind and decided to get out of Kilinochchi. We had to flee our home in order to be secure. No food, not shelter, nothing for several days. We were protected and helped by each other. My husband was killed. When we arrived in Vavuniya, we were very exhausted. We were given food and water by the government and some other organizations. Later, conditions in the camp became worse. Wounded people were dying, especially children and older people. My elder daughter who married at the age of 14 was separated from us and taken to another camp. I was left alone with the youngest daughter. My daughter had been disabled from the beginning. In February 2010, we came out of the camp and lived with a relative. We are living in a small house owned by them. My daughter is going to school. The hardest thing was just learning to survive here. After displacement, the workloads of many women have doubled. I am doing coolie work. I also make brooms and sell them. At the same time, I have to take care of the family and cook. At the beginning it was difficult for me to do everything alone. But now I am used to it. As I have to take the full responsibility for everything in my family, it gives a kind of satisfaction though it is challenging. Although I was never engaged in a job before migration, now I feel I can do much if there is an opportunity. I really want to go back to my village. I am fed up of the life here. I can do much better in our paddy field and garden.

Thangamma was unable to make sense of much of her past, clearly due to her present psychological condition. She is traumatized due to the hardships she underwent during the last stages of war. This does not mean that she did not have a happy past or that she had forgotten everything. Her account provides evidence that she had had a good and a happy family life. Like Thangamma, many women started the discussions with an emphasis on how good life had been in their former villages. Thangamma's experience of displacement may not be an ordinary one for people who live outside the war zone, yet many people who survived the war had similar experiences.

She was very worried about the abduction of her sons. Lund (2007), discussing countergeographies, identifies participation of the former LTTE child soldiers as a form of marginalization and exploitation. The LTTE's recruitment of child soldiers in the north and east meant that many children between the ages of 12 and 16 lost their future. They became victims of the war economy, marginalized, abused and rejected. As for many parents whose children were forcibly taken by the LTTE, losing her two boys was a great loss for Thangamma. At the same time her elder daughter got married at the age of 14, a common strategy employed by parents to avoid forced recruitment.

With the end of the protracted armed conflict in May 2009, many civilians from the northern districts of Kilinochchi and Mullaiteevu became like Thangamma victims of a major humanitarian crisis. With great difficulty people reached Vavuniya and took temporary shelter in IDP camps. For Thangamma, living in the camp with her children became insecure and uncomfortable. At the same time, when the conditions for many IDPs in the camp became unbearable, she decided to go with her children to a relative's place. Previously, women from Kilinochchi, as in many rural areas in Sri Lanka, had a secure, sheltered life in a traditionally conservative society, where they were confined to the kitchen and supported the income of the family by carrying out tasks that were either homebound or took place in other secure environments such as their paddy fields and farms. Thangamma also supported her family by supporting her husband in their family farm. However, conditions in Vavuniya put many women into a situation for which they were not prepared.

Thangamma commented that previously her husband did not permit her to work outside the household, even though she had done so prior to marriage. With the notion of 'man as the provider for the family and the woman taking care of the home' so deeply rooted in people's minds, it is mainly after displacement that many Kilinochchi women violated these traditional rules. Thangamma's account also tells that while much of the impact of conflict on gender roles is negative, there are also opportunities for change created by war, which may lead to the remaking of roles and opportunities, particularly for female-headed households in traditional societies. In an effort to survive during a conflict, women often engage in casual work and other economic activities, which may give them more control, autonomy and status at both household and community levels. In postcrisis development activities, it is essential to enhance and protect these opportunities and gains, whereas there may be a tendency to revert to tradition and or to impose new constraints. It is important to know how conflict and displacement have had an impact on traditional gender roles in order to respond accordingly. Thangamma's story reveals that displacement can provide spaces for exploring women's own agency. Thangamma's experiences provided her with the opportunities to explore her capabilities for finding new transformative strategies. As in Arulmani's case, Thangamma also wants to go back and work on her land. In order to sustain the agency of women like Arulmani and Thangamma, reconstruction and resettlement programmes should provide spaces for such women to look after themselves in the future.

Development-Induced Displacement

Sri Lanka is typically characterized as a patriarchal society (Risseeuw 1991, IWAID 1995, de Alwis 2002). Women's identities are conceptualized through their relations with men as daughters, wives and mothers (Schrijvers 1988). The foregoing descriptions of traditional Sri Lankan women show that they are not necessarily passive. The story of the teacher Menike reveals how women can work towards achieving their goals even under difficult circumstances. Menike teacher, as she is addressed, described how she was able to achieve most of her objectives in life through hard work and sacrifice even after forced displacement:

When we were asked to move to Mahaweli from Kothmale, we had three children. Two of them were going to school. I was living with my husband and children in my husband's family home. We had a plan to buy a plot of land in Kothmale. But when we were informed about the Mahaweli project, we decided to move there as we thought we may not be able to buy land in Kothmale. Living in Kothmale since childhood, my husband did not want to leave the place. He insisted on buying land somewhere in Kothmale. But I knew, with our poor salaries it would only be a dream. I thought if we came here [Maliyadevapura], we could work hard on the land and give our children a comfortable life. We also would own land. When, I explained all these things, with many difficulties, he agreed to go to System H of the Mahaweli project. As both of us were teachers, we could get a transfer easily to a school in Maliyadevapura. But I thought our children would not have good facilities to learn here. We decided to leave them with my husband's family. They were attending a good school in Kothmale. When we came to Maliyadevapura, life was much harsher due to an unfamiliar, warm environment, mosquitoes and the quality of water. Although it was extremely difficult for us to live in the dry zone, I made up my mind not to return to Kothmale. We were happy we had land where we could grow paddy and vegetables. My husband and I worked on the land. We got good profits when the years passed. We visited Kothmale during the weekend to see our children. Our children were good in education. Now my son is a surveyor, my daughter is a draftsman and the youngest daughter is a clerk. They all have a stable income. In 1999, my husband passed away. But I did not want to move to Kothmale. Instead, my younger daughter came to live with me. She is divorced and has one daughter. I am receiving a good pension. I still work in our paddy land. I am convinced that I managed to achieve my targets.

Scholars have noted that displacement generally leaves the displaced person uprooted with a sense of being abandoned (Lund 1981, 1993, Sørensen 1996, Azmi 2007). Leaving their native places also evokes a deep sense of loss and resentment because, when people are forced to leave, they leave behind a sense of identity, a culture, personal networks and community history. Although Menike teacher knew about the dry zone and faced many difficulties, she was more optimistic than her husband about the opportunities to improve their life by moving to Mahaweli. Menike teacher was not reluctant to face the challenges. Her plans for a better future for the whole family, in terms of money, education, landownership and ownership of a house, all materialized due to her strategic plans and hard work, although her decision challenged previously existing norms in a conventional upcountry Sri Lankan extended family where women had a muted voice at that time. Her account tells that, while being among the most vulnerable, women have also shown their strengths and capabilities in such internal displacement contexts and challenge the hindrances to their wellbeing.

Dayawathie now lives with her sons. She looks much older than her actual age. When I asked about her experience of displacement and what she was doing for a living, she explained:

> We were originally from Galle. I have two sons. My husband was a serious supporter of UNP [United National Party]. My husband did not have a permanent job. We had four children. We were very poor and lived in small house built in a reservation. We always had the threat of eviction. My husband came to know about the Mahaweli project through a friend. He talked to the MP [Member of Parliament) and told of his interest to migrate to Mahaweli settlements. The MP promised him a piece of land. Within six months we got a small plot of land. I was not very happy to move from my village. I relied on friends and relative in times of crisis. I thought I am taking a risk. But we have to get into the water to swim. So I decided to go. The whole family came here [Kongwewa]. During the first two or three months, life was extremely difficult. We did not know the neighbours. We did not find continuous jobs. My husband got a contract job in the Mahaweli construction site. It was not enough to cover our expenses. Then I decided to buy some dried chillies. I made chilli powder packets and sold them in the weekend market. My sons helped me in the market. I was able to cover the school expenses of my children. Then I expanded the business further and distributed the chilli powder packets to local shops. But then, within two years, I became ill and the doctor asked me to stop my business as it was not good for my health. I stopped that. I felt the need to find an alternative business as my husband's contract work was over. He became weak after the contract work. He also developed a kidney problem and was forced to stay at home for about three years and then passed away. Since then, I have had a very hard time. My elder son had to stop his education and he told he will help me. With the help of my relatives in Matara, I brought dried fish and sold in the weekend market. Now my sons are married and they are engaged in the vegetable business. I am still selling dried fish in the market. I think, we should not depend on our children when we become old. We should have an income. My sons do not want me to continue the business. But as long as I am healthy, I will do so.

For Dayawathie, too, leaving her village was not just a simple act of changing her place of residence. Unlike in other Mahaweli settlements, the places of origin of settlers in Kongwewa varied. They were from many different parts of the island. Before coming to the settlements, the majority were very poor and had been marginalized in the societies in which they lived. For most of them migration to Mahaweli settlements was the only option to find a better future. As outsiders, many settlers felt uncomfortable during the initial months. Dayawathie's husband's contract job provided security during the initial years of resettlement, while Dayawathie also supported the family by engaging in home-based self-employment. After her husband's death, she had to revitalize the links with her former village in order to start the dried fish business. Dayawathie's account challenges the notions of the dependent wife, both as an old woman and as an independent mother.

Discussion and Conclusion

I have attempted in this study to elaborate on the struggle of four internally displaced women to reestablish their livelihoods and I have highlighted the effectiveness and transformative nature of their agency in the process. The circumstances of displacements are usually worse for women than for men. Displacement creates changes in gender roles and women are forced into unaccustomed roles and responsibilities for which they were neither prepared nor trained. They are compelled to live in new situations, which are alien to the cultural and religious values that gave meaning to gender identities and gender roles. However, these women have given new meanings to practices of gender in new places.

Although the contexts of displacement varied, all four women have shown their strength by accepting new challenges in new places. They all had to undergo stress, uncertainty, hardships and deprivation during the initial period of displacement. They have stood up to different structures that were once considered barriers (gender, ethnicity, religion, age, culture and hierarchical power relations). Women generally lack community support due to displacement, as the communities are fragmented. When they are separated from the formal support system, they can become easy targets for harassment and abuse. When women lose their support networks, they become victims. However, Arulmani's and Thangamma's accounts reveal that, although they had to undergo many hardships and take on new roles and responsibilities, they sought their family's survival by confronting existing strong gender norms. Their actions exemplify transformative agency.

Similarly, the decision made by Menike teacher shows not only her eagerness to obtain a good life for her family but also a strategic long-term objective that provided her access to land and a home. Dayawathie's narrative similarly provides an example of both the effective and transformative nature of agency. It is clear from the above narratives that, amidst the destruction and trauma left by the war and development-induced displacement, women have shown an amazing capacity not just to get on with their lives, but also to participate fully in the process of redefining gender roles and relations in new environments. Through their struggles at different levels, these women have challenged the established patriarchal gender relations and social norms. The narratives also convey the changing nature of agency, identity and power in the interaction of IDP women with a variety of contexts, aiming to improve their own lives and negotiate challenging and changing realities. They formed new and diverse conceptions concerning their identity, agency and the transformative nature of agency.

The situation of displacement is an opportunity for renegotiating gender relations. While displacement modifies gendered worlds, this may not always occur in negative ways. Scholars and relief practitioners discuss the opening of new spaces specifically for women, allowing them to increase their independence and engage in roles that were formerly denied to them (Moser 2001, Hyndman and de Alwis 2003). War and development-related displacement in Sri Lanka have inevitably produced structural transformation for both women and men. For many women this has opened up new socioeconomic and political opportunities that challenge traditional gender hierarchies. While internal displacement has led to impoverishment and marginalization, the accounts of Arulmani, Thangamma, Menike teacher and Dayawathie show that it has also led to the transcending of social subordination and assisted transformation.

References

Azmi, F. 2007. Changing Livelihoods Among the Second and Third Generations of Settlers in System H of the Accelerated Mahaweli Development Project (AMDP) in Sri Lanka. *Norsk Geografisk Tidsskrift–Norwegian Journal of Geography*, 61, 1–12.

Azmi, F. 2008. *From Rice Barn to Remittances: A Study of Poverty and Livelihood Changes in System H of the Accelerated Mahaweli Development Project (AMDP), Sri Lanka.* Doctoral theses at NTNU, 2008:166. Department of Geography. Trondheim: Norwegian University of Science and Technology (NTNU).

Brettell, C.B. 2002. The Individual/Agent and Culture/Structure in the History of the Social Sciences. *Social Science History*, 26, 429–45.

Brun, C. 2008. *Finding a Place: Local Integration and Protracted Displacement in Sri Lanka.* Colombo: Social Scientists' Association.

CEPA 2009. *A Profile of Human Rights and Humanitarian Issues in the Vanni and Vavuniya – March 2009.* [Online: Centre for Policy Alternatives]. Available at: http://www.cpalanka.org/a-profile-of-human-rights-and-humanitarian-issues-in-the-vanni-andvavuniyamarch-2009/ [accessed: June 2010].

de Alwis, M. 2002. The Changing Role of Women in Sri Lankan Society. *Social Research*, 69, 675–91.

de Zoysa, D.A. 1995. *The Great Sandy River: Class and Gender Transformation among Pioneer Settlers in Sri Lanka's Frontier.* Amsterdam: Spinhuis.

El-Bushra, J. 2000. Gender and Forced Migration: Editorial. *Forced Migration Review*, 9, 4–7.

Giddens, A. 1984. *The Constitution of Society.* Berkeley, CA: University of California Press.

Hyndman, J. and de Alwis, M. 2003. Beyond Gender: Towards a Feminist Analysis of Humanitarianism and Development in Sri Lanka. *Women's Studies Quarterly*, 31, 212–26.

IDMC 2012. *Sri Lanka: A Hidden Displacement Crisis* [Online: IDMC Internal Displacement Monitoring Centre], Available at: http://www.internal-displacement. org/8025708F004BE3B1/(httpInfoFiles)/0F7746546306FCB3C1257AA800584 5A6/$file/srilanka-overview-oct2012.pdf [accessed: 18 November 2012].

IWAID 1995. *Women, Transition and Change*. Institute of Agriculture and Women in Development (IWAID). Colombo: Friedrich Ebert Stiftung and Gala Academic Press.

Kabeer, N. 1994. *Reversed Realities: Gender Hierarchies in Development Thought*. London: Verso.

Kabeer, N. 1999. Resources, Agency, Achievements: Reflections on the Measurements of Women's Empowerment. *Development and Change*, 30, 435–64.

Kabeer, N. 2003. *Gender Mainstreaming in Poverty Eradication and the Millennium Development Goals: A Handbook for Policy-Makers and Other Stakeholders*. London: Commonwealth Secretariat.

Long, N. 1984. Creating Space For Change: A Perspective on the Sociology of Development. *Sociologica Ruralis*, 24, 168–84.

Long, N. 1997. Agency and Constraint, Perceptions and Practices: A Theoretical Position, in *Images and Realities of Rural Life*, edited by H. de Hann and N. Long. Assen: Van Gorcum, 1–20.

Long, N. 2001. *Development Sociology: Actor Perspectives*. London and New York: Routledge.

Long, N. and Long, A. 1992. *Battlefields of Knowledge: The Interlocking of Theory and Practice in Social Research and Development*. London and New York: Routledge.

Lund, R. 1981. Women and Development Planning in Sri Lanka. *Geografisk Annaler*, 63B, 95–108.

Lund, R. 1993. *Gender and Place*, Volume 1: *Towards a Geography Sensitive to Gender, Place and Social Change*; Volume 2: *Examples from Two Case Studies*. Trondheim: Department of Geography, University of Trondheim.

Lund, R. 2000. Geographies of Eviction, Expulsion and Marginalization: Stories and Coping Capacities of the Veddhas, Sri Lanka. *Norsk Geografisk Tidsskrift– Norwegian Journal of Geography*, 54, 102–9.

Lund, R. 2007. At the Interface of Development Studies and Child Research: Rethinking the Participating Child. *Children's Geographies*, 5, 131–48.

Manchanda, R. 2004. Gender, Conflict and Displacement: Contesting 'Infantilisation' of Forced Migrant Women. *Economic and Political Weekly*, XXXIX(37), 4179–86.

Mason, J. 1996. *Qualitative Researching*. London: Sage.

Moser, C. 1993. *Gender Planning and Development: Theory, Practice and Training*. London: Routledge.

Moser, C.O.N. 2001. The Gendered Continuum of Violence and Conflict: An Operational Framework, in *Victims, Perpetrators or Actors? Gender, Armed Conflict and Political Violence*, edited by C.O.N. Moser and F. Clark. London: Zed Books, 30–52.

Muller, H.P. and Hettige, S.T. 1995. Mahaweli as a Field of Tensions: Irrigation Settlements in the Dry Zone of Sri Lanka, in *A Blurring of a Vision – The Mahaweli: Its Social, Economic and Political Implications*, edited by H.P. Muller and S.T. Hettige. Ratmalane, Sri Lanka: Sarvodaya Book Publishers, 1–22.

Peiris, G.H. 1996. *Development and Change in Sri Lanka: Geographical Perspectives.* New Delhi: Macmillan India Limited.

Raby, N. 1995. Redesigning a Settlement Agency in Sri Lanka to Promote Participation of Women in its Management, in *Development Displacement and Resettlement: Focus on Asian Experiences*, edited by H.M. Marthur and M.M Cernea. New Delhi: Vikas Publishing House, 190–203.

Rajagopalan, S. 2010. Gender Violence, Conflict, Internal Displacement and Peace Building. *Peace Prints: South Asian Journal of Peace Building*, 3, 1–15.

Rajasingham, D.S. 2007. Between Tamil and Muslim: Women Mediating Multiple Identities in a New War, in *Gender, Conflict and Migration*, edited by N.C. Behera. New Delhi –Thousand Oaks – London: Sage, 175–204.

Risseeuw, C. 1991. *Gender Transformation, Power and Resistance among Women in Sri Lanka.* New Delhi: Manohar Publications.

Schrijvers, J. 1988. Blueprint for Undernourishment: The Mahaweli River Development Scheme in Sri Lanka, in *Structures of Patriarchy: The State, the Community and the Household in Modernising Asia*, edited by B. Agarwal. London: Zed Books, 29–48.

Scudder.T. 2005. *The Future of Large Dams: Dealing with Social Environmental, Institutional and Political Costs.* London: Earthscan, 138–85.

Shanmugaratnam, N. 2001. *Forced Migration and Changing Local Political Economies: A Study from North Western Sri Lanka.* Colombo: Social Scientists' Association.

Shanmugaratnam, N., Lund, R. and Stølen, K.A. (eds) 2003. *In the Maze of Displacement: Conflict, Migration and Change.* Bergen: Høyskoleforlaget.

Siriwardane, S.S.A.L. 1981. *Emerging Income Inequalities and Forms of Hidden Tenancy in the Mahaweli H Area.* People's Bank Research Monograph. Colombo: Peoples' Bank.

Sørensen B.R. 1996. *Relocated Lives: Displacement and Resettlement within the Mahaweli Project, Sri Lanka.* Amsterdam: VU University Press.

Wanigaratne, R. 1987. Paddy Initial Diversification in the Mahaweli H Area: A Precondition for Poverty. *Upanathi: Journal of Sri Lanka Association of Economists*, 12, 36–75.

Walker, R. 2010. Violence, the Everyday and the Question of the Ordinary. *Contemporary South Asia*, 18, 9–24.

Zackariya, F. and Shanmugaratnam, N. 2002. *Stepping Out: Women Survival Amidst Displacement and Deprivation.* Colombo: Muslim Women's Research and Action Forum.

Chapter 16

Coping Capacity of Small-Scale Border Fish Traders in Cambodia

Kyoko Kusakabe

Introduction

Facing adverse natural calamity or a hostile policy environment, people at the margin – those who have little voice in the macropolicy environment, such as indigenous people, microentrepreneurs, unorganized workers and labourers – develop their own coping strategies for survival and to improve well-being. Their choice of strategies is determined by their coping capacity – what resources they are able to manage in order to safeguard their life and livelihoods. This chapter is inspired by the work of Ragnhild Lund on the coping capacity of the Veddhas in Sri Lanka (Lund 2000, 2003). I will take the case of small-scale border fish traders in Cambodia to illustrate how a policy of opening the border affected them and showed their ingenuity in coping with the change. At the same time, I argue that their limited coping capacity restricted their opportunity to take advantage of the change and pushed them into a further marginalized state.

Lund (2000) describes how the Veddhas, an indigenous population of Sri Lanka, have been subject to forced relocation throughout the nineteenth and twentieth centuries. She analyses the Veddhas in two locations: Henanigala and Dambane. The Veddhas in Henanigala were displaced to make way for a national park under a development programme. They were alienated from the forest and became hired labourers. Youths started to dress and speak like Sinhalese and even started to practice Buddhism. The Dambane Veddhas refused to move and went to court to fight the case. However, they were not given additional land or specific measures to serve their culture as promised. They became alienated from the jungle and started to engage in tourism activities. The Dambane Veddhas ended up worse off than Henanigala Veddhas and have depended heavily on external assistance. Through these cases, Lund (2000) argues that the coping capacity of the Veddhas had been eroded by the development policy of the Sri Lankan state.

In order to understand coping capacity, Lund considers 'how both external and internal factors, collective capabilities and individual characteristics interact on and influence the Veddhas' coping capacity' (Lund 2000: 103). This perspective on how agency becomes enabled provides an important dimension for understanding the situation of border fish traders when the border trade was officially opened between Cambodia and Thailand. In the following, I review the concept of coping capacity and apply it to the case of small-scale women border fish traders in Cambodia to explain the strategies they took and the limitations they faced.

The Concept of Coping Capacity

Wisner et al. (2004: 113) define coping as follows:

> Coping is the manner in which people act within the limits of existing resources and range of expectations to achieve various ends. In general this involves no more than 'managing resources', but usually it means how it is done in unusual, abnormal and adverse situations. Thus coping can include defence mechanisms, active ways of solving problems and methods for handling stress.

Similarly, UN/ISDR (2009) defines coping capacity as: 'The ability of people, organizations and systems, using available skills and resources, to face and manage adverse conditions, emergencies and disasters'. Hence, coping capacity concerns how far people are able to use the resources to maintain their livelihoods (Wisner 2001, Doocy et al. 2005). As Few et al. (2004) point out, coping capacity is inversely related to vulnerability. Wisner et al. (2004: 11) define vulnerability as 'the characteristics of person or group and their situation that influence their capacity to anticipate, cope with, resist and recover from the impact of a natural hazard'. Vulnerability can be measured by the bundles of resources that the individual or group command (Wisner et al. 2004). Access to resources is a crucial dimension in understanding coping capacity and vulnerability.

Understanding how access to resources is determined is important for the study of both the vulnerability and its inverse, the coping capacity of a group of people (Bagchi et al. 1998, Eriksen et al. 2005). We need to analyse a specific person's situation, how access to resources has changed and what the person did in the face of adversity in order to gain a full grasp of vulnerability. We need to understand the situation both at the microlevel, such as relationships with others, and at the macrolevel, such as the wider political set up in the region and nation (Wisner et al. 2004). Analysing events and responses of specific people allows us to understand how people are able to cope with adversities.

In analysing coping capacities of Veddhas, Lund (2000) focuses on events and processes as well as on the specific situation in which individual Veddhas find themselves. She examines the events they experienced – their eviction and expulsion from the forest, and the development programme that in practice facilitated the process of marginalization. She also examines individual characteristics that define differential access to resources and individual livelihood choices in specific conditions. She includes family and community support, reflecting social relations. The Veddhas' knowledge was deeply related to the forest. When the state prohibited them to use and enter the forest, Veddhas had to change their livelihood from one that was heavily reliant on the forest to one that does not utilize forest resources. The relationship between the Veddhas and the Sri Lankan state determined the extinguishing of Veddhas' entitlements to the forest and hence restricted their use and maintenance of indigenous knowledge.

In this chapter, I follow Lund's approach and analyse events, processes and the specific situation of small-scale border traders in order to highlight their capacities and coping strategies following the opening of the border between Cambodia and Thailand. In the case of women border fish traders, their coping capacity is gendered in that their responses to changes in border trade reflect the gender norms and ideologies of society. Women traders were able to flourish when a border conflict led to the official closure of the border, since women were not seen as combatants and enjoyed comparative mobility compared to men. However, when the border was officially reopened, women traders, who had much less capital than large enterprises that were often managed by men, had difficulty competing with large traders. Small-scale women traders were also vulnerable to demands from corrupt officials. Thus structures of domination (Wisner et al. 2004) between the Cambodian or Thai border authorities and women traders as well as between large enterprises and women traders put the women traders to great disadvantage.

Border Fish Traders in Cambodia

In the present study, I discuss the border between Aranyaprathet, Thailand, and Poipet, Cambodia. Cambodia has a large freshwater fishery based upon the Tonle Sap Lake and Mekong River. A study of border fish traders on the Thai–Cambodian border was conducted in 2003 (Kusakabe et al. 2008). Eighty-six traders (37 Thai traders and 49 Cambodian traders) were interviewed in Poipet commune, Au Chorov District, Banteay Meanchey Province, Cambodia and in Klong Luek Commune, Aranyaprathet District, Sa Kaew Province, Thailand.[1] In 2002, freshwater fish production contributed 7–12 per cent of the total gross domestic product (GDP) of Cambodia (Chea and McKenney 2003) and employed over one million people. The fishery's production from wild freshwater fish was recorded at 385,000 tons in 2001 (MAFF 2004).

Freshwater fish production has always been an important economic activity in Cambodia, except during the Pol Pot regime from 1975 to 1979, when commercial fisheries were banned. After the regime collapsed in 1979, there was an exodus

---

1   The study was conducted by the Asian Institute of Technology (the author), Department of Fisheries, Thailand (Ubolratana Suntornratana and Napaporn Sriputinibondh) and the Cambodian Centre for Study and Development in Agriculture (CEDAC), a nongovernmental organization (Prak Sereyvath). For further details of this research, see Kusakabe et al. (2008). We are grateful for the generous support of the Swiss National Centre of Competence in Research North–South, Joint Area of Case Studies, Partnership Actions for Mitigating Syndromes of Global Change. Ragnhild Lund served as an external advisor to the project.

of refugees from Cambodia to the Thai border, but there were no official trade relations between the two countries. The border area was still a battlefield between the government military and Khmer Rouge. During this conflict era, Cambodian women traders travelled in government military trucks to trade fish in Thailand. It was easier for women to engage in border trade, since as noncombatants they were less controversial in accompanying the soldiers with their fish. Since it was still rare to get Tonle Sap fish in Thailand, the fish fetched good prices.

As the fighting subsided, it became much easier to trade fish and border trade flourished. Since large traders were absent, the market was dominated by small-scale women border traders. After peace was restored, the state-owned fish export company Kampuchea Fish Import and Export Company (KAMFIMEX) opened an office at the border town of Poipet in 1990, and required all traders to go through KAMFIMEX and pay export fees when they exported fish to Thailand. However, KAMFIMEX eventually lost its power to control the trade, as more and more traders started to cross the border without paying export fees and by 1997 KAMFIMEX was unable to control the flow.

When the Poipet–Aranyaprathet border became an official border crossing, customs and quarantine offices were set up on both sides of the border. Alongside the official structure, many informal checkpoints started to crop up. According to Chea and McKenney (2003), traders in 2002–2003 had to make 27 payments to 15 institutions at 16 locations between Tonle Sap Lake and the border market on the Thai side of the border, and 69 per cent of their potential profit went to such fees. More than 50 per cent of the payments were without any official basis. Despite the absence of KAMFIMEX, fish traders faced even more barriers with demands to pay various fees.

With the official opening of the border, large companies started to export fish. With such competition small-scale border fish traders, predominantly women, faced many difficulties. First, they had difficulty in procuring fish. In order to secure a fish supply, large traders provided advance payment to fishers. Smaller traders lacked capital to provide advance payments and hence were not able to buy fish directly from fishers. This disadvantaged them, since they could then only buy fish from other, larger traders. Second, they were more vulnerable to 'informal payments' – that is, arbitrary fees collected by various officers along the way to the border. Without refrigeration facilities, small traders could not afford to waste time negotiating with the authorities for payment, which resulted in higher rates to avoid spoilage. Such low negotiating power made it difficult for small traders to predict costs for these payments and plan their businesses. Third, at the Thai–Cambodian border, fish needed to be reloaded into carts as trucks were not allowed to pass the border gate. In order to get their fish through the border by cart, traders had to pay transporters to bring their fish to the other side. Since the amount of fish that they dealt with was small, the cost of transportation per unit increased for small traders. At the time of our study in 2003, large exporters paid 3 baht (USD 0.075) per kg to transport fish from the landing site to the border town of Poipet in Cambodia, and 2 baht (USD 0.05) per kg to transport it across the border to Rong Kluer Market on the Thai side. The payment was higher for small-scale traders than large traders. Small-scale traders paid transporters 2.5 baht (USD 0.063) per kg, although for microtraders with only a washing tub full of fish (30–40 kg) no payment was required. Hence the 'opening' of the border became an impediment for business opportunities for many of the small fish traders. Similar 'closure' of opportunities for local people at the border was identified by Walker (2000) in northwestern Laos.

Fish Traders' Coping Capacity

Small-scale border fish traders' opportunities for business have been restricted by the government's policy of liberalizing border trade. The participation of larger businesses in the trade has squeezed them out of procuring fish directly from fishers. A trade environment that favours economies of scale has increased the operation costs of small border traders. Women traders had little capital to offer to fishers for fish or to buy and transfer fish in bulk. The opening of the border and the difficulties that women small traders with little capital had in procuring fish and paying informal fees left them at a disadvantage. In the following sections, I focus on relational aspects of their coping capacity, and how the traders used their resources not only to survive but also to resist domination by the state.

Community Support

Community support was forthcoming during collective bargaining between about 60 traders and transporters on the one hand and KAMFIMEX on the other. Although the leader was a male transporter and many women traders declined to join, fearing that they would not be able to do fish business if they are blacklisted by KAMFIMEX, there were still many women traders who joined in to get their voices heard. Fish traders as well as fish transporters were not happy about the extra charge that was imposed by KAMFIMEX in 1990. KAMFIMEX had no licence to control the export of fish after 1993, but nonetheless continued to collect payments. There was only one road to the border crossing from inner Cambodia, and KAMFIMEX owned a storage facility on this road. It was therefore able to force most of the traders to stop to make payment. In February 2002, 58 people (33 women and 25 men) gathered thumbprints (instead of signatures for those who could not write their names) and protested against KAMFIMEX. Of the 58, four were transporters and the others were small-scale traders and exporters. Small-scale traders are those who trade 200–500 kg per day on average, although there is a large seasonal fluctuation in the amount of fish traded. The chronology of events was provided by the leader of the protest. The leader, a man, was a transporter and a native of Poipet. While employed with KAMFIMEX, he had been able to inspect their legal documents and realized that KAMFIMEX was not authorized to collect any payments at all from traders. Traders and transporters brought their complaints to the sub-district, district and provincial authorities. Through a year-long process, the traders and transporters were successful and were able to force KAMFIMEX to stop taking these payments. During the protest against KAMFIMEX, the leader of the protest advised small traders not to pay any unauthorized fees to collectors. He advised traders to give only small 'cigarette fees' and not pay the amounts demanded. 'I told them that I will protect you, so you do not need to pay.' These protests, he argued, have reduced informal payments along the road. This male leader acted as a 'protector' for women traders. In one sense, it is a reflection of gender roles in general, but, in another sense, women traders mobilized support from these male transporters by relying on such gender roles.

However, with the 'official' opening of the border and when the market was open to competition, individual small traders could do very little. Collective action against KAMFIMEX was possible when there was a clear partner to negotiate with, but with many players coming into the border trade it was more difficult to organize collective action, especially given the diverse background of traders. Some traders were seasonal traders and others were full-time, so not all felt the same degree of importance to maintain their trade. After the border was opened and the border trade flourished, the relative visibility of small traders has been greatly reduced.

Local Knowledge and Identity

Among the most important knowledge that the traders have is knowledge about their customers. Since many traders have been trading for over a decade, they have developed good relations with their regular Thai customers, whom they call *mooi*. They are loyal to each other. Their Thai customers buy from particular Cambodian traders even when there is a lot of fish elsewhere in the market, and the Cambodian traders will make sure that they sell to their regular Thai customers even when fish is scarce and other traders offer better prices. Such relationships are important for small fish traders in order to stabilize their sales, bearing in mind that they may have to make unpredictable informal payments to authorities. However, to rely on Thai traders and not to cultivate connections and support among Cambodian traders limits the negotiating power that Cambodian border fish traders have. Cambodian traders do not have access to cold storage, unlike the Thai traders, as the cold storage facilities are on the Thai side of the border. Hence Cambodian traders need to sell their fish quickly, then and there. This is a clear disadvantage when selling fish. Relying on the Thai buyers rather than looking for support among the Cambodian traders might be making their status more precarious than before.

Despite such limitations, traders' knowledge about and their linkages with their customers can allow them to circumvent temporary closures of the border by the state. From time to time, whenever there is a bilateral conflict between Cambodia and Thailand, the gate on the Aranyaprathet–Poipet border is closed. Women traders will then call their Thai counterparts and transport their fish to another smaller informal border crossing that it is not officially closed. Thus, by mobilizing their network, they are able to overcome the state policies that inhibit their trade. Male officers and male transporters of goods at the border call the Thai side of the border *din ke* ('their land'), referring to the border market that is located on the Thai side of the border. On the other hand, for women traders, whether the market is physically on the Thai side or the Cambodian side is not a problem. For them, the limit to their mobility is not the international border. Their 'territory' is where their customer is, and the 'border' is their reach of human relationships.

Another way that the traders cope with the changing trade environment is through their identity as mothers and as Khmer. They negotiate by using their image as mothers and poor women as leverage for negotiation with officers. Traders said that when facing authorities for payment, they will 'beg them' (*som ke*) and ask for pity from the officer, saying that they are poor women and they have to feed their family. One woman trader said that when customs officers asked for a very high fee, she tried to annoy them by crying out loud and accusing them of taking money from poor women like herself until the officer finally agreed to let her go at a lower fee. The women traders said that men are not capable of such a strategy ('men will soon start to fight'), and felt that such behavior can be tolerated in women.

Border fish traders also use their national identity to create better linkage with their customers. They were comfortable shifting their identity from one of difference to one of similarity, a phenomenon we did not see among male transporters. The differences between traders and transporters in the shifting use of identities may, however, not be because of gender differences but because of role differences – men are transporters and need to bargain less with the Thais. This practice was especially strong when the Cambodian women traders talked to Thai women traders. They emphasized the similarities between Thais and Cambodians, and that there was no difference between them – they were from the same 'nationality' (*chat*). The emphasis on identities was made when the Cambodian women traders were negotiating with them over price or when we were interviewing them in front of Thai traders. In negotiating with a Thai woman trader, a Cambodian woman in the border town of Ban Laem said: 'We are both from the same nationality [*chat diao kan*], so give me a discount.' In the border market at Rong Kluer, both Cambodian and Thai women were saying that Cambodians and Thais are the same ('so we should help each other, and you should give me a good deal!'). One Thai woman trader, when interviewed, said: 'I am Thai 100 per cent'. Then she went on to explain how she used to be in Koh Kong in Cambodia before the Khmer Rouge regime, doing border trade with her husband (a Cambodian national), then fled to Thailand. 'Oh, so you are actually a Cambodian?' we asked, and she got irritated and told us that Thais and Cambodians are the same, and it does not matter which nationality one is. Thus, women traders, in order to negotiate better with their clients, like to keep their 'nationality' boundaries flexible and as open as possible so that they can include others into their classification, thus adding legitimacy to their demands for preferential treatment. For women traders, state-designated nationality boundaries are irrelevant. What matters is their relationship with their customers.

Conclusion

The chapter has analysed the coping capacity of small-scale women border fish traders in Cambodia. Following Lund (2000), I have tried to understand the context-specific situation of these traders and what they are doing in the particular situation in which they find themselves. I have demonstrated not only how the traders are coping but how coping capacity may be strengthened or weakened. Women traders were adversely affected by the policy of officially opening the border, which has increased competition with large enterprises and put them at a disadvantage in terms of capital and their power relations with the authorities. They have been targeted for 'informal payments', which are detrimental to their business, and since such payments are unpredictable they are not able to foresee the cost for their businesses. However, women traders have not retreated in the face of these difficulties. They were successful in collective action against the export fees collected by KAMFIMEX. However, once the border was open and many larger players came in, it was not possible for the women to organize united action. Community support diminished and so did their coping capacity.

On the other hand, women border traders have used their relations with their customers when government orders have closed the border. They have utilized a fluid identity in order to construct solidarity with different people to establish linkages across the border, defying the authorities' efforts to demarcate the border and establish national identities. They have circumvented the state and taken themselves out of its reach by defying the border closure and rejecting classification into a single national identity. When the government has followed a policy of opening the border, female small-scale border fish traders were put at a disadvantage. The analysis has shown that their coping strategies were sufficient to help them to adapt, but were not enough to change the larger market structure at the border.

References

Bagchi, D. K., Blaikie, P. Cameron, J., Chattopadhyay, M., Gyawali, N. and Seddon, D. 1998. Conceptual and Methodological Challenges in the Study of Livelihood Trajectories: Case Studies in Eastern India and Western Nepal. *Journal of International Development*, 10, 453–68.
Chea, Y. and McKenney, B. 2003. *Fish Exports from the Great Lake to Thailand.* Working Paper 27. Phnom Penh: Cambodia Development Resource Institute.

Doocy, S., Teferra, S., Norell, D., and Burnham, G. 2005. Credit Program Outcomes: Coping Capacity and Nutritional Status in the Food Insecure Context of Ethiopia. *Social Science and Medicine*, 60, 2371–82.

Eriksen, S.H., Brown, K. and Kelly, P.M. 2005. The Dynamics of Vulnerability: Locating Coping Strategies in Kenya and Tanzania. *Geographical Journal*, 171, 287–305.

Few, R., Ahern, M., Matthies, F. and Kovats, S. 2004. *Floods, Health and Climate Change: A Strategic Review*. Tyndall Centre Working Paper No. 63. Norwich: Tyndall Centre for Climate Research.

Kusakabe, K., Prak, S., Suntornratana, U. and Sriputinibondh, N. 2008. Gendering Border Spaces: Impact of Open Border Policy Between Cambodia–Thailand on Small-Scale Women Fish Traders. *African and Asian Studies*, 7, 1–17.

Lund, R. 2000. Geographies of Eviction, Expulsion and Marginalization: Stories and Coping Capacities of the Veddhas, Sri Lanka. *Norsk Geografisk Tidsskrift– Norwegian Journal of Geography*, 54, 102–9.

Lund, R. 2003. Representations of Forced Migration in Conflicting Spaces: Displacement of the Veddas in Sri Lanka, in *In the Maze of Displacement: Conflict, Migration and Change*, edited by N. Shanmugaratnam, R. Lund and K.A. Stølen. Kristiansand: Høyskoleforlaget, 76–104.

MAFF 2004. *Cambodia* [Online: Ministry of Agriculture, Forestry and Fisheries]. Available at: www.maff.gov.kh/fi sheries.html [accessed: August 2004].

UN/ISDR 2009. *Coping Capacity* [Online: PreventionWeb]. Available at: http://www.preventionweb.net/english/professional/terminology/v.php?id=472 [accessed: 9 February 2011].

Walker, A. 2000. Regional Trade in Northwestern Laos: An Initial Assessment of the Economic Quadrangle, in *When China Meets Southeast Asia: Social and Cultural Change in the Border Regions*, edited by G. Evans, C. Hutton and K.K. Eng. Singapore: Institute of Southeast Asian Studies, 122–44.

Wisner, B. 2001. *Notes on Social Vulnerability: Categories, Situations, Capabilities and Circumstances* [Online: Environmental Studies Program, Oberlin College]. Available at: http://www.radixonline.org/resources/vulnerability-aag2001.rtf [accessed: 1 November 2011].

Wisner, B., Blaikie, P., Cannon, T. and Davis, I. 2004. *At Risk: Natural Hazards, People's Vulnerability and Disasters*. London: Routledge.

Chapter 17

Spontaneous Frontier Migration in Sri Lanka: Conflict and Cooperation in State–Migrant Relations

Berit Helene Vandsemb

Introduction

Until the twentieth century, migration frontiers (implying migration into uninhabited or sparsely populated parts of the country) were a common feature of the political landscape. They were considered essential for accommodating the fast-growing population of many states, for instance in Sri Lanka where succeeding governments have adopted a policy of sponsored planned migration to rural areas since the 1930s (Moore 1985, Peiris 1996). Most of these settlement schemes have been located in frontier areas claimed by the Tamil minority, and hence are seen as one of the major causes of the ethnic conflict in Sri Lanka. However, the focus here will be on the less studied spontaneous frontier migration taking place in the wake of the government-sponsored settlement schemes. Increasing population density, landlessness and unemployment in southern Sri Lanka, combined with government land policies and the prevalent ideology of smallholder agriculture, have led to spontaneous out-migration from the south into the remote forested areas in the southeast. In this chapter, I examine the extent to which the interests of the spontaneous migrants coincide or clash with those of the Sri Lankan state. I will look into the state interests related to deforestation, rural development and political control as well as the migrants' interests related to land, home and livelihood. Such state–migrant relations are a key dimension in understanding alternative geographies of development. Everyday politics matter (Rigg 2007) and these spontaneous migrants' actions have influenced state policies in Sri Lanka. A spontaneous migrant is someone who chooses his destination and time of departure without the intervention of any institution. I will show how the frontier at Tanamalwila represents a space of opportunity and a place to live for in-migrants and how the frontier has changed from spontaneous settlements to more established villages. According to Tuan (1977: 6), 'what begins as undifferentiated space becomes place as we get to know it better and endow it with value.' Thus, place is a distinctive type of space that is defined by the lived experiences of people. When people invest meaning in a portion of space and become attached to it in some way, it becomes place.

Frontier migration and settlement have been part of most societies' demographic and economic history. The frontier has often been viewed by the core as a region of great resource potential, and the core has sought to incorporate the frontier into its domain in order to control the resources and to achieve various state goals. Frontier land settlement has been a popular development policy in a number of developing countries, especially in Latin America and South and Southeastern Asia (Findley 1988, Shrestha 1990, Brown and Sierra 1994). It has been viewed as a strategy for population redistribution from densely populated areas, and a strategy of agricultural development, easier to execute than land reforms implying redistribution of land. There is often also a sociopolitical consideration – to avoid potential revolt among the landless rural population by moving them to the frontier. In its historical origin the frontier was not a legal or political concept, but rather a manifestation of the spontaneous tendency for growth on the margin of the inhabited world. The frontier refers to the division between the settled and uninhabited parts of a country (Johnston et al. 2000: 282). According to Kristof (1959: 281):

> ... the nature of frontiers differs greatly from the nature of boundaries. Frontiers are characteristic of rudimentary socio-political relations – relations marked by rebelliousness, lawlessness and/or absence of laws. The presence of boundaries is a sign that the political community has reached a relative degree of maturity and orderliness, the stage of law-abidance.

Both frontiers and boundaries (or borders) are the manifestation of sociopolitical forces, but while frontiers are the result of spontaneous or at least ad hoc solutions and movements, boundaries are fixed and enforced through a more rational and centrally coordinated effort after a conscious choice is made. However, borders do not need to be territorial constructs for them to constitute lines of separation or contact. This may occur in real or virtual space, horizontally between territories, or vertically between groups and/or individuals (Newman and Paasi 1998). Borders exist in almost every aspect of society, categorizing people into those who belong to the group and those who do not. The essence of a border is to separate 'self' from the 'other' – to act as a barrier to protect 'us, the insiders' from 'them, the outsiders'. Thus, the bordering process creates order through the construction of difference between groups and/or territories.

I have studied the spontaneous frontier migration process to Tanamalwila, where poor Sinhalese peasants have moved from traditional villages in the interior southwestern part of the island and settled in spontaneous agricultural settlements to create a place to live. My first fieldwork in the area was in 1985–86, while the main periods of data collection for this study were in 1989, 1990, 1991 and 2004. The study is based on data collected both at the frontier (Tanamalwila Division in Moneragala District) and in the major area of out-migration (Weeraketiya Division in Hambantota District). I have applied an extensive multi-method strategy including survey questionnaire, qualitative interviews, narratives, observation, maps, air photos, photos, documents and secondary literature. The long time perspective enabled my study of two different realities in the same geographic location at the frontier; both the new spontaneous settlement and the more mature village that has developed over time (Vandsemb 2007).

Sri Lanka and the State Project of Rural Development

Sri Lanka is divided into two geographical regions – the wet zone in the southwest and the dry zone in the north and east – which have repercussions on the pattern of settlement and development. The wet zone, covering approximately one third of the country, is more developed and more densely populated than the larger dry zone. These regional differences have been central to Sri Lanka's policies on rural development and population distribution. Rural migration is an outcome of the national policies concerning economic development, poverty alleviation strategies, and regional and rural development, and in contrast to many other developing countries, the predominant internal migration pattern in Sri Lanka has been rural to rural movement.

Rural development policies have played a major role in Sri Lanka since before independence in 1948, as the majority of the poor live in rural areas. Landless labourers, sharecroppers and marginal peasants together constitute the majority of rural residents in Sri Lanka. From early in the twentieth century, indigenous leaders advocated a self-determined national development strategy, where the underlying philosophy was rural development and the preservation of the peasantry. There was a great 'preoccupation with the general upliftment (*sic*) of the majority of Sri Lankan people, namely, the peasants in their rural habitats' (Sessional Paper 1990: 132). The strategy contained a romantic vision of rebuilding the splendour of an ancient culture, which had been neglected through centuries of foreign rule. The image of the Sinhalese as an essentially peasant nation was intrinsic to this historical myth. This nationalist rural development ideology has remained strong since independence. Indeed, the presence of nationalist ideologies in rural development has grown since the late 1970s (Moore 1985, Spencer 1990, Woost 1993). However, the political demand for rural development has not come from the poor peasants themselves. Moore (1985) points out that the elite articulated the plight of the peasants in such a way as to give moral support to its own claims to rule. The political elite has been accepted as the custodian of the smallholder population, and therefore also as the legitimate spokesperson for peasants' interests. Sri Lanka's first Prime Minister, D.S. Senanayake, is legendary for his efforts in promoting colonization and irrigation development in the dry zone, and his successors have followed in his footsteps.

The state is the biggest landowner in Sri Lanka. In 1985, more than 80 per cent of the land remained either as state lands or under some form of government control (Sessional Paper 1990). During the 1990s, a number of state-owned plantations were privatized but the state is still by far the biggest landowner as it owns about

60 per cent of all land (World Bank 2002). At least up to the 1920s, land alienation was in favour of large estates, but from around 1930 onwards there have been new trends in agrarian policy. The focus of agricultural development shifted from the wet zone to the dry zone and from plantation crops to rice, and alienation of land to the peasantry became the primary land policy. There are several reasons for this shift in policy. Universal franchise was introduced in 1931 and, as peasants constituted the majority of the electorate, priority was given to addressing the welfare needs of the peasantry and combating the widespread poverty in rural areas by supporting self-employment in agriculture. The stability of the country was seen to rest upon the creation of a dynamic peasant sector that could increase food production and thus reduce the dependency on food imports. State-aided peasant resettlement schemes became an alternative to land reform, which helped to preserve the existing property relations.

Sri Lanka's policy of spatial redistribution of its population has mainly occurred through induced migration to new settlement schemes in the dry zone. By the late 1960s, about a hundred such settlements had been developed with approximately 100,000 peasant families being settled there (Nelson 2002). The settlement schemes are connected to the expansion of irrigable land for paddy cultivation, both by building new irrigation facilities and by restoring ancient village ponds. During the 1970s and 1980s, the government pursued major irrigation settlement projects, notably the massive Mahaweli programme. This was planned to resettle 150,000 families (Mahaweli Authority 1982) and up to the end of 1993 had involved the resettlement of about 115,000 families (Nelson 2002). The motivation for these resettlement programmes was to reduce land pressure and unemployment in the wet zone and to increase the production of rice and other food products in order to reduce import costs. Further, it was part of the nationalist rural development ideology of creating a class of land-owning farmers and restoring the ancient civilization of the dry zone: 'This transmigration of our people to their historical homelands is of enormous societal significance ...' (Mahaweli Authority 1982: 3). Most of the settlement schemes, mainly encompassing Sinhalese peasants, were established in northeastern and eastern Sri Lanka, areas bordering and sometimes overlapping the Tamil-speaking areas. The land policy did not address the contradictions arising from such policies in a multiethnic society. Consequently, these resettlement schemes are seen as one of the major causes of the conflict between the Sinhalese government and the Tamil minority.

In the wake of this induced migration to settlement schemes, there has been a spontaneous rural migration process taking place, both to the fringes of the settlement schemes and to other sparsely populated rural areas, such as Tanamalwila. There is a lack of data on this spontaneous movement; however, most of these spontaneous migrants have become encroachers on state land and engage in *chena* (slash-and-burn) cultivation. In 1979, an island-wide survey of encroachments on state lands revealed that about 6 per cent of the total area of Sri Lanka consisted of more than 600,000 illegal encroachments. From 1979 to 1985, the number of encroachments increased by 15 per cent (Sessional Paper 1990). Most of these encroachments

---

were under *chena* cultivation in the dry zone. However, a substantial amount of these encroachments were cultivated by nonmigrants.

A policy of encroachment regularization was adopted by the Government from 1979 (Sessional Paper 1990). The following conditions were applied in rural areas: the encroacher had to be landless; the encroached land had to be under cultivation; and the maximum extent to be regularized per encroacher was two acres of highland and one acre of paddy land. The government put a ban on *chena* cultivation by stopping the issuing of *chena* permits; instead they issued annual permits for specific crops to be grown on the same land. Despite the ban, people continued to cultivate *chena* illegally. However, people became more reluctant to open up new land by burning forest; instead they cultivated old plots, which involved burning grass and bushes only, resulting in reduced fallow periods and decreased soil fertility.

To strengthen the rural development work, District Integrated Rural Development Programmes (IRDPs) were introduced by the government in the late 1970s. Moneragala IRDP (MONDEP), funded by the Norwegian Agency for Development Cooperation (NORAD), was launched in 1984 (Dale 1992). The aim was to improve the standard of living of the local population, and marginalized farmers and encroachers were among the target groups (Sørbø and Zackariya 1995). MONDEP planned, funded and coordinated both sectoral projects and area development projects, which were implemented by various agencies, and Tanamalwila Division was included in the second area development project. The MONDEP development efforts have mainly stressed the enhancement of local infrastructure such as roads, health and education facilities, improved drinking water supply, and electricity networks, as Moneragala District had been lacking much of the basic infrastructure found in most other parts of the country. MONDEP has also emphasized social mobilization, with the help of 'social mobilizers' in rural areas to encourage participation. Furthermore, as part of the government's rural development policy from the late 1970s, Rural Development Societies were promoted and established in most villages. These are democratically organized and constitute the main base for local self-help activities – assisted by the government (Dale 1992). However, factionalism along political lines is commonly seen in these rural organizations, which hinders participation by all local people (Abeyrama and Saeed 1984).

Spontaneous Frontier Migration to Tanamalwila

Until the 1990s, most studies on internal migration in Sri Lanka, with a few exceptions (e.g. Abeysekera 1984, Crooks and Ranbanda 1981), were based on census data concentrating on general patterns of migration and the settlement schemes (e.g. Abeysekera 1981, Gunawardena 1982, Indraratna et al. 1983, Kearney and Miller 1987), while other studies dealt with the experiences of people resettled in the settlement schemes (e.g. Lund 1978, 1993; Sørensen 1996) or forced migration and displacement due to the ethnic conflict (e.g. Schrijvers 1999, Hasbullah 2001, Brun 2003). Few studies have focussed on the spontaneous frontier migration process, although some have dealt with the conditions of *chena* cultivation and village life in spontaneous migrant settlements (e.g. Vitebsky 1984, Baker 1998).

In my work (Vandsemb 2007), I have used the concept of frontier migration to refer to the migration into remote forested areas in the interior of southeastern Sri Lanka. The study area is located in Tanamalwila Division, which has attracted migrants coming for *chena* cultivation because of low population density (33 per sq. km in 1981) and the consequent availability of land. The slack enforcement of laws prohibiting *chena* cultivation and the policy of regularization of encroachments have further increased the attraction of the area. From 1971 to 1981, the population of Tanamalwila Division grew at an annual rate of 11 per cent (MONDEP 1986). Less than 10 per cent of the total area in Tanamalwila Division is agricultural land, while most of the area is considered to be forest and scrubland (Department of Census and Statistics 2003, Land Use Planning Division 2004). Both traditional villagers and in-migrants cultivate *chena* in the forest and scrubland. Settlement schemes have had little influence on Tanamalwila as there are only three irrigation settlement schemes and these comprised rehabilitation of existing traditional villages. The traditional irrigation facilities were renovated and the irrigated land was divided among the people who already lived there. In addition to these, a few smaller settlement schemes were established in Tanamalwila in the early 1970s, but these schemes were not a success:

> I came here 20th August 1974 together with 77 men from Batapola (the west coast); we were given land here by the government. No one was here. The families came later. Within six years 75 of these families gave up and went away because there were no facilities here and the government did not help us. Later new people came to cultivate *chena* on these lands. (Male pioneer migrant 1990)

This spontaneous frontier migration process does not represent a transition in livelihood and lifestyle, although the work load, labour-sharing and gender roles related to *chena* and paddy cultivation are different. Rather, it represents a way of enabling people to continue with a rural livelihood and a peasant way of life, which many of these migrants prefer. The migrants in Tanamalwila Division were mostly young nuclear families whose members hailed mainly from large smallholding peasant families who cultivated paddy and coconut in Weeraketiya Division. In my sample (358 households), the average age of the head of household was 36 years while the average age of the spouse was 31 years, and the average size of household was 4.2 members (Vandsemb 2007). The socioeconomic changes in home villages that impelled this out-migration were factors such as increasing population density, decreasing size of landholdings, increasing landlessness, increasing unemployment and under-employment, and stagnation in the development of agriculture and alternative means of livelihood. This situation has intensified both inequality and poverty, and Weeraketiya Division represents a classic setting of rural out-migration (cf. Connell et al. 1976). The out-migrants belong to a group of people who are economically and politically marginalized and who feel that their livelihood alternatives are constrained as they lack education and skills for work outside agriculture. The frontier migration has enabled these young couples to construct their own households and become less dependent upon their parents.

What kind of frontier is Tanamalwila? Newman and Paasi (1998: 189) point out that: 'the political frontier was differentiated from the settlement frontier, the former affected by the existence of the international boundary, the latter constituting the, as yet, uninhabited region lying within the state territory and representing the spatial margin of the state's ecumene'. It is my contention that this area can be seen as both a settlement frontier as well as a political frontier. Furthermore, I also apply the term development frontier as the area represents the margins of state-implemented rural development efforts. Although state–migrant relations are intrinsic to the whole process of settlement, development and political control, I will deal with these three different notions of frontier separately.

Settlement Frontier: From a Space of Opportunity to a Place to Live

When in-migration and settlement started in the area about 40 years ago, the pioneers entered a densely forested area with a few tiny, traditional villages spread out with long distances from each other. A map in Brohier's book (1935) shows the emptiness of the area, marked as *chena* and jungle except the three tiny villages of Niyandagala, Suriya Ara (later abandoned) and Angunakolapelessa. The pioneer migrants settled in the forest far away from these traditional villages:

> The whole area was jungle and elephants attacked us when we started *chena*
> cultivation. I started to clear the jungle ... and built a hut in a tall tree to
> guard the *chena* against elephants, wild boars, wild buffaloes and monkeys.
> (Male pioneer migrant 1991)

Thus, this was a push forward into largely uninhabited areas, i.e. unknown spaces, where the migrants challenged state authority by encroaching on state land to cultivate *chena*. Typically, the men arrived first; they built a small hut, cleared the land and started cultivation before they brought their families or got married and brought their wives. However, the area was not altogether uninhabited when the migrants arrived. Although the original villagers were few, dispersed, and economically and politically marginalized, they did constitute a small population and the area in which they lived could not be called a wilderness. In general, the original inhabitants had a positive view of migrants arriving into their neighbourhood instead of having forest and wild animals as their closest neighbours. Their view may be interpreted as welcoming an expansion of civilization. However, they did not look upon themselves as uncivilized – on the contrary, they expressed disapproval of the migrants' bad habits:

> People from the same village used to help each other, but now people are not
> so helpful because people from many different places have different ideas ..
> People here are not making problems; people are not like that here. People in
> Weeraketiya are fighting with coconuts and coconut leaves and bringing that
> attitude here. They are no good, but they are not making any problems for me.
> (Female original inhabitant 2004)

The frontier was a zone of cultural encounter (Kristof 1959) between the
migrants coming from different areas with male-dominated paddy cultivation,
and the original villagers who were dependent on *chena* cultivation based on
a more equal gender division of labour. In ethnic and religious affiliation, the
frontier migrants in Tanamalwila have a common background as they are all
Sinhalese Buddhists. Yet, they constitute a socially heterogeneous group as they
originated from many different places, which sometimes led to conflicts because
people from different parts of Sri Lanka have prejudices against each other.
Original villagers and migrants from up-country districts tend to look upon the
'southerners' (the majority of the migrants) as people who drink excessively,
fight and in general cause problems.

Gender differentiates migrants' experiences of life at the frontier. In general,
men seemed to be more optimistic than women about the opportunities in
Tanamalwila. Men find a more attractive situation because of land availability and
income possibilities in cultivation. Women also see the possibilities in land, but
they emphasize the hard work, the drudgery and the social isolation:

> That day, when I saw this area, I was feeling alone. There were no people here.
> After some time, I saw some people but they were not nice. I thought I could
> not go and talk to them. I thought these people were living very difficult lives,
> cultivating *chena* and earning money from the *chena*. I saw some 3–4 houses
> around in the jungle. Later, I went to those places to talk and be friends with
> them. They welcomed me. I can remember that like today. (Female migrant 2004)

In general, at the early stages of settlement, the migrants seemed to know or
want to know very little about each other. Spontaneous settlement involved a
mixture of houses situated along tracks and houses spread out in the landscape
at a distance from one another. Initially, there was a very low level of interaction
among the migrants and between the migrants and the original villagers. Many
migrants kept close contact back to their home villages as they chose a prolonged
translocal migration strategy, where the men migrated while the women and
children stayed on in the home village. Nevertheless, this close contact with their
original home villages has faded with the maturing of the new settlements as
former translocal households have become permanent settlers in Tanamalwila.
For most migrants today, the visits to the home village have become limited to
special occasions or ceremonies.

The frontier migrants moved from their home place into fairly unknown spaces and created a new home place at the frontier. Thus, place-making is part of the frontier migrants' livelihood strategy. Place connotes 'ownership' or a connection between person and location, and implies notions of privacy and belonging (Cresswell 2004). To be able to get a land deed, the frontier migrants had to demonstrate their commitment and belonging to the place they cultivated. They did this by place-making activities such as shifting from *chena* to permanent cultivation, creating a home and taking part in community activities. To create a home they built permanent houses to reconstitute their families in one place. Outside the house they established a garden around their home with trees, bushes and ornamental flowers. At village level they took part in community organizations and communal work, co-operated with their neighbours and developed social networks. The migrants have started to perceive their new place as a proper village or home:

> During the last five years this has become a proper village. After the second generation is growing up and getting married to each other this is a proper village. Earlier people were from many different places, now they become related to each other. My daughter is married to the grandson of an original villager. They are our neighbours. (Male migrant 2004)

> It is only now that this is a real village. For the first time I did not have to go to my home village for the coming New Year festival as my mother has recently moved in to live with us. (Female migrant 2004)

It may be interpreted as a sign of an established settlement when the grandparents leave the home village and come to settle down with the migrants. Yet, only a few families have become extended by the arrival of the grandparents. A number of migrants have become grandparents themselves. Over the years, the migrants have experienced a growing attachment to the place. Their sense of place or feeling of home is developed through the interaction of the state (land policy and rural development policy) and themselves (building permanent houses and marrying neighbours). The land deed, the permanent house and the village infrastructure are important material aspects of their feeling of home, but immaterial factors such as social networks and especially marriage and blood lines seem to be decisive. Although most marriages still occur with partners from the origin area, there are signs of a new trend of 'love marriages'. When the second generation grows up and gets married in the neighbourhood, people become related to one another and their scepticism to the 'other' decreases. However, the new village is still loosely integrated socially compared with the villages of origin because of the scattered settlement pattern and the fact that people originate from many different places. Those with relatives in their neighbourhood seem to concentrate their social life around these relatives and mix less with other villagers. Those with no relatives around are 'forced' to mingle with their neighbours; otherwise they will remain socially isolated in the village.

According to Tuan (1991), naming is the creative power to give a certain character to things and naming is one among several place-making activities the migrants in Tanamalwila have been involved in. Names give meaning to a place – the villagers have given some of the new settlements names that confirm their ties to their area of origin. The frontier has changed in people's minds from a vast undifferentiated space of forest to specific places – the new settlements with familiar names. Today, it appears that a 'civilizational' process (i.e. people instead of forest) has been going on as there are a number of new villages established in the area.

Development Frontier: State–Migrant Relations

With the heavy influx of new people in Tanamalwila, both the migrants and the original villagers were expecting more development to come in the future. Their expectations are closely connected to the state policies of settlement schemes and rural development in the dry zone. Tanamalwila Division is among the least developed regions in Sri Lanka while the migrants' home areas are more developed, especially regarding infrastructure. Hence, a conceptualization of the settlement frontier as a development frontier involving the division between more developed and less developed areas seems appropriate.

For the migrants, access to land is fundamental to their development of a livelihood at the frontier. The value of land is quite distinct from other types of capital and resources. Ownership of land gives social status, represents a store of wealth, and its value as an asset is continually rising in a market economy as it becomes scarce and is more intensely used. Land also provides security against loans that can be used to further improve livelihoods (Ellis 2000). The migrants learned from each other how they could ensure their access to land:

> Gamini [her youngest son] and I stayed together with my eldest son for about two years. Then people told us that if we were living together like that we would not get more land. Therefore, Gamini and I moved to encroach on a new plot of land and Gamini built our first house. (Female migrant 1990)

Settlement frontiers are usually places at the periphery of state control (Kristof 1959, Newman and Paasi 1998), and to a certain extent this has also been the case in my study area. The early migrants were not subject to any state control at the frontier. They encroached upon government land, burned the forest and started to cultivate *chena*. After a couple of years, when land fertility decreased, they would move on to a new plot and burn the forest. In addition to *chena* cultivation, there was also illegal logging. As a result the area covered with forest has been dramatically reduced (Storm-Furru and Vandsemb 1989). During the 1980s the government started to become concerned about the serious deforestation of the area, and various initiatives were taken. The police were told to stop illegal logging and to enforce the ban against burning original forest. However, this effort was not very effective as the police had to be sent in from distant places, and they were not too keen to go into the jungle. As a response to government policy, the migrants organized in order to strengthen their claim to land:

> People wanted to stay here, but the government was pressing for them to leave and the police came and asked them to leave because they did not have legal permits to cultivate the land. People wanted to get land deeds so they started the Village Development Society to approach the government and solve the problems they were facing, especially those concerning land deeds. The first thing they did was to mark out the villages. Earlier the whole area was called Suriya Ara; now they defined the village boundaries and named the new villages.
> (Male migrant 2004)

By cooperating with the government in deciding the boundaries of the new villages, the migrants demonstrated their willingness to become sedentary cultivators, i.e. the transformation from lawlessness to orderliness (Kristof 1959). The government also implemented policies to stabilize and control *chena* cultivation. By giving the migrants entitlement to the land they were cultivating, the government wanted to transform *chena* cultivation into sedentary agriculture. This policy was more successful than policing in curbing *chena* cultivation. MONDEP saw the regularization of encroachments in Tanamalwila as an essential strategy to develop the area. 'To elevate these encroachers to the status which makes them entitled to institutional assistance and help them devoid from harmful agricultural practices and improve their lands and their living conditions, the regularization of encroachments is considered as an essential need' (MONDEP 1986, 59). In 1990, 53 per cent of my sample had some kind of entitlement to their land, with the greater part of these having had their encroachments regularized recently. Another 15 per cent had been selected to get their land regularized while the remainder were hoping to get it sometime in the future.

Entitlement to land gave people the possibility of improving their standard of living by obtaining housing loans from credit institutions. Around 1990, most houses were constructed of wattle-and-daub; however, a number of settlers had applied for housing loans and had started to make bricks to build permanent houses. In 2001, there were 387 permanent houses in the study area compared to 231 impermanent houses (Tanamalwila Divisional Secretariat 2002). During the years since the migration started, there has been development of infrastructure in the frontier area, some of it established by the migrants themselves but most of it funded and implemented by MONDEP and the state. The first common meeting place was built by the migrants themselves in 1988; a small wattle-and-daub building functioned as a temple and pre-school. A year later a small government office and a health clinic were built by MONDEP. During the following years, a large number of latrines and improved wells were built and village access roads were improved. There was no school when I was there in 1990. The villagers collected money from various sources and built the first school building in 1996; later the building was extended with the help of an international NGO. Today, there is also an electric power line built through the village and the houses along the main road have been connected to electricity.

In the early phase of in-migration, many migrants were sceptical of state intervention since by their actions they challenged the legitimacy of state power in the area by violating the laws on *chena* cultivation, and they were sometimes punished for this. On the other hand, the migrants fulfilled the state goals of colonizing the dry zone and in that way helped the state's realization of boundary assertion – i.e. gaining state control over the frontier. By letting the migrants stay at the frontier, the state also achieved a certain reduction of potential revolt in the areas of origin, which are well known source areas for the rebellious political party *Janatha Vimukthi Peramuna*. Hence, successive governments have taken a lenient attitude towards the encroachers and treated them more as pioneers than offenders. State–society relations in Tanamalwila Division have developed and changed since the in-migration started four decades ago. There has been a transition from a frontier of spontaneous settlements with rudimentary state–society relations to mature villages with closer ties between the state and society. Through MONDEP the state has gained increasing influence and control in the in-migration areas. Today, most migrants live in permanent houses and they cultivate their entitled land on a permanent basis. Furthermore, the establishment of village organizations and village boundaries in the new settlements shows that the community has reached a relative degree of maturity and orderliness under state control. However, a certain lack of state control is still evident. Faulty household and election lists, illicit cultivation of cannabis, brewing of illicit liquor, and recent encroachments on reserved land such as water catchments and national parks provide evidence of clashing interests between the state and certain groups among the settlers.

Political Frontier: A Division between Sinhalese and Tamil Nationalisms

The frontier at Tanamalwila is not a territorial boundary between two distinct areas but it can still be conceptualized as a political frontier with an ethnic boundary – a boundary between Sinhalese and Tamil areas and between Sinhalese and Tamil nationalisms. Sinhalese and Tamil nationalist historical narratives share a common assumption of the two groups being mutually exclusive. They both embrace notions of ethnically pure territory and identity, despite the history of mixed settlement, intermarriage and bicultural communities in most parts of the island (Rajasingham-Senanayake 2002).

The settlement schemes or the colonization of the dry zone in Sri Lanka can be seen as part of a political strategy to build the Sinhalese state. The policy of establishing new settlements, mainly of Sinhalese peasants, in the dry zone was profoundly accelerated after independence in 1948 with the growth of Sinhalese nationalism. The political rhetoric included the notion of restoring the glorious ancient Sinhalese civilization based on irrigated peasant agriculture in the dry zone

(Woost 1993). Kearney and Miller (1987: 116) point out that the formerly nearly uninhabited areas of the dry zone had once served as a broad belt of demarcation between the Sinhalese and the Tamils, but colonization changed this situation.

The notion of the frontier as a political–ethnic boundary is clearly most relevant for the northeastern parts of Sri Lanka where most of these new settlement schemes were established, often near Tamil-dominated areas, on land claimed to be ancestral Tamil homeland by Tamil nationalists. Tanamalwila has never been a Tamil area or an area of mixed settlement – on the contrary, it was part of the ancient Sinhalese kingdom of Ruhuna (de Silva 1981). Yet, Tanamalwila can be seen as part of this political frontier as colonization is helping to extend the area under Sinhalese state control. Between Tanamalwila and the Tamil dominated east coast there is a large national park that is virtually a no-man's land. Large parts of it were closed for years due to the fear that Tamil guerrilla soldiers might have been hiding there. An attack in 2008 on a police post in Tanamalwila, where three police men were killed by suspected Liberation Tigers of Tamil Eelam (LTTE) militants (BBC 2008), reinforced this fear. Thus, this is clearly a border dividing 'us' from 'the other'. As Newman and Paasi (1998) point out, borders do not need to be territorial constructs for them to constitute lines of separation. In this way, Tanamalwila gained a strategic meaning as a defensive line keeping the enemy out. To function as a defensive line, the area has to be controlled by the state. Seen in this light, spontaneous settlement in Tanamalwila by Sinhalese peasants is in the interest of the state as the state needs to develop the frontier to demonstrate its political presence in the zone of conflict (cf. Newman 2006). Thus, MONDEP has been an active agent in this process of bordering by the Sinhalese state. The permanent settlement of migrants has supported the state's need for political control. Coinciding interests of the state and the migrants have enhanced the Sinhalese state's control of the political frontier.

Concluding Remarks

The process of spontaneous settlement in Tanamalwila has been a process of both conflict and cooperation in state–migrant relations. Initially, at the settlement frontier, there was a clash of interests between the spontaneous migrants and the state. Although the state is seen as the custodian of poor peasants, these marginalized migrants challenged state authority by encroaching on state land and burning the forest to cultivate *chena*. They took 'development into their own hands' by migrating, opening up new land and establishing new settlements. The state tried to prevent in-migration and the consequent deforestation by sending in the police, a strategy that proved to be less than effective. There are, however, also concurrent state–migrant interests as the migrants move away from an area with a great potential for revolt against the state. Thus, one may claim that it is in the state's interest that the migrants leave their home area, but they should not engage in *chena* cultivation. Coinciding state–migrant interests are also found at the development frontier. The state offers cooperation instead of conflict by providing land deeds, housing loans and infrastructure. The migrants respond to this by shifting to sedentary agriculture, building permanent houses and establishing social organizations. The state is playing its role as the custodian of the marginalized farmer and the migrants in return are transforming the spontaneous settlements into more mature villages. This shows that the state has been more successful in solving the clash of interests by rewarding legal activity than in trying to police illegal activity. The spontaneous settlements may be regarded as 'cheap settlement schemes', fulfilling the government's goals of population redistribution, reducing unemployment, and increasing agricultural production in addition to securing state control over the frontier. Thus, this may be seen as a state–nonstate border where the influence of the state has increased over the years since the migration started. Finally, the migrants' search for a livelihood and a place to live have supported the need of the Sinhalese state for political control over areas bordering the Tamil dominated east coast. After the civil war came to an end in May 2009, the state has gained control over the whole island, yet Tanamalwila is still a frontier between Sinhalese-dominated and Tamil-dominated areas.

References

Abeyrama, T. and Saeed, K. 1984. The Gramodaya Mandalaya Scheme in Sri Lanka: Participatory Development or Power Play? *Community Development Journal*, 19, 20–31.

Abeysekera, D. 1981. *Regional Patterns of Intercensal and Lifetime Migration in Sri Lanka*. Honolulu: East–West Population Institute.

Abeysekera, D. 1984. Rural to Rural Migration in Sri Lanka, in *Rural Migration in Developing Nations*, edited by C. Goldscheider. Boulder, CO: Westview Press, 109–207.

Baker, V.J. 1998. *A Sinhalese Village in Sri Lanka: Coping with Uncertainty*. Fort Worth: Harcourt Brace College Publishers.

BBC 2008. *Policemen Killed in Thanamalwila*. [Online. BBC, 21 January]. Available at: http://www.bbc.co.uk/sinhala/news/story/2008/01/080121_thanamalwila_police.shtml [accessed: 23 January 2008].

Brohier, R.L. 1935. *Ancient Irrigation Works in Ceylon, Part III: Western, Southern and the Eastern Areas of the Island*. Colombo: Ceylon Government Press.

Brown, L.A. and Sierra, R. 1994. Frontier Migration as a Multi-Stage Phenomenon Reflecting the Interplay of Macroforces and Local Conditions: The Ecuador Amazon. *Papers in Regional Science*, 73, 267–88.

Brun, C. 2003. *Finding A Place: Local Integration and Protracted Displacement in Sri Lanka*. Dr. polit. thesis, Department of Geography. Trondheim: Norwegian University of Science and Technology (NTNU).

Connell, J., Dasgupta, B, Laishley, R. and Lipton, M. 1976. *Migration from Rural Areas: The Evidence from Village Studies*. Delhi: Oxford University Press.

Cresswell, T. 2004. *Place: A Short Introduction.* Oxford: Blackwell.

Crooks, G.R. and Ranbanda, H.A. 1981. *The Economics of Seasonal Labour Migration in Sri Lanka.* Colombo: Agrarian Research and Training Institute.

Dale, R. 1992. *Organization of Regional Development Work.* Ratmalana: Sarvodaya Book Publishing Services.

Department of Census and Statistics 2003. *Census of Agriculture 2002 (Smallholding Sector) Moneragala District (Preliminary Report).* Colombo: Department of Census and Statistics.

de Silva, K.M. 1981. *A History of Sri Lanka.* Delhi: Oxford University Press.

Ellis, F. 2000. *Rural Livelihoods and Diversity in Developing Countries.* Oxford: Oxford University Press.

Findley, S E. 1988. Colonist Constraints, Strategies, and Mobility: Recent Trends in Latin American Frontier Zones, in *Land Settlement Policies and Population Redistribution in Developing Countries,* edited by A.S. Oberoi. New York: Praeger, 271–316.

Gunawardena, K.A. 1982. Some Recent Changes in the Pattern of Internal Migration in Sri Lanka. *Progress,* 2, 7–13.

Hasbullah, S.H. 2001. *Muslim Refugees: The Forgotten People in Sri Lanka's Ethnic Conflict,* Volume 1: *Introduction.* Colombo: Research and Action Forum for Social Development.

Indraratna, A.D.V.d.S., Codippily, H.M.A., Abayasekera, A.W.A.D.G. and Abeykoon, A.T.P.L. 1983. Migration Related Policies: A Study of the Sri Lanka Experience, in *State Politics and Internal Migration,* edited by A.S. Oberai. London: Croom Helm, 79–135.

Johnston, R.J., Gregory, D., Pratt, G. and Watts, M. (eds) 2000. *The Dictionary of Human Geography.* 4th Edition. Oxford: Blackwell.

Kearney, R.N. and Miller, B.D. 1987. *Internal Migration in Sri Lanka and its Social Consequences.* Boulder, CO and London: Westview Press.

Kristof, L.K.D. 1959. The Nature of Frontiers and Boundaries. *Annals of the Association of American Geographers,* 49, 269–83.

Land Use Planning Division. 2004. *Tanamalwila D.S. Division Land Use Map.* Moneragala: Kachcheri.

Lund, R. 1978. *Prosperity to Mahaweli: A Survey on Women's Working and Living Conditions in a Settlement Area.* Colombo: People's Bank.

Lund, R. 1993. *Gender and Place,* Volume 1: *Towards a Geography Sensitive to Gender, Place and Social Change;* Volume 2: *Gender and Place: Examples from Two Case Studies.* Trondheim: Department of Geography, University of Trondheim.

Mahaweli Authority. 1982. *Ganga: Randeniyagala Story and Mahaweli Settlements.* Colombo: Ministry of Lands and Land Development and Ministry of Mahaweli Development.

MONDEP. 1986. *BASIS: Balaharuwa, Sittarama, Suriya Ara Area Development Project,* Volume II. Moneragala: MONDEP office, Kachcheri.

Moore, M. 1985. *The State and Peasant Politics in Sri Lanka.* Cambridge: Cambridge University Press.

Nelson, M.D. 2002. *Mahaweli Programme and Peasant Settlement Development in the Dry Zone of Sri Lanka.* Peradeniya: University of Peradeniya.

Newman, D. 2006. Borders and Bordering: Towards an Interdisciplinary Dialogue. *European Journal of Social Theory,* 9, 171–86.

Newman, D. and Paasi, A. 1998. Fences and Neighbours in the Postmodern World: Boundary Narratives in Political Geography. *Progress in Human Geography,* 22, 186–207.

Peiris, G. H. 1996. *Development and Change in Sri Lanka.* Delhi: Macmillan India Limited.

Rajasingham-Senanayake, D. 2002. Identity on the Borderline: Modernity, New Ethnicities, and the Unmaking of Multiculturalism in Sri Lanka, in *The Hybrid Island: Culture Crossings and the Invention of Identity in Sri Lanka,* edited by N. Silva. Colombo: Social Scientists' Association, 41–70.

Rigg, J. 2007. *An Everyday Geography of the Global South.* London and New York: Routledge.

Schrijvers, J. 1999. Fighters, Victims and Survivors: Constructions of Ethnicity, Gender and Refugeeness Among Tamils in Sri Lanka. *Journal of Refugee Studies,* 12, 307–33.

Sessional Paper No. III. 1990. *Report of the Land Commission – 1987.* Colombo: Department of Government Printing.

Shrestha, N. 1990. *Landlessness and Migration in Nepal.* Boulder: Westview Press.

Spencer, J. 1990. *A Sinhala Village in a Time of Trouble: Politics and Change in Sri Lanka.* Delhi: Oxford University Press.

Storm-Furru, I. and Vandsemb, B.H. 1989. Agricultural Intensification in the Dry Zone of Sri Lanka: Impacts on Farming Practices, Welfare and Women's Work. *Norsk Geografisk Tidsskrift,* 43, 55–65.

Sørbø, G., and Zackariya, F. 1995. *Moneragala District Integrated Rural Development Programme: Review of Socio-Cultural Aspects.* Bergen: Chr. Michelsen Institute.

Sørensen, B.R. 1996. *Relocated Lives: Displacement and Resettlement within the Mahaweli Project, Sri Lanka.* Amsterdam: VU University Press.

Tanamalwila Divisional Secretariat. 2002. *Tanamalwila Division Economic Profile 2002.* Tanamalwila: Tanamalwila Divisional Secretariat.

Tuan, Y.-F. 1977. *Space and Place.* London: Arnold.

Tuan, Y.-F. 1991. Language and the Making of Place: A Narrative-Descriptive Approach. *Annals of the Association of American Geographers,* 81, 684–96.

Vandsemb, B.H. 2007. *Making a Place and (Re)constructing a Life: The Role of Gender in Spontaneous Frontier Migration to Tanamalwila, Sri Lanka.* Doctoral theses at NTNU, 2007:28. Department of Geography. Trondheim: Norwegian University of Science and Technology (NTNU).

Vitebsky, P. 1984. *Policy Dilemmas for Unirrigated Agriculture in South Eastern Sri Lanka: A Social Anthropologist's Report on Shifting and Semi-Permanent Cultivation in an Area of Moneragala District.* Report prepared jointly for Ministry of Agricultural Development and Research, Sri Lanka and Overseas Development Administration, U.K. Centre of South-Asian Studies. Cambridge: University of Cambridge.

Woost, M.D. 1993. Nationalizing the Local Past in Sri Lanka: Histories of Nation and Development in a Sinhalese Village. *American Ethnologist*, 20, 502–21.

World Bank. 2002. *Sri Lanka Poverty Assessment.* Report No. 22535-CE. Washington, D.C.: The World Bank.

Chapter 18
Researching Forced Migration at the Interface of Theory, Policy and Practice

Cathrine Brun and Ragnhild Lund[1]

Introduction

Forced migrants displaced because of war, disasters and development are commonly considered to be 'out of place', in an abnormal state of being and on the margins of societies. Since 1995, the Research Group on Forced Migration at the Norwegian University of Science and Technology (NTNU) has analysed realities and representations of forced migrants at different times and places. Our research has challenged stereotypical common knowledge concerning forced migrants by critically examining how categorizations are made, and how refugees are labelled in research and in humanitarian policy and practice. In this chapter, we use the work of the research group to analyse the state of the art regarding forced migration research conducted in the disciplines of geography and development studies although with an interdisciplinary orientation. Through a process of deconstructing our findings and knowledge production, we discuss the challenges and perspectives we have faced and how different geographical sub-disciplines may have gained from our work. Three interconnected areas are analysed: the relationship between policy, practice and research; methodologies, with particular focus on action oriented research; and theoretical contributions.

Researching Forced Migration – Border Geographies

Working in the field of forced migration, we often find ourselves in the dilemma described by Landau and Jacobsen (2003) as the 'dual imperative'. The dual imperative implies that research should be both academically sound and policy-relevant. According to Landau and Jacobsen, most forced migration research seeks to explain the behaviour, impact and problems of the displaced with the intention of influencing agencies and governments to develop more effective responses. In geography this dual imperative has been much discussed in relation to critical geographical thought and the relationship between academia and the 'real world'. Castree (2002) describes this as an anxiety that has been experienced

---

1 Equal authorship.

more strongly in geography than in other social sciences, and it may be possible to draw parallels between geography and forced migration studies in this context. Castree (2002: 103) describes this discussion as 'border geography': 'a critical geography that can transcend the real but nonetheless permeable divide, separating academics from myriad non-academic constituencies'. Castree's point is that it is not necessary to worry about the policy relevance of our research. Producing knowledge in itself is enough, and our responsibility can be identifiable more to the academic environment than to the world outside university. In the same vein, Bakewell (2008) states that, by trying to be policy-relevant in forced migration studies, we tend to overlook people who are not covered by the policy definitions. He advocates more policy-irrelevant research that enables us to see beyond the policy categories and thus produce new knowledge about people and processes overlooked by the policy-makers.

While both Castree and Bakewell provide interesting arguments in the discussion of our responsibilities as researchers of forced migration, they do not clarify what constitutes the dual imperative and how problematic the relationship is between being policy-relevant and academically sound. According to Landau and Jacobsen (2003), there is a tendency that, as work becomes more and more academically sophisticated, it becomes ever more irrelevant to practitioners and policy-makers. Based on the experiences from our research, we would contradict this statement. One example is related to how research can potentially theorize how subjects are categorized and researched. 'Refugees', 'internally displaced people' and 'development-induced displacement' are all categories constructed under specific historical circumstances with specific protection aims. In many studies these categories have been taken for granted. In our work we have challenged these categories and the associated theories and methodologies through basic research, the results of which may advance assistance and protection for forced migrants (e.g. Birkeland 2000, Lund 2000a, 2000b, 2002a, 2002b, 2003, 2004, Brun 2003). While researchers should go beyond policies and not take them for granted, we find it crucial to think critically about the way policies work and their impact on the people affected by forced migration (Brun 2008, Blaikie and Lund 2010).

A missing dimension in much of the current discussion on the responsibilities of researchers in forced migration studies and geography is the relationship with our research subjects and/or research participants. We are working with displaced populations, researching their experiences of violence and listening to people's stories of being forced to leave their homes and losing family members, livelihoods and their houses. This is extremely difficult if one does not think one's research can challenge the causes of their displacement and help alleviate suffering. Coming close to our research subjects requires us as researchers to see ourselves as part of the field we are researching (Brun 2009). Herein lies a key dimension of the way we see the dual imperative. On the one hand, our responsibility as researchers of forced migration is to academia, including our students and general knowledge production. On the other hand, we have a responsibility to the people and places affected by forced migration, which we engage in and become part of as researchers (Lund 2012).

Researching forced migration often implies working with people on the margins of societies. They are often located in societies to which they are defined as not belonging. They may be living in the border lands between warring parties and thus in conflicting spaces. Malkki (1997a) discusses how she has been criticized for studying groups on the margins and often under extreme circumstances in refugee camps, because the knowledge she has produced was not 'academically sound' as people who are displaced are not in their 'natural environment'. Similarly, Lund (2002a) was met with scepticism when studying the Veddhas in Sri Lanka, a group that was classified as virtually extinct.

We would argue, however, that researching the margins – as transgressive in a society – may challenge existing knowledge. We agree with Cresswell (1996: 26), who states:

> ... transgression is important because it breaks from 'normality' and causes questioning of that which was previously considered 'natural', 'assumed' and 'taken for granted'. Transgressions appear to be 'against nature'; they disrupt the patterns and processes of normality and offend the subtle myths of consensus.

The concept of transgression is pursued in two distinct ways in this chapter: as a research theme and as a type of research. Regarding research theme, we critically examine the research topics we have chosen to study over the years and how they may signify various situations 'on the margin' or redundancies, i.e. notions of transgressions, and their importance for understanding processes of forced migration. Regarding type of research, we try to identify where we belong, or rather where these kinds of geographies may be placed in our discipline. We describe what has characterized our theoretical and methodological approaches and what kinds of representations we have been able to construct on the basis of our research. Transgression, which literally means going beyond accepted boundaries of taste, convention or the law (*Collins English Dictionary* 2003), may here be defined as working at the interface of various subdisciplines of geography, working in an *ad hoc* manner and by 'crossing borders' theoretically and methodologically.

Working at the Interface of Contemporary Geographies

Most of our studies can be placed at the interface of various subdisciplines of geography, most notably cultural, political and development geography. All of our studies are concerned with discussions of policy. We identified two major continuums.

In the *policy–politics–culture continuum*, emphasis has been put on studying cultural and political dimensions such as identity politics, ethnicity and belonging, the meaning of various histories of displacement, and how such dimensions transform political processes and group conflicts over time and at different places. This continuum particularly addresses issues relevant to cultural and political geographies. It has been important for the research group to contextualize its findings as well as recognize the significance of history in the process of conflict and displacement. A major concern has been to identify how culture produces and constitutes meanings, and how meanings may be deconstructed. We are preoccupied with culture-related concepts across all the subdisciplines of human geography, as we have pursued issues such as identity (Azmi and Lund 2009), making home (Brun and Lund 2008) and the right to return (Tete 2004, 2009, 2011). We have seen it as a major imperative that culture is understood as related to wider transformations, such as human mobility worldwide. Everyday experiences and individual consequences of displacement are linked to wider political processes, and our cultural studies stand in close relation to politically and policy-oriented research. Following Flint (2003), we relate to politics with both a big 'P' and a small 'p', meaning that political geography is concerned with institutional arrangements regarding the state and its foreign relations (big 'P') as well as politics that challenge existing institutions by using non-traditional political venues (small 'p'). We are particularly concerned with the power of structural oppression during war and displacement, how people contest spaces and places, and how displacement leads to reterritorialization and redefinition of place. Continuous research by Brun (2003, 2008) on the northern Muslims in Sri Lanka has contributed to understanding how people change their connections with different places during different stages of war and in the peace process. Lund (2000b) has studied the indigenous population, the Veddhas, in Sri Lanka and their territorial expulsion and cultural marginalization from British colonial times up to the recent war situation.

The *policy–development continuum* relates to how social and economic transformations are articulated during and after crises, how displaced people strategize, how power is articulated and how displacement takes place over time. The inclusion of conflict in development studies coincides with studies of the political economy of conflict in addition to the emerging field of humanitarianism. Policy is at the core of our work in the context of development studies and development geography.

Development geography is a broad field that encompasses cultural and political geographies. Although there is no single definition of development geography, common to all directions is that they critically examine the relationship between the state, the market and civil society. Inspired by alternative development approaches, which emphasize the role of agency and civil society, our research group has focused on refugees as active participants in situations of deprivation, displacement and eviction. Inspired by poststructural and postcolonial approaches, a major focus has been to identify power relations, how power–geometries work and are articulated at various levels and regions in and between the Global North and South, and how this focus has produced multiple representations of the displaced.

Methodologies

A common denominator of our work as geographers has been 'the field', the arena where knowledge-making takes place (Dewsbury and Naylor 2002, Lorimer 2003, Crang and Cook 2007), and fieldwork is a shared practice for most projects in our group. Methodologically our work is qualitative and interpretative. Qualitative methods range from various types of interviews, text analysis, visual methods and participant observation to action research. The main body of our work is concerned with the human consequences of forced migration, particularly the experiences and strategies of forced migrants. Ethnographic fieldwork helps us understand how people go about their day-to-day life. Its actor-oriented approach has proved crucial for understanding, but also challenging, the notion of forced migrants as capable actors. Some of the master's theses by students belonging to our research group have used individual narratives and life histories (Negera 2002, Tete 2009, Heer 2005). Some of the students come from war-affected areas and therefore could make use of their local knowledge and local language. Negera (2002), for example, based his analysis of Oromo refugees purely on individual narratives, which provided extremely strong testimony of the atrocity of persecution during war. Tete (2004, 2009, 2011) used individual life history interviews of Liberian refugees in Ghana and displaced people in Sri Lanka to explore situations of vulnerability and insecurity in refugee settlements and camps.

However, ethnographic standards may not always be possible to achieve. One issue is the time needed to understand people's experiences and strategies. In some conflict situations, curfews may restrict a researcher's possibilities to move around and stay with people in camps. For example, Birkeland (2000) experienced how general academic standards regarding methodology may not be possible to follow in situations of war and violence, when there may be insecurity, restricted access to people and territories, and the impossibility of raising certain questions due to sensitivity and the risk of putting people in danger. She concludes that more important than ensuring the quality of one's data is to follow principles of 'do no harm' in situations of conflict. Hence, ethical concerns become more important than scientific ones. This accords with David Smith's (1997) quest for a moral turn in geography, emphasizing normative issues and the ethics of studying the *other*, postcolonialism, disabilities etc. Debates on professional ethics have focused on how geographers work in the field: what the appropriate methods are, how one should behave, what is considered good conduct, how interviews and texts should be analysed, and how research results should be disseminated. Our self-reflective and critical evaluations determine our success or failure as researchers. Lund (2002a) argues that the typical fieldwork situation concerns how the contrast between informant and fieldworker as representatives of different cultures becomes substituted by differences in positioning and power. As an outsider and a researcher, one may be in a superior position in terms of status and able to choose among different types of information. However, the informant controls what he or she wants to convey and the researcher receives different information from one informant to another.

In developing knowledge, many of us have tried to orientate ourselves in a complex global discourse rather than trying to identify what is local and then derive knowledge about a place, a people, a region or nation (Lund 2002b). Forced migration, especially, is a phenomenon tied to the global political and economic scene. Long (2001: 230–31) says that new migratory research concerns:

> ... the nature and development of particular transnational networks of people and places. This demands an understanding of the interlocking of 'localised', 'transnational', 'nomadic' and 'hybrid' experiences and also how these constituent elements transmute into new 'globalised' cultural identifications associated with 'migrants on the move' ... The other face of contemporary population movements is the displacement of people ...

Geographical representations are subject to different interpretations depending on whether one follows the mapping approach or the text approach. What is global and local and what is spatial and nonspatial become subject to different interpretations and theoretical definitions.

While the majority of work conducted in the research group has followed qualitative interpretative methodologies, an emerging trend as we have come closer to our research subjects and more integrated in the field is to engage more actively in action research as an approach and methodology.

Constructing Knowledge about Forced Migration and Geography

Some closely interconnected discourses characterize several of the projects conducted in the research group. These may be summarized in three crosscutting themes of research:

1. The relations between agency and place and how research has contributed to both essentialize and deessentialize either;
2. The relations between power, boundaries, identities and flows, and how these may inform about processes of displacement and coping;
3. How relations between representations, categories and labelling influence the knowledge produced.

Essentializing Place or Agency?

The relationship between people and place is a core issue in studies of forced migration as in the discipline of geography (Massey 1993, 2005, Stepputat 1994, 1999, Kibreab 1999, Warner 1999, Brun 2001, Turton 2005). However, there seems to be an increasing distance between the researchers' construction of knowledge and the policies applied to deal with people who have fled from their places of origin to settle elsewhere. Questions relating to where forced migrants belong, how they are treated at the place of refuge and when their displacement should end are closely connected to the fundamental question of the relationship between people and place.

Several authors have shown how refugees have been represented and viewed with a basis in an essentialist notion of place (Malkki 1995, Sørensen 1996, Brun 2001). To be granted the status and rights of a refugee, one has to cross an internationally recognized state border. From an essentialist viewpoint, refugees challenge the 'national order of things' (Malkki 1995). Only when refugees either return to their origin or 'naturalize' in the host society is order reestablished. However, this is no longer a prevalent view in academic studies. Based on poststructuralist and postcolonial ideas, and influenced by the cultural and spatial turns, it is argued that states of movement, homelessness and displacement are 'normal' elements, not transgressive ones, because people are more mobile than ever and understandings of nativeness are thus difficult to maintain. It has been concluded that even though people are forced to flee, they are not torn loose from their culture, they do not lose their identity and they do not become powerless (Allen and Turton 1996, Gupta and Ferguson 1997, Malkki 1997b).

Our research group has reacted to these polarized views on the state of refugees by analysing the changing power, status and meanings of places for people on the move. Displacement changes people's relationship with specific places and creates new practices to deal with these changes. For example, we have found that people develop strong attachments both to their place of origin and place of refuge (Syvertsen 2004, Tete 2004, 2009, 2011, Aase 2005). However, while research shows that translocal strategies have become common among forced migrants, policies are still very much oriented towards a linear state–place relationship. For example, so-called 'durable solutions' expect that one either integrates, resettles or returns, while the reality for many forced migrants is that they develop translocal strategies where connections are nurtured with both the places they fled from and the places where they live as displaced, in order to make the most out of the situation in terms of livelihoods and for safety in case the conflict that caused their displacement resumes (Brun 2008). Researchers no longer regard displaced people as passive victims in an abnormal state of being, but as active agents who are able to develop strategies to continue functioning socially (Birkeland 2003, Hoem 2008, Refstie 2008, Serwaija 2008, Oltedal 2009). However, members of the research group have also cautioned against a tendency to essentialize agency (Shanmugaratnam et al. 2003). Although the majority of the research projects have relied on actor-oriented perspectives, the group has pointed to marginalization as an effect of displacement and passiveness as a form of survival (Lund 2002b). Hence, because forced migrants are both victims and actors of change, we need to examine, problematize and understand the limits of agency. Cresswell's (1996: 23) critical discussion of the role of intentionality and resistance is useful for rethinking the role of agency in research on forced migration:

Transgression, in distinction to resistance, does not, by definition, rest on the intention of actors but on results – on the 'being noticed' of a particular action. The question of intentionality remains an open one ... To have transgressed in this project means to have been judged to have crossed some line that was not meant to have been crossed. The crossing of the line may or may not have been intended. Transgression is judged by those who react to it, while resistance rests on the intentions of the actors(s).

Boundaries, Identities and Flows

There is no natural relationship between people and places, but rather people and places must be understood through people's relationships to institutions at multiple locations. Additionally, the process of migration, whether forced or voluntary, creates, strengthens and even breaks down boundaries between different groups of people, and social categories are being constructed, negotiated and reconstructed.

Much of our research has considered various dimensions of boundaries, identities and flows in an attempt to analyse processes of forced migration. We have pursued a grounded approach to the wider processes of globalization and change (Brun 2005). Regarding globalization as a contingent and constructed process formed out of the specificities of the people and places involved in shaping globalization processes, a grounded approach starts from the lives of a variety of people with diverse relationship to globalization (Nagar et al. 2002, Rigg 2007). Through a grounded approach we have attempted to explore the range of social locations (gender, class, ethnicity, race and sexuality) and the multiple ways in which globalization is lived, created and articulated.

The majority of our research has considered situations of civil war where cleavages between people are revitalized, with fatal consequences. In the Balkans, the whole set of thoughts around reconstruction, recovery and return has been challenged because people cannot move back to where they lived before due to ethnic tensions and the creation of new boundaries. However, people identify themselves as belonging to their previous home place (Skotte 2004, Heer 2005). This is also the case in Sri Lanka, where both the war and the tsunami have prevented return for various political and 'developmental' reasons (Blaikie and Lund 2010).

Other research in the group has shown how people negotiate and challenge boundaries and borders. Sagmo (2004) analyses how Burmese students in exile use networks as a tool in the fight for political change in Burma, including underground networks within Burma, networks crossing the Thai–Burmese border and a strong international network linked via the Internet. Fossan (2005) shows how Karen refugees on the Thai–Burmese border negotiate their immobility in the refugee camps by challenging the boundaries defined by the Thai refugee administration regime.

Multiple strategies involving networks with a number of different places throughout the world have become important for many displaced people. For some, networks can be used to escape the situation in the affected country and try to obtain refugee status abroad. To send remittances from relatives abroad is another global connection that has provided a lifeline for many internally displaced people, indicating the strong relationship between internal displacement and international migration (Van Hear 2002, Brun 2005, Azmi and Lund 2009). However, such strategies not only enable people to pursue their livelihoods but also may sustain tension and conflicts. Relationships between refugees (those who have crossed an international border) and internally displaced persons (IDPs) show the reality of the power–geometries of time–space compression (Massey 1993). International migrations are embedded in broader social, economic and political processes (Sassen 1993, Giles and Hyndman 2004). The increasingly strict migration regimes of states and regions have made the costs of international migration unattainable for many of the poorer civilians affected by civil war. These politics of mobility have an impact on gender, class and ethnicity, and are closely related to citizenship (Brun 2003, Giles and Hyndman 2004).

Not only people but also displacement and flows of money, skills and materials move across boundaries. In geography, Hyndman (2000) challenges the political and cultural assumptions of current humanitarian practices and exposes the distancing strategies that characterize present operations. She provides an insightful analysis of humanitarianism on the ground in examining the policies and practices of its organization at various levels. From our group, Khasalamwa (2009, Khasalamwa-Mwandha 2012) has studied the policies and practices of nongovernmental organizations (NGOs) in Sri Lanka and Uganda and shows how similar interventions may not bring about similar outcomes in different places. This demonstrates the importance of taking local practices and contexts into consideration when formulating policies and practices for humanitarian actions (Lund et al. 2011).

Representations, Categories and Labelling

The field of refugee studies developed from a label: the 'refugee'. This is a politically constructed category formed to deal with the refugee crisis in Europe after World War II (Zetter 1988, Hein 1993). Considerable criticism has been levelled towards studies that have uncritically taken up these categories formulated by policy organizations and used them as a starting point for research (Malkki 1995, Chimni 2009). Since its inception in 1995, our group has taken part in challenging these narrow policy categories and today we can clearly identify a broader field of forced migration studies examining a much wider repertoire of forced migrants.

Our starting point was to challenge the refugee category and study groups of forced migrants falling outside it, primarily those displaced within the borders of their own country. Although this is no longer the main perspective, we started with studying situations of internal displacement caused by environment, development projects and civil war. Often two or more causes acted together in causing displacement. For example, in Angola, Birkeland (2000) has studied how – in a country where it is believed that 4.1 million people were displaced because of war – many people had clearly learnt how to live with the war, but when environmental degradation in the same area made access to food difficult people had no choice but to move. Later, much of our group's work has concentrated on how various causes of displacement (e.g. conflict, disaster and development) interplay in the same area and for the same populations, making the situation more complex for the displaced and for the regimes dealing with displacement. Lund (2000b) found in her research on the Veddhas that there is a close relationship between development projects and the Sri Lanka government's strategy of expanding and controlling territories to make a buffer zone between the Tamil-dominated north and east and the Sinhalese-dominated south at the time of the civil war on the island.

Both Lund's and Birkeland's examples show that causes are complex and that there is a case for understanding the relationship between root causes and triggers of displacement. An important lesson from these studies is that placing people in a single category may obscure understanding of the causes of displacement and thus risk the formulation of irrelevant strategies for ending displacement.

Our projects have studied how categories emerge, how the meaning of categories change, and what the implications of these categories are for the people identified with them (Fadnes 2008, Refstie 2008). One issue is the way constructed categories such as 'internally displaced people' and 'host communities' become social categories in a society. Brun's (2008) research on local integration processes in protracted situations of internal displacement shows how the categories of 'IDPs' and 'hosts' were constructed first by the Muslim discourse of hospitality and later by the humanitarian discourse on the necessity of labelling people 'internally displaced persons'. During protracted displacement, these categories became integral to people's identities, claims for the rights to territories and resistance to integration. Nonetheless, the way whereby labels are formed, and guidelines and laws formulated on their basis, do not necessarily ensure protection of and assistance to the people displaced (Fadnes 2008, Refstie 2008, Tete 2009).

Contingent representations may exist and can be illustrated by a quote from Lund's (2003: 101) conclusion concerning different representations of forced migration related to the Veddhas:

> First, representations of forced migration in science were found to be hegemonic and should be subject to reconsideration and redefinition. Secondly, representations of forced migration may be embodied and gender biased and thus not equally relevant to both genders. Third, representations of internal displacement are often short-term and a-historic and therefore make it difficult to analyse the complex root causes of migration. Fourth, the displacement of indigenous populations shows how forced migration is an ongoing social process ... Finally, representations of forced migration indicate lack of official recognition of ethnic minorities and displaced.

From Representations to Practice – How Knowledge Travels

While our work up to 2004 was mainly concerned with collecting data and writing about displacement, the focus changed after this as the research matured, alongside institutional changes and changing realities in the cases we were studying. Working for ten years with forced migration, we had accumulated knowledge, established contacts with practitioners and interest organizations, and secured positions in the university to continue our work. Close relations in the field with people categorized as forced migrants as well as with organizations and institutions working with forced migration made us part of the field and responsible to it. This was particularly the case in Sri Lanka, where major changes were taking place. The peace process had led to plans for massive repatriation of IDPs to their home places in the north, and the World Bank was involved in planning housing schemes for the returnees. The plans demonstrated that there was limited acknowledgement of previous experiences of housing schemes for IDPs before and during the war. The plans were top-down with limited participation by the returnees and civil society organizations.

There is an increasing trend of action research and participatory research in geography (Kindon et al. 2007) and refugee studies (Doná 2007, Mackenzie et al. 2007). Against the background of substantial knowledge on displacement in Sri Lanka, we aimed for more action-oriented approaches to assisting in recovery after war. We started a collaboration programme with a Norwegian NGO to examine how the housing plans could be made more participatory and more grounded in past experiences. When the collaboration was just about to start, however, the Indian Ocean tsunami of 2004 devastated the coastline of the country and all attention on housing was directed towards recovery in the tsunami-affected areas. This again reshaped our research agenda. We started to examine the relationship between reconstruction and recovery with respect to housing, identity and homemaking (Brun and Lund 2008, Azmi and Lund 2009). We explored previous resettlement and housing policies with respect to lessons learnt and 'building back better' (Brun and Lund 2009, Khasalamwa 2009). We also started exploring intervention issues such as gender, livelihoods (FONT 2008) and how knowledge travels in humanitarian organizations, especially through action research (Brun and Lund 2010).

We were granted funding by the Research Council of Norway for a project examining the relationship between practitioner knowledge and academic knowledge. Much of our work became directed towards the way knowledge travels between various actors in the humanitarian field, including us as researchers, how we can intervene to influence the way knowledge travels, and how humanitarian work can be strengthened by engaging with the different sets of knowledges available to different actors in the field.

In the action research, the main aim was capacity building. We organized various activities: workshops focusing on gender, conflict mediation and livelihood development; developing the NGO's livelihoods approach through strategic action planning (SAP); offering visits to and collaboration with other NGOs and community-based organizations (e.g. the Self-Employed Women's Association (SEWA) of India); and offering a Diploma in Humanitarianism. We also conducted research on several tsunami recovery projects in order to provide critical feedback on performance and make recommendations. This collaboration was a unique opportunity to collect information repeatedly over very short time intervals in close partnership with development agencies, allowing us to understand their production of knowledge and practices, and to identify changes over time in the communities concerned (FONT 2008).

We also developed 'real time research' (RTR), defined as a process of data collection, analysis and dissemination of knowledge that can inform specific humanitarian interventions and policy formulation as they occur over time in a particular location. It is a way of exploring the interfaces of knowledge production between research and practice, as well as the capabilities of the humanitarian actor under investigation. Hence, RTR is an aid to monitor a project's progress based on livelihoods indicators rather than an evaluation or solely an academic exercise. Methodologically, such an approach has some limitations. For researchers it creates ethical dilemmas related to sampling and choice of sites, possible NGO bias, differential access to actors and limited contact hours with local stakeholders (FONT 2008). However, action research methodology has been taken one step further in other projects, whereby the researcher has worked more closely with the affected IDPs and consequently succeeded in mobilizing for change (Refstie 2008, Refstie and Brun 2012).

The development of our RTR methodology has been a long and elaborate process of interaction and joint decision-making between an international NGO that we collaborated with, its district staff, local stakeholders (i.e. people affected locally and community-based organizations) and researchers. The RTR methodology has been used to: establish monitoring routines for the organization; understand the changing contexts of post-tsunami projects; understand the relevance and impacts of the organization's projects; understand community responses; and hold discussions on what is being done within the organization as well as among stakeholders. RTR has also helped us understand and analyse how knowledge travels and how organizations work with knowledge transfer today (Brun and Lund 2010, Attanapola et al. 2013).

We have found that knowledge is generally perceived to be transferred as a linear process 'whereby untransformed knowledge acts as a technical solution to a given development "problem"' (McFarlane 2006: 288). Alternative discourses, such as feminist, postcolonial and critical geography, examine how knowledge is always situated (Haraway 1988, Rose 1997), is partial, multiple and relational, and consequently also embedded in power relations. Knowledge that 'sits' in organizations is always coproduced in hybrid entities (Attanapola et al. 2011). We have found that ideas do not necessarily travel smoothly from one level to another within an organization. The way international NGOs are structured means that there is neither acknowledgement of such knowledge transfer and creation nor space for it to take place, causing increasing distance between local and theoretical knowledge (FONT 2008).

Conclusions

Where does our knowledge production situate the NTNU Research Group on Forced Migration, and what does it say about current research in geography, forced migration and development studies? First, the group has contributed to the rapidly expanding field of forced migration studies. Our work at the interface of contemporary geographies has also contributed to theoretical advances within our respective disciplines. Theoretical and methodological processes are subject to continuous negotiation and renewal of discourse, and are cumulative.

By transgressing fields, we realize the need to address and readdress conventional disciplines and identify how they can change. Such positioning has sensitized us to our different capabilities. It addresses the significance of collaborating with practitioners to develop knowledge through mutually responsible partnerships; and collaborating with researchers from contingent fields to develop and assess how knowledge is constructed and travels among partners and fields. We have learnt how knowledge is always situated, partial and relational, and is embedded in power relations among various stakeholders and partners.

For the group, working in teams and valuing the contribution of junior researchers are important for gathering experiences from forced migration crises throughout the world. Thus, our knowledge is not only derived from isolated case studies, but forms parts of a larger whole. The continuous focus on action-oriented perspectives has meant that the various research projects have kept people in focus. Our work demonstrates how research on forced migration must be grounded in local realities and practices, and in ways that reveal how policy processes are embedded in the analyses. This focus coincides with an emerging trend in geography and development studies to include policy processes, while forced migration studies increasingly link refugee and displacement issues to discourses of long-term development and social change.

Developing our own methodology and entering into the field of action research meant that to begin with most of our work was practical. This has given us increased insight into how policies are implemented and how academics can make a contribution together with the people affected by disasters and forced migration. Such a shift towards 'new' participatory and empirically grounded research may provide additional inputs into various subdisciplines of geography, such as development geography, political geography and cultural geography, as well as forced migration studies in general. More specifically, we aim through the research to continue to unravel marginalization and voice change.

References

Aase, T.F. 2005. *Marginalization of an Ethnic Minority in Northern Thailand.* Master's thesis. Trondheim: Department of Geography, NTNU.

Allen, T. and Turton, D. 1996. *In Search of Cool Ground.* Oxford: James Currey.

Attanapola, C., Brun, C. and Lund, R. 2011. Working Gender After Crisis: Partnerships and Disconnections in Sri Lanka After the Indian Ocean Tsunami. *Gender, Place and Culture*, 20, 70–86.

Azmi, F. and Lund, R. 2009. Shifting Geographies of House and Home – Female Migrants Making Home in Rural Sri Lanka. *Journal of Geographical Science*, 57, 33–54.

Bakewell, O. 2008. Research Beyond the Categories: The Importance of Policy Irrelevant Research into Forced Migration. *Journal of Refugee Studies*, 21, 432–53.

Birkeland, N. 2000. Forced Migration and *Deslocados* in the Huambo Province, Angola. *Norsk Geografisk Tidsskrift–Norwegian Journal of Geography*, 54, 110–15.

Birkeland, N.M. 2003. Peace in Angola: IDPs on their Way Home?, in *Researching Internal Displacement: State of the Art: Conference Proceedings, Trondheim, Norway, 7–8 February 2003*, edited by C. Brun and N.M. Birkeland. Acta Geographica–Trondheim Series A, No. 6. Trondheim: Department of Geography, NTNU, 329–36.

Blaikie, P. and Lund, R. (eds) 2010. *The Tsunami of 2004 in Sri Lanka: Impacts and Policy in the Shadow of Civil War*. Abingdon and New York: Routledge.

Brun, C. 2001. Reterritorializing the Relationship Between People and Place in Refugee Studies. *Geografiska Annaler*, 83B, 15–25.

Brun, C. 2003. Local Citizens or Internally Displaced Persons? Dilemmas of Long Term Displacement in Sri Lanka. *Journal of Refugee Studies*, 16, 376–97.

Brun, C. 2005. Women in the Local/Global Fields of War and Displacement. *Gender, Development and Technology*, 9, 57–80.

Brun, C. 2008. *Finding a Place: Local Integration and Protracted Displacement in Sri Lanka*. Colombo: Social Scientists' Association.

Brun, C. 2009. A Geographers' Imperative? Research and Action in the Aftermath of Disaster. *Geographical Journal*, 175, 196–207.

Brun, C and Lund, R. 2008. Making a Home During Crisis: Post-Tsunami Recovery in the Context of War in Sri Lanka. *Singapore Journal of Tropical Geography*, 29, 274–87.

Brun, C. and Lund, R. 2009. 'Unpacking' the Narrative of a National Housing Policy in Sri Lanka. *Norsk Geografisk Tidsskrift–Norwegian Journal of Geography*, 63, 10–22.

Brun, C. and Lund, R. 2010. Real Time Research: Decolonising Practices – or Just Another Spectacle of Researcher-Practitioner Collaboration? *Development in Practice*, 20, 812–26.

Castree, N. 2002. Border Geography. *Area*, 34, 103–12.

Chimni, B.S. 2009. The Birth of a 'Discipline': From Refugee to Forced Migration Studies. *Journal of Refugee Studies*, 22, 11–29.

Collins English Dictionary. 2003. Transgression. Glasgow: HarperCollins, 1710.

Crang, M. and Cook, I. 2007. *Doing Ethnographies*. London: Sage.

Cresswell, T. 1996. *In Place/Out of Place: Geography, Ideology, and Transgression*. Minneapolis, Mn: University of Minnesota Press.

Dewsbury, J.D. and Naylor, S. 2002. Practicing Geographical Knowledge: Fields, Bodies and Dissemination. *Area*, 34, 253–60.

Doná, G. 2007. The Microphysics of Participation in Refugee Research. *Journal of Refugee Studies*, 20, 210–29.

Fadnes, E. 2008. *Between Promise and Fufilment: An Assessment of the Discrepancy Between the Colombian Legal Framework on Internal Displacement and the Realization of Rights*. MPhil thesis. Trondheim: Department of Geography, NTNU.

Flint, C. 2003. Dying for a 'P'? Some Questions Facing Contemporary Political Geography. *Political Geography*, 22, 617–20.

FONT 2008. *The FONT Programme: Building Capacity for Reconstruction and Recovery: A Review of Activities 2005–2008*. Trondheim: NTNU/FORUT.

Fossan, K. 2005. *Facing Immobility, Practicing Mobility: Karen Refugees on the Thai Border*. Master's thesis. Trondheim: Department of Geography, NTNU.

Giles, W. and Hyndman, J. 2004. Introduction: Gender and Conflict in a Global Context, in *Sites of Violence: Gender and Conflict Zones*, edited by W. Giles and J, Hyndman. Berkeley: University of California Press, 3–23.

Gupta, A. and Ferguson, J. (eds) 1997. *Culture, Power and Place: Explorations in Critical Anthropology*. Durham: Duke University Press.

Haraway, D. 1988. Situated Knowledges: The Science Question in Feminism and the Privilege of Partial Perspective. *Feminist Studies*, 14, 575–99.

Heer, J. 2005. *The Serbs of Vukovar and their Experience of Place*. Master's thesis. Trondheim: Department of Geography, NTNU: Trondheim.

Hein, J. 1993. Refugees, Immigrants and the State. *Annual Review of Sociology*, 19, 43–59.

Hoem, I.B. 2008. *Women and Children at Risk: A Case Study of Urban IDPs from War Affected Northern Uganda in Kampal*a. MPhil thesis. Trondheim: Department of Geography, NTNU.

Hyndman, J. 2000. *Managing Displacement: Refugees and the Politics of Humanitarianism*. Minneapolis, MN and London: University of Minnesota Press.

Khasalamwa, S. 2009. Is 'Build Back Better' a Response to Vulnerability? Analysis of the Post-Tsunami Humanitarian Interventions in Sri Lanka. *Norsk Geografisk Tidsskrift–Norwegian Journal of Geography*, 63, 73–88.

Khasalamwa-Mwandha, S. 2012. *Spaces of Recovery: An Exploration of the Complexities of Post War/Disaster Recovery in Uganda and Sri Lanka*. Doctoral theses at NTNU, 2012: 356. Trondheim: Department of Geography, NTNU.

Kibreab, G. 1999. Revisiting the Debate on People, Place, Identity and Displacement. *Journal of Refugee Studies*, 12, 384–410.

Kindon, S., Pain, R. and Kesby, M. (eds) 2007. *Participatory Action Research Approaches and Methods: Connecting People, Participation and Place*. London: Routledge.

Landau, L. and Jacobsen, K. 2003. The Dual Imperative in Refugee Research: Some Methodological and Ethical Considerations in Social Science Research on Forced Migration. *Disasters*, 27, 185–206.

Long, N. 2001. *Development Sociology: Actor Perspectives*. Abingdon: Routledge.

Lorimer, H. 2003. Telling Small Stories: Spaces of Knowledge and the Practice of Geography. *Transactions of the Institute of British Geographers NS*, 28, 197–217.

Lund, R. 2000a. Geographical Research on Forced Migration and Development. *Norsk Geografisk Tidsskrift–Norwegian Journal of Geography*, 54, 89.

Lund, R. 2000b. Geographies of Eviction, Expulsion and Marginalization: Stories and Coping Capacities of the Veddhas, Sri Lanka. *Norsk Geografisk Tidsskrift– Norwegian Journal of Geography*, 54, 102–9.

Lund, R. 2002a. Methodological Choices and Dilemmas in Contested Spaces: Researching Minorities and the Displaced, in *Geographical Methods – Power and Morality in Geography: Proceedings of the Annual Conference of the Norwegian Geographical Society, Trondheim, Norway, April 5th–6th 2002*, edited by G. Setten and S. Rudsar. Acta Geographica–Trondheim Series A, No. 2. Trondheim: Department of Geography, NTNU, 3–23.

Lund, R. 2002b. Ethics and Fieldwork as Intervention, in *Geographical Methods – Power and Morality in Geography: Proceedings of the Annual Conference of the Norwegian Geographical Society, Trondheim, Norway, April 5th–6th 2002*, edited by G. Setten and S. Rudsar. Acta Geographica–Trondheim Series A, No. 2. Trondheim: Department of Geography, NTNU, 214–19.

Lund, R. 2003. Representations of Forced Migration in Conflicting Spaces: Displacement of the Veddas in Sri Lanka, in *In the Maze of Displacement: Conflict, Migration and Change*, edited by N. Shanmugaratnam, R. Lund and K.A. Stølen. Kristiansand: Høyskoleforlaget – Norwegian Academic Press, 76–104.

Lund, R. 2004. *Forced Migration, Resource Conflicts and Development: Final Report: Globalisation and Marginalisation, Multi- and Interdisciplinary Research on Development Paths in the South*. Oslo: Research Council of Norway.

Lund R. 2012. Researching Crisis – Recognizing the Unsettling Experience of Emotions. *Emotions, Space and Society*, 5, 94–102.

Lund, R., Khasalamwa, S. and Tete, S.Y.A. 2011. Beyond the Knowledge–Action Gap: Challenges of Implementing Humanitarian Policies in Ghana and Uganda. *Norsk Geografisk Tidsskrift–Norwegian Journal of Geography*, 65, 63–74.

McFarlane, C. 2006. Knowledge, Learning and Development: A Post-Rationalist Approach. *Progress in Development Studies*, 6, 287–305.

Mackenzie, C., McDowell, C. and Pittaway, E. 2007. Beyond 'Do No Harm': The Challenge of Constructing Ethical Relationships in Refugee Research. *Journal of Refugee Studies*, 20, 299–319.

Malkki, L. 1995. Refugee and Exile: From 'Refugee Studies' to the National Order of Things. *Annual Review of Anthropology*, 24, 495–523.

Malkki, L. 1997a. News and Culture: Transitory Phenomena and the Fieldwork Tradition, in *Anthropological Locations*, edited by A. Gupta and J. Ferguson. Berkeley: University of California Press, 86–101.

Malkki, L. 1997b. National Geographic: The Rooting of Peoples and the Territorialization of National Identity Among Scholars and Refugees, in *Culture, Power, Place*, edited by A. Gupta and J. Ferguson. Durham, NC: Duke University Press, 52–74.

Massey, D. 1993. Power Geometry and a Progressive Sense of Place, in *Mapping the Futures*, edited by J. Bird, B. Curtis, T. Putnam, G. Robertson and L. Tickner. London: Routledge, 59–69.

Massey, D. 2005. *For Space*. London: Sage.

Nagar, R., Lawson, V., McDowell, L. and Hanson, S. 2002. Locating Globalization: Feminist (Re)readings of the Subjects and Spaces of Globalization. *Economic Geography*, 78, 257–84.

Negera, K. 2002. *Identity and Forced Migration: Causes of Migration and Challenges: Case of the Oromo Settlers in Norway*. MPhil thesis. Trondheim: Department of Geography, NTNU.

Oltedal, G. 200. *'Her kan vi leve': En studie av bosettings- og integrasjonsprosesser blant flyktninger i Time kommune*. Master's thesis. Trondheim: Department of Geography, NTNU.

Refstie, H. 2008. *IDPs Redefined – Participatory Action Research with Urban IDPs in Uganda*. MPhil thesis. Trondheim: Department of Geography, NTNU.

Refstie, H. and Brun, C. 2012. Towards Transformative Participation: Collaborative Research with 'Urban IDPs' in Uganda. *Journal of Refugee Studies*, 25, 239–56.

Rigg, J. 2007. *An Everyday Geography of the Global South*. London: Routledge.

Rose, G. 1997. Situating Knowledges: Positionality, Reflexivities and Other Tactics. *Progress in Human Geography*, 21, 305–20.

Sagmo, T.H. 2004. *'The New Light of Burma' – ABFSU's Fight for Democracy in Burma*. Master's Thesis in Geography. Trondheim: Department of Geography, NTNU.

Sassen, S. 1993. Rethinking Immigration. *Lusitania: A Journal of Reflection and Oceanography* 5, 97–102.

Serwaija, E. 2008. *A Childhood Lost? A Case of Gulu Support the Children Organisation in Northern Uganda*. MPhil thesis. Trondheim: Department of Geography, NTNU.

Shanmugaratnam, N., Lund, R., Stølen, K.A. 2003. Introduction: Conflict, Migration and Change, in *In the Maze of Displacement: Conflict, Migration and Change*, edited by N. Shanmugaratnam, R. Lund and K.A. Stølen. Kristiansand: Høyskoleforlaget – Norwegian Academic Press, 8–23.

Skotte, H. 2004. *Tents as Concrete: What Internationally Funded Housing Does to Support Recovery in Areas Affected by War: The Case of Bosnia-Herzegovina.* Doctoral theses at NTNU, 2004: 61. Trondheim: Department of Urban Design and Planning, NTNU.

Smith, D. 1997. Geography as Ethics: A Moral Turn? *Progress in Human Geography*, 21, 583–90.

Sørensen, B.R. 1996. *Relocated Lives: Displacement and Resettlement within the Mahaweli Project, Sri Lanka.* Amsterdam: VU University Press.

Stepputat, F. 1994. Repatriation and the Politics of Space: The Case of the Mayan Diaspora and Return Movement. *Journal of Refugee Studies*, 7, 175–85.

Stepputat, F. 1999. Dead horses? *Journal of Refugee Studies*, 12, 416–19.

Syvertsen, I.K. 2004. *To Swing or Not to Swing, That is the Question: Akha Identity in a Changing Society: A Case Study of Three Akha Villages in Northern Thailand.* MPhil thesis. Trondheim: Department of Geography, NTNU.

Tete, S. 2004. *Narrative of Hope? Displacement Narratives of Liberian Refugee Women and Children in the Gomoa-Budumburam Refugee Camp in Ghana.* MPhil thesis. Trondheim: Department of Geography, NTNU.

Tete, S.Y.A. 2009. Whose Solution? Policy Imperatives Vis-À-Vis Internally Displaced Persons' Perceptions of Solutions to their Situation in Sri Lanka. *Norsk Geografisk Tidsskrift–Norwegian Journal of Geography*, 63, 46–60.

Tete, S.Y.A. 2011. *Protracted Displacement and Solutions to Displacement: Listening to Displaced Persons (Refugees and IDPs) in Ghana and Sri Lanka.* Doctoral theses at NTNU, 2011:287. Trondheim: Department of Geography, NTNU.

Turton, D. 2005. The Meaning of Place in a World of Movement: Lessons from Long-Term Field Research in Southern Ethiopia. *Journal of Refugee Studies*, 18, 258–80.

Van Hear, N. 2002. Sustaining Societies Under Strain: Remittances as a Form of Transnational Exchange in Sri Lanka and Ghana, in *New Approaches to Migration: Transnational Communities and the Transformation of Home*, edited by K. Koser and N. Al-Ali. London and New York: Routledge, 202–23.

Warner, D. 1999. Deterritorialization and the Meaning of Space: A Reply to Gaim Kibreab. *Journal of Refugee Studies*, 12, 411–16.

Zetter, R. 1988. Refugees and Refugee Studies – A Label and an Agenda. *Journal of Refugee Studies*, 1, 1–6.

PART V
Conclusion

Chapter 19

Researching Alternative Development: An Autobiographical Discussion with Ragnhild Lund

Michael Jones

Introduction: Approaches to Histories of Knowledge

During the course of the twentieth century, scholarship became increasingly reflexive. Beginning with an increasing preoccupation with the general history of scientific endeavour during the first half of the century, there developed a growing interest in the history of individual disciplines during the second half of the century. From disciplines writing accounts of their own history, there emerged during the last two decades of the century an increasing emphasis on the role and influence of the individual researcher on what is investigated at any particular time. Hence two broad approaches to histories of knowledge can be discerned. The first consists of histories of ideas and the social context in which these ideas were produced. The second consists of personal histories of practitioners in different fields in the form of biographies and autobiographies and illustrating how personal life stories both reflect and affect the development of knowledge within and beyond particular disciplines. Although general historical accounts and biographical or autobiographical accounts are often considered as two distinct genres, they are complementary. Every historical narrative is influenced by the particular perspective of its author, while every personal narrative is influenced by the time in which it is written. Broad historical accounts provide a context for individual thoughts and actions, while personal accounts provide the building blocks for the development of general ideas and practices.

The present chapter adopts the latter approach. It presents an account of the Norwegian geographer Ragnhild Lund's life and career in the form of a narrative based on interviews conducted with her by the present author. As both researcher and practitioner, Ragnhild Lund has been among the pioneers of feminist approaches in development studies and has been in the forefront of arguments concerning alternative development, providing alternative understandings of the geography of development in opposition to mainstream development theory.

The chapter starts by introducing the autobiographical approaches that developed in part as a response to the feminist critique of conventional accounts

Figure 19.1   Ragnhild Lund
Photo courtesy of Department of Geography, NTNU

of the history of science, followed by a short presentation of the place of autobiography in geography and development studies. The interview with Ragnhild Lund is preceded by a positioning of the relationship between the interviewer and the interviewed. The autobiographical discussion then takes the form of long extracts from the interview, in which Ragnhild tells of her career as a geographer and her approach to researching alternative development. Finally there comes a presentation of Ragnhild's published work and how this is reflected in the chapters of the present book.

Autobiographical Approaches and Feminism

Conventional accounts of scientific history and practice have in particular been challenged by feminist ideas, which criticize such accounts for often being presented as 'objective and disembodied' (Aitken and Valentine 2006: 2). The Canadian sociologist Dorothy E. Smith (1988, 1990, 2005) provides a critique of the presentation of knowledge as abstract, objective and universal, and examines knowledge formation in relation to her own lived experiences as a woman situated in a patriarchically dominated system of power. Similar arguments for recognition of the situatedness of science have been advanced by the US-American biologist and social scientist Donna J. Haraway (1988, 1991) and in geography by Gillian Rose (1993). Such criticisms of the detachment of social science from lived realities have led to what has been termed a 'biographical turn' in social science (Chamberlayne et al. 2000).

The growing number of women entering academic research has brought 'distinctive understandings of and perspectives on the world ... based upon women's places, experiences and training' (Nash 1994: 55). Discussing fieldwork, Kim England (1994: 85–7) notes that this is affected by the biography of the researcher in two ways: first, the researcher is always positioned by personal characteristics such as gender, age, ethnicity or sexual identity, and these both inhibit and enable particular insights; second, every research situation is imbued with power relations. The feminist critique of universality and objectivity draws attention to the ways in which social relations, including gender dynamics, affect the research process (Gilbert 1994: 90–91). Hence there is a need to integrate the personal into the research process. Therefore, understanding the lives and experiences of individual researchers is an important part of the history of ideas.

Autobiography in Geography and in Development Studies

In the late 1970s, the geographers Anne Buttimer and Torsten Hägerstrand at Lund University in Sweden initiated the International Dialogue Project, in which senior colleagues were invited to relate their personal histories, experiences and views as practising geographers. One of the products of this work was a series of brief essays in which seven North European geographers – including one woman, Aadel Brun Tschudi – were asked to give their personal reflections on the significance of particular places for their creativity (Buttimer 1983a: 65–90). In her book, *The Practice of Geography* (1983b), Buttimer presented the autobiographies of twelve European and North American geographers. Two of them were women, including Tschudi (1983), who described how her birth and upbringing in China as the daughter of Norwegian missionaries influenced her later research orientation in the field of development geography. In 1988 Hägerstrand and Buttimer edited a book providing twelve autobiographical presentations from retired colleagues in the Nordic countries, of which only one was a woman and only one had undertaken some research in the Global South.

The imbalance between men and women has begun to be addressed in recent autobiographical collections. *Placing Autobiography in Geography*, edited by Pamela Moss (2001a), contains autobiographical accounts by five men and four women. Two of the accounts discuss research in the Global South. One of them is by a female writer, Robin Roth (2001). She explains her motivation for getting involved in international development, based on a critique of conventional development discourse and praxis, which she sees as perpetuating colonial relationships and privileging the knowledge of what she calls the 'Minority World' (the Global North) over that of the 'Majority World' (the Global South).

Moss (2001b, 2001c) discusses three ways in which biography and geography come together. First, this approach to the history of the discipline concerns not only the geographical literature as such but also the people who have produced it. Second, the method allows for reflexivity and positioning on the part of the researcher. Third, in extension of this, it focuses analytically on the personal element in how information is collected, thus affecting the research process and its content. She notes that geographers became interested in autobiography as an approach to disciplinary history in the mid-1970s and early 1980s. In the 1990s, attention was on reflexivity and positioning. In data collection, however, the focus has been mainly on the personal life histories of those being researched while the importance of the researcher's autobiography has been neglected.

Recently, the inclusion of autobiography as an integral part of understanding a discipline has reached textbooks. In *Approaches to Human Geography*, edited by Stuart Aitken and Gill Valentine (2006), nine contemporary geographers – five men and four women – were invited to explain how events in their lives have influenced their approaches to theory and practice in their academic careers. The autobiographies indicate some of the personal factors that influence how geographers 'come to know the world', showing how their work 'has been shaped by their academic context, place, and personal experiences'. It is argued that personal writing is an important means of challenging 'the disembodied and dispassionate nature of much academic writing' (Aitken and Valentine 2006: 169). One of the nine autobiographical accounts relates to the Global South: this is by Richa Nagar (2006), an Indian woman who moved to the USA and did field work in Tanzania on the complexities of race, religion, caste, class, and gender in the South Asian immigrant community in Dar es Salaam before shifting to the field of women's studies and working with NGO activists in India.

This brief account suggests that the main emphasis in autobiographical approaches has been male-dominated and centred on research undertaken in the Global North. Still only a few of the autobiographical accounts of geographers present the careers of women, although the number is growing, and only a few – whether of women or men – are concerned with research in the Global South. While scarcely redressing the problem, the following autobiographical discussion with Ragnhild Lund provides a small contribution towards changing the picture.

The Interviewer and the Interviewed

An argument for the scholarly importance of autobiographical discussions is that it takes consequence of the fact that knowledge is socially situated. In arguing this, I therefore give here a short account of the academic cooperation I have had with Ragnhild Lund over more than 30 years.

My contact with Ragnhild began in 1980 when she came as a research fellow to the Department of Geography at the then University of Trondheim. I had been appointed senior lecturer at the department in 1975, shortly after it was founded. Ragnhild came from the University of Bergen, where she had completed a master's degree in 1979 with a dissertation on women's working and living conditions in a new settlement area, Mahaweli in Sri Lanka. The focus on women in developing countries has since been a central feature of Ragnhild's research. Already at the end of the 1970s, before coming to our department, she was involved in the establishment of the women's network DAWN (Development for a New Era), which continues to play a constructive and critical role internationally and to provide inspiration for female researchers worldwide. As a research fellow, Ragnhild organized a field course in Sri Lanka for our master's students, in which I and other colleagues also took part in. This was my first experience of the Global South, and stimulated my interest, although my research continued in another direction and I did not specialise in development studies. However, one result of the field course was a special issue of the Norwegian Journal of Geography, containing articles by both Norwegian and Sri Lankan researchers, which Ragnhild and I edited (Jones and Lund 1983).

Ragnhild was appointed senior lecturer in the Department of Geography in 1984. Between 1986 and 1991, she led together with social anthropologist Merete Lie an interdisciplinary project on female workers in industry established by Norwegian firms in Malaysia (Lund and Lie 1989, Lie and Lund 1991, 1994). From 1990 to 1999, together with colleague Axel Baudouin, she had a central role in initiating and coordinating a cooperation project for research and education between the Department of Geography in Trondheim and that at the University of Peradeniya, Sri Lanka. This was the first such cooperation project initiated under the Norwegian National Programme for Development, Research and Education. The project enabled the exchange of researchers and students. The research focus was on the livelihood conditions of settlers in the Mahaweli irrigation area.

In the meantime the Department of Geography in Trondheim set up a doctoral degree programme. Ragnhild successfully defended her thesis, titled *Gender and Place* (Lund 1993), in April 1994 as the first candidate to complete a doctorate in the department. As her supervisor I remember Ragnhild as highly motivated and very independent, and my task was limited to reading and commenting on a largely completed text on the basis of my general geographical knowledge. The most memorable supervisory meeting we had was in Bangkok at Christmas time in 1992. Ragnhild had at that time recently taken leave from her position in Trondheim to work for two-and-a-half years as associate professor in rural development planning at the Asian Institute of Technology (AIT).

In 1995 Ragnhild was appointed professor in geography in what was shortly to become the Norwegian University of Science and Technology (NTNU) in Trondheim. She took the initiative for and has subsequently been coordinator of the Faculty of Social Science's first international master's programme, established in 1997. Many of her colleagues, including myself, have found it stimulating to teach and supervise students from Asia, Africa and Latin America.

As editor of the *Norwegian Journal of Geography* from 1999, I have highly appreciated the contribution that Ragnhild has made since then as guest editor for four special issues of the journal. The first presented results from a project on forced migration that Ragnhild had initiated (Lund 2000a). The second was on feminist geographies (Berg and Lund 2003), a field in which Ragnhild has been a Norwegian pioneer. The third took up new faces of poverty in Ghana (Lund et al. 2008a). The fourth was concerned with Sri Lanka's postcrisis reconstruction in the wake of the catastrophic tsunami of 2004 and the still then ongoing civil war (Lund and Blaikie 2009). Except for the issue on feminist geographies, all reported on externally financed research projects. In 2011 Ragnhild joined the editorial team of the journal as editor with special responsibility for development studies.

This brief account only touches on a few of the large number of projects, publications, consultancies and positions that Ragnhild has had in the field of development studies. Her work was acknowledged when she was awarded the university's internationalization prize for 2011.

In January 2011 I undertook a two-hour long videoed interview with Ragnhild. This was one of several autobiographical conversations I have undertaken with colleagues in the department (Jones 2012: VII–VIII), inspired by the work of Anne Buttimer (1983b, also Hägerstrand and Buttimer 1988). The interviews are intended as potential sources of departmental and disciplinary history. They aim to illuminate factors that have influenced the academic choices made by individual researchers and which have left their stamp on the department's scholarly profile at different times and hence on the development of geography as a discipline in Norway. The interviewee is asked to prepare for the interview by reflecting beforehand on his or her career and academic activities. The interviewer's role is to keep the conversation going by asking questions.

The interviewee is asked to reflect over factors influencing the choice of geography as a discipline and the particular career followed as a geographer. Relevant influences may include upbringing, family, schooling, colleagues, social contacts and networks. The interviewee is further asked to discuss the scholarly ideas and geographical thinking that have influenced the research topics and problems addressed and the methods adopted during his or her career. Other questions take up teaching, supervision and the dissemination of research results. Of particular interest is the social context, such as engagement in academic or political debates. For geographers, it is also relevant to ask if there are particular places that have influenced their academic activities. The interview is loosely structured at the outset, allowing the interviewee flexibility in deciding the content of the interview. It is primarily a conversation that presents a life history with an emphasis on the scholarly dimension.

There are certain pitfalls and ethical questions arising in such interviews. The interview situation is inevitably influenced by the personal relationship between the interviewer and the interviewee. When the interviewer and interviewee are close colleagues, they will generally have a degree of common understanding of the discipline's history and context, with the result that certain things may be left unsaid as being taken for granted (Adriansen and Madsen 2009). There is a further danger that an autobiographical presentation may emphasize the successes of one's career, whereas failures and unsuccessful research projects may not be communicated. Some ethical issues in the transcription and publication of personal narratives are raised by Valentine (1998), such as whether the interviewee should be given the opportunity to correct a transcript (which then departs from the original interview and opens the possibility of censoring), and who owns the published product of what is essentially a cocreation.

In the present case, the transition from a videotaped oral conversation to a written text occurred as follows. The interview with Ragnhild was conducted in Norwegian, converted to an audiotape and transcribed by a professional transcriber. I am responsible for selecting the extracts, which I have translated into English and edited. Ragnhild was given the opportunity to read through the resultant text and to suggest any necessary adjustments. In this way the result is a jointly created dialogue.

The Interview and Autobiographical Discussion

MJ: First, Ragnhild, could you mention some distinguishing characteristics of your career as a geographer?

RL: Initially I did not think I would have a career as a geographer or even a university career. My choice of subject was influenced because I thought I would be a teacher in a sixth-form college. I began here [in Trondheim] by studying English and later moved to Bergen ... I took French and then I wanted to have a social science ... I have always liked geography and I liked it at school ...The influence it had was an interest for places and other places beyond the immediate city area. I had a grandmother who taught me to read maps ...

MJ: Where did you travel?

RL: In the family? My father was a commercial traveller in North Norway so summer holidays for us meant taking either the train or the coastal steamer north to Tromsø or Bodø and then driving south back to Trondheim by car ... We were always told it was important to learn to get to know one's own country before travelling to other places in the world. It was not until I was 18 that I went abroad, when I travelled to England to learn English at a language school ...

MJ: How was it to study geography at the University of Bergen?

RL: There were two directions in geography that attracted me. One was what later became called resource geography, emphasizing the natural environment and environmental change, and the other was development research, which was just starting up within geography in Bergen. A new senior lecturer had just been employed, who came from Sweden and was called Jan Lundqvist. It was he who eventually became my supervisor in development studies.

MJ: How did you experience taking a master's degree in geography?

RL: I was not always present when I was a master's student because I worked for NORAD [Norwegian Agency for Development Cooperation]. Against the background of my undergraduate assignment in geography (a literature study on the position of women in Tanzania), I was asked to write a report for NORAD, which was in fact my first research task. I had to interview case workers and experts within development regarding how the projects they had worked with had benefited the position of women in society (Lund 1977) ... On the basis of my experience with NORAD ... and because a Norwegian embassy was to be established in Colombo, I received strong encouragement from development researchers to do work in Sri Lanka ... We were granted funding from NORAD. There were three master students working on the project, two from Sri Lanka and one from Norway ... I went to Sri Lanka and did fieldwork together with the two lads, their local supervisor and Jan Lundqvist ...

MJ: In which area in Sri Lanka did you do fieldwork?

There was a large development project at that time, the settlement project in Mahaweli, which practically covers the whole North Central Province of Sri Lanka. It was a very controversial project because it meant that poor landless Sinhalese farmers moved quite far north into what were earlier Tamil areas ...

MJ: Did you notice tension between the ethnic groups while you were doing fieldwork, or was it before it became visible?

We only heard that there had been some protest marches in the nearby town ... It was not especially visible for us at the time – even though we could read about it in the newspaper.

MJ: What did you do when you had completed your master's thesis [in 1979]?

RL: I worked on a report for a working committee appointed by the Norwegian Research Council for the Social Sciences and Humanities to examine the significance of social science research in Norway for developing countries. I worked on this as secretary for the sociologist Ingrid Eide from the University of Oslo (Eide and Lund 1980) ... I reckon her as my mentor in women's research ...

I had an assignment from CMI [Chr. Michelsen Institute for development research in Bergen] to contribute to the establishment of a women's network for developing countries ... with Devaki Jain, a well-known social scientist in India ... We invited women activists and researchers from Latin America, Africa and

Asia to Bangalore, where Devaki Jain lived then, and we established the DAWN group ...

I was then encouraged to apply for a research fellowship from the Research Council ... I was awarded a fellowship ... and was then asked to come to Trondheim ... Since both my husband and I are from Trondheim, we thought it was good also for many private reasons to return here ...

MJ: You organized and led a field course to Sri Lanka in 1981?

RL: I have led three master's field courses to Sri Lanka, and this was the first ... I remember we struggled with the programme. We experienced both flooding and a traffic accident, so we got delayed and had to improvise ... The programme we set up primarily studied the changes that were occurring in the south of Sri Lanka, which was the big development project area for Norwegian assistance at that time ... Hambantota and later Monaragala had rural integrated development programmes ... Then we drove into the highland region and to Kandy ... via the old Survey Department near Nurelya ... There was a former British club ... where we stayed overnight ... We drove long distances then. We drove to Anuradhapura from Kandy. I am very glad that I saw this area so early, because it is very changed now today. During the whole of the war it was a high security zone with checkpoints and soldiers with loaded weapons and it is a much destroyed area today – not directly destroyed by the war but by neglect and lack of maintenance ...

MJ: After the field course in 1981 we edited a special issue of Norsk Geografisk Tidsskrift, the Norwegian Journal of Geography, where researchers from both Sri Lanka and Norway contributed.

RL: That is correct. I remember specially that there was an agricultural researcher who contributed an article on Dry Zone agriculture and the possibilities of development in the Dry Zone (Perera et al. 1983). It was quite a pioneering article – because such things were only studied much later ... We had visited his agricultural research station during the field course. He had a doctoral degree from Russia, and had a different approach to development problems than we had at that time ...

MJ: Which other Asian countries besides Sri Lanka have you worked in?

RL: I have worked in several countries in various capacities ... I came in contact with Merete Lie ... who was a social anthropologist and we were keen to study the removal of industry from Norway to Asia. We are still studying this in that we have investigated footloose industry that moves from land to land and uses labour and competence where it is most profitable to do so at any time ... We were in Malaysia with our families and did fieldwork, and wrote a book that we called *Renegotiating Local Values* (Lie and Lund 1994). This dealt with how female workers were recruited from agricultural areas to industry, and what effects it had on local societies ... We have followed how one firm moved its production on to China and studied its motives for establishment there, working conditions and social change ... We edited a book in which our study is one of the chapters (Lie and Lund 2008). We called this book *Making it in China*. Half the book examines how Norwegian researchers see the removal of industry to China and the other half how Chinese researchers look on Nordic industry that comes to China ...

I was asked ... if I would be interested in a job at the Asian Institute of Technology in Bangkok ... This was in 1992 and I thought it would be interesting to try something new so I applied and got the job. I was there for two-and-a half years with my family. We established a very close relationship with Thai society through that experience. But it was challenging in very many ways because the students came from the whole Asian region – from countries that did not at that time have master's education at their universities ... And it was challenging for me who was to teach development theory and so judge the actuality of development studies ... They were oriented towards problem-solving – because they had a mission to go back to their countries, not to go into research and university work, but into planning ...

MJ: Could you tell about the international master course you built up [in Trondheim]?

RL: The international Master's Course in Development Studies was established after I returned from Thailand, when the Dean of the Faculty asked me to look into this together with colleagues from other social sciences. The result was initially a master's programme for a trial period of three years. It was so successful that it became permanent and has now had its ... twentieth anniversary ... It was originally called Master in Social Change. But the students, especially those from Africa, wanted it to refer to development studies. So now it is called Master in Development Studies Specializing in Geography ... We have had between 12 and 20 students each year ... Now up to five Norwegian students can take it ...

Otherwise the department has got more and more used to having international students. To begin with many were hesitant to teach in English, but now it is quite normal for most. We have to take account of international dimensions such as the students' background and upbringing, and also to the knowledge they bring with them. They bring a somewhat different type of geographical knowledge and development knowledge than we expect Norwegian students to have. Norwegian students are quite good on theory ... while the students who come from Africa and Asia and also some from Latin America ... have more empirical and experience-based knowledge. But unfortunately it is also often repetitive knowledge ... Our tradition has become more and more oriented towards critical thinking and self-reflection on what you do and why, positioning and ethics.

MJ: Is there good interaction between the Norwegian and foreign students?

RL: I do not think there is such good interaction as there should be. Because they have a different experience, the foreign students are not so used to working in groups, for instance. They have been trained in individual performance. Norwegian students are familiar with groups and cooperation. But language is a barrier in practice, such as when taking a break over sandwiches at lunch time there is very little communication between them and the foreign students ... The foreign students who manage best are where there are several from the same country ... It is also a question of money. The foreign students save as much as they can ... and do not want to take part in common activities that cost something, for example. This is something we have not recognized – an activity may cost only a hundred crowns, but a hundred crowns to them is a huge amount. So there are such cultural and structural factors that make integration somewhat difficult. ...

MJ: A criticism that has been raised against the increasing turn to English in the universities is that it can be at the cost of the development of Norwegian as an academic language. Do have any opinions about this?

RL: Yes, I believe this is correct ... It is very seldom that I publish anything in Norwegian. I find it much easier to use English as a scientific or working language. I think this is a loss, but I think also perhaps that it is a loss academically that we adopt uncritically the Anglo-Saxon scientific ideal. Before, when we published more in Norwegian and had debate forums in Norwegian, then Norwegian and Nordic issues were more to the fore... I think we had a more critical self-initiated scientific view ...

MJ: Can you tell about different administrative tasks you have had?

RL: The first administrative task I had, not so long after I was appointed to a permanent position in the department, was as head of department and I had to learn how the university functioned ... I had my second child while I was head of department ... and only had a short maternity leave before going back to the position as head of department ... My husband took paternity leave, and was one of the first in Trondheim to do so ...

I have been a member of the Research Council's programme steering committee on gender and development, and in a national committee for development research. Later ... I got the task of research director for the university ... I spent most of my time in identifying and working on the university's interdisciplinary programmes ... The programme I have most followed up afterwards has been the globalization programme. I have also been one of the leaders of a programme on conflict and migration ...

Then I had an interesting task working for Worldagroforestry, one of the 15 agricultural centres of the CGIAR consortium [Consultative Group on International Agricultural Research] ... in Nairobi. I was first an ordinary member of the Executive Board and later became Chair of Programmes. In that period I travelled around in the world, especially Africa. I inspected and checked – I had responsibility for more than 2000 projects in Latin America and Africa ... We were responsible that everything went well and were continually controlling and evaluating. I learnt a lot. Among other things I met in that capacity the President of India ...

I have also been leader of the Asia network in Norway. For two periods I was leader. I learnt a lot, especially from researchers at the University of Oslo, which has broader Asia research than at our university ...

MJ: How have your personal attitudes and views affected your research?

RL: I am convinced that our attitudes and views affect our choice of what we want to do and become interested in. For example I have mentioned earlier that as children we learnt to read maps. I found a map on internet a while ago, which shows where you find indigenous peoples in Orissa state in India ... Indigenous peoples are called tribal peoples in India. The area shown in green on the map shows a high concentration of indigenous people. It is in these areas today that mining and plantations are coming in as new economic activities under the current neo-liberalistic policies in India ...Over 50 per cent of the population are tribal. When the mining companies move in – it can be Indian companies or multinationals – ... then people are forced to move out of these areas. This map was for me a means of identifying a problem ...

So we are making a comparative study in Laos, India and China on this problem. One becomes inspired by the visual presentations one sees. And one becomes inspired by trends of the time. The feministic understanding I gained in the 1970s has influenced me the whole time even though it has developed along the way. So I believe that you become best in what you are interested in. I always say to my master students not to study something you are not interested in, because you will not succeed. I believe that motivation is very important.

MJ: There must have been some ethical questions you discussed in your study of Norwegian firms in developing countries?

RL: Yes, we discussed many ethical questions and also research ethics – concerning how we should do that study, but there was also a debate on whether there was exploitation of female workers in industry. It was very unpopular, for example, when we found that the Norwegian firms in Malaysia at that time openly said that they did not need to bother about Norwegian laws and protection of workers' rights because they had now become international. Such discussions refer to corporate social responsibility today and concern ethics on a very high level.

It was also an ethical question when we found, for instance, that the industrial company we studied in depth was the one that paid the lowest wages of all the international firms there. That they were Norwegian firms did not mean that they worked in an ethically acceptable manner ...

There were also ethical dimensions about how we cooperated and how we should share the research results and how we should relate professionally to one another. I remember a long discussion we had on whether it was necessary for good colleagues to be good friends to do good research. And although we became good friends, we said then that it was not a requirement. The most important thing was that we could rely on one another. This has followed me all along in my choice of partners for cooperation. Trust is very important. And generosity in sharing one's knowledge, and not being afraid to share it ... Of course informing and gaining the consent of respondents in the field goes without saying.

MJ: Tell about your project on forced migration.

RL: We decided to apply to the Research Council for a grant to study internal displacement ... because of conflict and war. At that time all the literature on conflict was about refugees who had crossed national borders. We identified a niche where we studied people forced to flee within their own country, which in many ways was a bigger refugee problem that those crossing form one country to another. This has become even more topical today because Europe closes its borders and so do Australia, the USA and Canada more or less. Canada, which had been very liberal, has now come with a more restrictive immigration policy ...

I found it important to study other ethnic groups [than Tamils, Singhalese and Muslims] that had been made invisible in the conflict in Sri Lanka. Hence I chose to study the Veddhas, an indigenous group that were marginalized ... and found documentation that showed they had been subject to displacement over a long period in history ... Their displacement became accelerated after a major development programme in the 1980s and a national park was established (Lund 2000b) ...

Displacement is not a new phenomenon, but has taken place historically the whole time. The reasons for migration often lie in development projects. In academic language this is called development-induced displacement ... as opposed to conflict-induced displacement ...

We established a research group on forced migration. This was in 1997 [the group received funding from the Research Council of Norway in 1997 although it was actually set up in 1995] and we have had other projects in extension of this activity. We developed a very interesting cooperation with a nongovernmental organization based in Sweden and Norway and doing a lot of work in Sri Lanka ... We developed a method that we called 'real-time research', where we collected information on how their projects developed, and we entered into a dialogue with the users of research and discussed the results we had come to (Brun and Lund 2010, 2014) ... Later we applied to the Research Council and received funding for a project that we called 'Beyond the Action-Knowledge Gap'. We wanted to study the meeting point between practitioners and researchers in connection with handling crises ...

We are also doing a study of young people. Youngsters in Sri Lanka are soon a majority of the population, and underemployed ...

MJ: Is the study of children and youth a new direction for you?

RL: I had a sabbatical and spent five months at the San Diego State University, where I met Stuart Aitken ... Then I invited him as adjunct professor to our department ... He was partly at the Centre for Child Research and partly at the Department of Geography. He persuaded me to take part in a writing workshop on child research. So I came to write an article, or rather a chapter in a book which we called *Global Childhoods* ... I wrote a chapter where I tried to see how the discussion of participation has been in child research compared with development research, and where there were synergies and convergences (Lund 2008).

MJ: You have also worked in other parts of the world apart from Sri Lanka and Asia. Haven't you also worked in Ghana?

RL: Yes, I was invited by colleagues to take part in a project on health and poverty ... My contribution to the project was to give it a title. There had come out a World Bank report that spoke of the new types of poverty, where the poor were women, or refugees, or migrants from the countryside to the city. We formulated a project that we called 'The New Faces of Poverty'. There were several sub-projects – on livelihoods and female employment, and on health, especially HIV/AIDS. What I remember best from the fieldwork we did in Ghana was when I travelled together with a Ghanaian colleague to an orphanage for those whose parents had died of HIV/AIDS. We came to a village where the main occupation must have been making coffins for there was one carpenter's workshop after the other as we drove into the village ... Many people were displaced when the Volta Dam was built. Farmers became lorry drivers. They drove across the border and on to Nigeria, and many came back with HIV and AIDS. And I was so surprised when my colleague began to cry. It was unexpected because he is an AIDS specialist and has worked for many years with the problems of HIV/AIDS. However, he had mainly worked with statistical material. So he became very concerned about the children and the next generation – and how the local society handled the crisis with all the orphans. This was described in the literature as a ticking bomb – because so many millions of children became orphans and the national state and local society were unable to handle it. But it was a theoretical discussion. We saw that they coped. They had a traditional system in Ghana whereby everybody was taken care of, either by the queens, or queen mothers, or the chiefs. The formal state apparatus provided health services in Ghana, but at the same time the traditional system functioned. We wrote an article in *GeoJournal* with the title 'Queens as Mothers', in which we examined the need for care from a health perspective (Lund and Agyei-Mensah 2008) ... We interviewed the children in groups and individually and used a lot of photography and observation ... We edited a theme issue of the *Norwegian Journal of Geography* ... in which most of the Ghanaian researchers in the research group wrote articles (Lund et al. 2008a).

MJ: Do you have any general reflections on what you have achieved through these many different projects?

RL: I have tried everything from questionnaires to a combination of methods and now in cooperation with aid organizations to working more with action research ... Theoretically I have developed from being a researcher working at the interface between development studies and feminist geography to be more and more ... poststructuralist ... and to be more and more actor-oriented ... This corresponds with much that is written on power, and with an increased focus on power relations in development research. Also here there is an increased focus on actors and processes rather than the formal structures. At the basis of development research as we originally learnt it was the relationship to the national state and the way institutions should strengthen the national state. But on the agenda today are completely different concepts, such as globalization, transnationalism, phenomena and processes and relations and interactions that are shaped across national boundaries ...

A second thing I have achieved is to gain broad competence on Asia and on other themes than gender. There are war and conflict, there are capabilities and livelihoods and there is postwar reconstruction ... This broad competence shows in my teaching. I am able to be much freer in communicating my own experiences, and to give empirical insights into development studies more than I could do before ... I think that it is underestimated that the most important task for an academic is not research for research in itself, but to build up knowledge for others. I think this is also the most important thing in supervising students. You teach them method and theory and different perspectives that they take with them into their working life ... I think that what we produce and develop of knowledge together with young people, who will be the next generation, is underestimated in our way of thinking about academia ... in which individual production counts more than collective production. I have been concerned with working together with people in other disciplines and problem-oriented teams, and I feel that many structural constraints make interdisciplinary work difficult ... The university loses its ability to see science and knowledge as a whole, and knowledge as a common good in society. We forget this dimension in our own scientific meriting and counting of publications ... I get very indignant over how Eurocentric publishing practice has become ... Things published in Sri Lanka or India ... and read by many .... do not count before they are published by a recognized publisher. An article ... published in Colombo counts for nothing here. Many of the publications that you have for ideological reasons reported back to those you have worked with are not recognised in our scientific meriting system ... Asia is becoming a much more important global actor than ever. If we think provincially about our research we will lose in the long run.

Unravelling Marginalization and Voicing Change

The autobiographical discussion with Ragnhild Lund reflects the path of alternative development thinking. Her work has received impulses from feminism, poststructuralism and action research. She has moved from studies of the place of women in the development context to examination of the effects of globalization and investigation of responses to forced migration due to conflict and other crises. Throughout, her work has focused on the poor and disadvantaged as active agents of change in their struggle against marginalization and she has listened to and given voice to their concerns. These concerns are reflected in the chapters of the present volume (Brun and Blaikie 2014).

Ragnhild's early engagement in gender studies is seen in her first written work, a report on the significance of Norwegian development aid projects for the position of women (Lund 1977). A later report on the significance of social science research for developing countries was written together with Ingrid Eide (Eide and Lund 1980), who later in the 1980s became director of the Division for Women in Development of the United Nations Development Programme (UNDP). Eide (2014), in her chapter in the present volume, describes how during her tenure at UNDP an international consensus to focus on women in the development context was established.

Ragnhild's concern with women's working and living conditions in the Mahaweli settlement area in Sri Lanka, beginning with her master's thesis (Lund 1979), broadened into a deep theoretical and practical understanding of the place of women in development planning (1981, 1983, 1993). Studies of the livelihood conditions of settlers in Sri Lanka were pursued further in several doctoral theses at the Department of Geography in Trondheim (Seneviratne 2003, Vandsemb 2007, Azmi 2008). Two chapters in the present volume by Ragnhild's former students take this work further. Berit Helene Vandsemb (2014) examines the experiences of poor farmers who migrated to a relatively sparsely populated 'migration frontier' in southeastern Sri Lanka, and we hear the voices of these spontaneous migrants over a 15-year period. In Fazeeha Azmi's (2014) study of internal displacement and women's agency in two Mahaweli settlements, we hear the voices of four women who through in-depth interviews tell how they have coped with development-induced displacement in one settlement and war-induced displacement in the other.

Ragnhild's work with Merete Lie on the experiences of female workers recruited by foreign industrial companies in Malaysia (Lund and Lie 1989, Lie and Lund 1991, 1994) inspired a similar study by another of Ragnhild's PhD students presenting the accounts of female workers in an export processing zone in Sri Lanka, with a focus on health (Attanapola 2005). In Chamila Attanapola's (2014) contribution to the present volume, in-depth interviews with three women workers give voice to their agency in coping with human rights violations in the export processing zone. Using a similar approach, another PhD student has presented accounts by craftswomen in self-employment in Odisha (Orissa), India (Acharya and Lund 2002, Acharya 2004).

The research on forced migration has gone through various stages since it began in the mid-1990s (Lund 2000a, Shanmugaratnam et al. 2003, Lund and Blaikie 2009). Ragnhild has supervised several doctoral theses on forced migration (Brun 2003, Skotte 2004, Azmi 2008, Tete 2011, Khasalamwa-Mwandha 2012, see also Lund et al. 2011). Cathrine Brun (who was Ragnhild's first doctoral student) and Ragnhild Lund present the work of the project in the present volume (Brun and Lund 2014). This work has inspired chapters in this volume in different directions. The consequences of war in Sri Lanka were not only internal displacement; Jennifer Hyndman (2014), who has held an adjunct professorship in the Department of Geography in Trondheim, presents in her chapter harrowing accounts by women who suffered the violent atrocities of the war as well as telling of their struggle for justice. A colleague of Ragnhild at AIT, Kyoko Kusakabe (2014), was inspired by her account of the coping capacity of displaced indigenous Veddhas in Sri Lanka (Lund 2000b) to write about the coping capacity of small-scale women border traders in fish as they have faced changing regimes of border control in Cambodia. The chapter by Ragnhild's former student at AIT, Smita Mishra Panda (2014), who now works in India, presents documentation from civil society organizations of the struggle of *Adivasi* (tribal) women in Odisha against displacement, loss of rights and dispossession of their livelihoods by national and international mining corporations (see also Lund and Panda 2011). Two contributions critically analyse discourses linking migration to climate change: Bernadette Resurrección (2014) (a colleague at AIT) critiques the discourses on climate change, migration and security as presented in international documents and advocacy literature, while Haakon Lein (2014) (a colleague in Trondheim) shows that there is little evidence for the view that climatic events are a significant cause of migration from Bangladesh to India.

Together with Trondheim colleague Stig Jørgensen, Ragnhild worked between 2003 and 2008 in the project 'The New Faces of Poverty' (Lund et al. 2008a) in cooperation with geographers in Ghana. Ragnhild contributed to studies of women workers in stone chip production (Lund et al. 2008b), slum-dwellers' livelihoods (Owusu et al. 2008) and children orphaned due to HIV/AIDS (Lund and Agyei-Mensah 2008). The last-mentioned study contributed to theoretical reflections on what it means to be a 'participating child' (Lund 2008). The principal partner at the University of Ghana was Samuel Agyei-Mensah, who had earlier completed a doctorate in Trondheim (Agyei-Mensah 1997). The chapter in this volume by Agyei-Mensah and colleague Charlotte Wrigley-Asante (2014) provides a historical overview of changing gender relations in the light of political ideologies and development issues. George Owusu, whose PhD thesis Ragnhild supervised (Owusu and Lund 2004, Owusu 2005), has contributed a chapter giving an overview of Ghana's housing policy as it relates to the urban poor (Owusu 2014). Although both chapters deal with Accra, they link the local with the global by showing the impacts of globalization.

The continual renegotiation of local values that occurs in the meeting of the global and the local – as well as the national and the local – permeates Ragnhild's research, whether concerning those affected by development schemes, female workers in global industrial concerns, indigenous groups or slum-dwellers. Ragnhild's studies of the ramifications of Norwegian investment in Malaysia and more recently China (Lie and Lund 2008) have a certain parallel in the chapter in the present volume by Ragnhild's colleague in Bergen, Arnt Fløysand, who together with Jonathan Barton investigates the complex outcomes of Norwegian foreign direct investment in the salmon industry in Chile, based on interviews with managers, workers, local authorities and civil society organizations (Fløysand and Barton 2014). Ragnhild's people- and place-centred perspective has provided inspiration for another former adjunct professor in Trondheim, Stuart Aitken, who in a chapter together with three coauthors presents a method for participatory mapping as a means of engaging local knowledge in a national nature reserve in China (Aitken et al. 2014).

Ragnhild has contributed to ideas of alternative development both empirically and theoretically, and both through basic research and applied research. In 1989 she coedited a theme issue of the Norwegian Journal of Geography on 'Development Geography', introducing a collection of articles within two fields: rural problems in the Third World and the relocation of Western industry (Lund 1989). In theoretical writings she argues for the integration of ecological and socioeconomic explanations in development geography, cross-fertilization of ideas between feminist geography and development studies, sensitivity in development studies to the particularity of place, and the need for development geography not only to search for explanations but to be of practical relevance (Lund 1986, 1992, 1993, 1998). Perspectives from feminist geography (Berg and Lund 2003) suffuse her work as well as many of the chapters in the present volume. A particularly original contribution is the chapter by Bergen colleague Vibeke Vågenes (2014), who in a critique of universalist approaches to female empowerment argues for the need to understand gender relations and power inequalities in their own social and cultural context; she presents a study based on interviews of women in eastern Sudan showing how gender segregation allows women to influence their own life situations within female domains outside the orbit of male domination.

Understanding local diversity and people's own concerns is at the centre of research on alternative development. A major challenge is making this knowledge useful in practical action and in influencing policy. Recent action-oriented research that Ragnhild has been engaged in has focused on home-making and housing in the context of postwar reconstruction (Brun and Lund 2008, 2009, Lund and Azmi 2009) and on crisis management (Boano and Lund 2011, Palttala et al. 2012). The challenge of bringing the ideas of alternative development into policy-making is discussed in the present volume by Piers Blaikie, who worked closely with Ragnhild as adjunct professor for a period at the Department of Geography in Trondheim. Blaikie (2014) is concerned with linking the researcher's experience with policy and proposes a research and communication strategy for a useful and engaged political ecology, in which emancipation and justice are central. Hans Skotte (2014) (for whose doctoral thesis, completed in 2004, Ragnhild served as cosupervisor) presents in his chapter a method of mutual teaching and learning based on the experiences of students engaging in practical architectural and urban planning projects in the Global South. Such approaches involve emotional engagement with the subjects of investigation. In an important article, Ragnhild uses her own experience of doing ethnographic research in Sri Lanka on the displacement and recovery of people affected by war and the tsunami. She discusses how emotions have an impact on the research process in crisis situations and how sensitivity to the emotions of both the researchers and the research participants can make research more reflexive and ethical (Lund 2012).

Conclusion

As a contrast to the more well-established approach to histories of knowledge that presents the development of ideas in different disciplines and the social context in which these ideas were produced, this chapter has focused on a second approach consisting of personal histories and how these both reflect and affect the development of disciplines. Adopting the autobiographical approach, this chapter has presented an account of the Norwegian geographer Ragnhild Lund's life and career in the form of a narrative based on interviews conducted with her by the present author. The extracts from the interviews present her life experiences and career as a geographer who has specialized in development studies in the Global South. The chapter has presented her education and work, her scholarly ideas and geographical thinking, influences on her ideas, and ways in which her work has influenced others. As both researcher and practitioner, Ragnhild Lund has been among the pioneers of feminist approaches in development studies and has been in the forefront of arguments based on alternative understandings of the geography of development in opposition to the historical mainstream.

References

Acharya, J. 2004. *Gendered Spaces: Craftswomen's Stories of Self-Employment in Orissa, India.* Doctoral theses at NTNU, 2004:74. Department of Geography. Trondheim: Norwegian University of Science and Technology (NTNU).

Acharya, L. and Lund, R. 2002. Gendered Spaces – Socio-Spatial Relations of Self-Employed Women in Craft Production, Orissa, India. *Norsk Geografisk Tidsskrift–Norwegian Journal of Geography*, 56, 207–18.

Adriansen, H.K. and Madsen, L.M. 2009. Studying the Making of Geographical Knowledge: The Implications of Insider Interviews. *Norsk Geografisk Tidsskrift–Norwegian Journal of Geography*, 63, 145–53.

Agyei-Mensah, S. 1997. *Fertility Change in a Time and Space Perspective: Lessons from Three Ghanaian Settlements.* Dr. polit. thesis. Department of Geography. Trondheim: Norwegian University of Science and Technology (NTNU).

Agyei-Mensah, S. and Wrigley-Asante, C. 2014. Gender, Politics and Development in Accra, Ghana, in *Alternative Development: Unravelling Marginalization, Voicing Change*, edited by C. Brun, P. Blaikie and M. Jones. Farnham: Ashgate, 117–33.

Aitken, S. and Valentine, G. (eds) 2006. *Approaches to Human Geography.* London – Thousand Oaks –New Delhi: Sage Publications.

Aitken, S., An, L., Wandersee, S. and Yang, Y. 2014. Renegotiating Local Values: The Case of Fanjingshan Reserve, China, in *Alternative Development: Unravelling Marginalization, Voicing Change*, edited by C. Brun, P. Blaikie and M. Jones. Farnham: Ashgate, 171–90.

Attanapola, C. 2005. *Unravelling Women's Stories on Health: Female Workers' Experiences of Work, Gender Roles and Empowerment Relating to Health in Katunayake Export Processing Zone, Sri Lanka.* Doctoral theses at NTNU, 2005:218. Department of Geography. Trondheim: Norwegian University of Science and Technology (NTNU).

Attanapola, C. 2014. Ignored Voices of Globalization: Women's Agency in Coping with Human Rights Violations at an Export Processing Zone in Sri Lanka, in *Alternative Development: Unravelling Marginalization, Voicing Change*, edited by C. Brun, P. Blaikie and M. Jones. Farnham: Ashgate, 135–54.

Azmi, F. 2008. *From Rice Barn to Remittances: A Study of Poverty and Livelihood Changes in System H of the Accelerated Mahaweli Development Project (AMDP), Sri Lanka.* Doctoral theses at NTNU, 2008:166. Department of Geography. Trondheim: Norwegian University of Science and Technology (NTNU).

Azmi, F. 2014. Impacts of Internal Displacement on Women's Agency in Two Resettlement Contexts in Sri Lanka, in *Alternative Development: Unravelling Marginalization, Voicing Change*, edited by C. Brun, P. Blaikie and M. Jones. Farnham: Ashgate, 243–57.

Berg, N.G. and Lund, R. (eds) 2003. Feminist Geographies. *Norsk Geografisk Tidsskrift–Norwegian Journal of Geography*, 57, 129–83.

Blaikie, P. 2014. Towards an Engaged Political Ecology, in *Alternative Development: Unravelling Marginalization, Voicing Change*, edited by C. Brun, P. Blaikie and M. Jones. Farnham: Ashgate, 25–37.

Boano, C. and Lund, R. 2011. Disasters, Crisis and Communication, in *Developing a Crisis Communication Scorecard*, edited by M. Vos, R. Lund, Z. Reich and H. Harro-Loit. Jyväskylä Studies in the Humanities 152. Jyväskylä: Jyväskylän Yliopisto, 49–152.

Brun, C. 2003. *Finding a Place: Local Integration and Protracted Displacement in Sri Lanka*. Dr. polit. thesis. Department of Geography. Trondheim: Norwegian University of Science and Technology (NTNU).

Brun, C. and Blaikie, P. 2014. Alternative Development: Unravelling Marginalization, Voicing Change, in *Alternative Development: Unravelling Marginalization, Voicing Change*, edited by C. Brun, P. Blaikie and M. Jones. Farnham: Ashgate, 1–22.

Brun, C. and Lund, R. 2008. Making a Home During Crisis: Post-Tsunami Recovery in a Context of War, Sri Lanka. *Singapore Journal of Tropical Geography*, 29, 274–87.

Brun, C. and Lund, R. 2009. 'Unpacking' the Narrative of a National Housing Policy in Sri Lanka. *Norsk Geografisk Tidsskrift–Norwegian Journal of Geography*, 63, 10–22.

Brun, C. and Lund, R. 2010. Real-time Research: Decolonising Research Practices – or Just Another Spectacle of Researcher–Practitioner Collaboration? *Development in Practice*, 20, 812–26.

Brun, C. and Lund, R. 2014. Researching Forced Migration at the Interface of Theory, Policy and Practice, in *Alternative Development: Unravelling Marginalization, Voicing Change*, edited by C. Brun, P. Blaikie and M. Jones. Farnham: Ashgate, 287–304.

Buttimer, A. (ed.) 1983a. *Creativity and Context: A Seminar Report*. Lund Studies in Geography, Ser. B. Human Geography No. 50. Lund: The Royal University of Lund, Department of Geography/CWK Gleerup.

Buttimer, A. 1983b. *The Practice of Geography*. London and New York, Longman.

Chamberlayne, P., Bornat, J. and Wengraf, T. 2000. Introduction: The Biographical Turn, in *The Turn to Biographical Methods in Social Science: Comparative Issues and Examples*, edited by P. Chamberlayne, J. Bornat and T. Wengraf. London and New York: Routledge, 1–30.

Eide, I. 2014. Implementing International Consensus on Women in Development: Context, Policy and Practice, in *Alternative Development: Unravelling Marginalization, Voicing Change*, edited by C. Brun, P. Blaikie and M. Jones. Farnham: Ashgate, 87–97.

Eide, I. and Lund, R. 1980. *Samfunnsvitenskapelig forskning av betydning for utviklingsland: En utredning*. Oslo: NAVF.

England, K.V.L. 1994. Getting Personal: Reflexivity, Positionality, and Feminist Research. *The Professional Geographer*, 46, 80–89.

Fløysand, A. and Barton, J.R. 2014. Foreign Direct Investment, Regional Development and Poverty Reduction: The Sustainability of the Salmon Industry in Chile, in *Alternative Development: Unravelling Marginalization, Voicing Change*, edited by C. Brun, P. Blaikie and M. Jones. Farnham: Ashgate, 55–71.

Gilbert, M.R. 1994. The Politics of Location: Doing Feminist Research at 'Home'. *The Professional Geographer*, 46, 90–96.

Hägerstrand, T. and Buttimer, A. (eds) 1988. *Geographers of Norden: Reflections on Career Experiences*. Lund Studies in Geography, Ser. B. Human Geography No. 52. Lund: The Royal University of Lund, Department of Geography/Lund University Press.

Haraway, D.J. 1988. Situated Knowledges: The Science Question in Feminism as a Site of Discourse on the Privilege of Partial Perspective. *Feminist Studies*, 14, 575–99.

Haraway, D.J. 1991. *Simians, Cyborgs and Women: The Reinvention of Nature*. London: Free Association Books.

Hyndman, J. 2014. 'No More Tears Sister': Feminist Politics in Sri Lanka, in *Alternative Development: Unravelling Marginalization, Voicing Change*, edited by C. Brun, P. Blaikie and M. Jones. Farnham: Ashgate, 155–67.

Jones, M. 2012. Om faghistorie og selvbiografi: Innledningsessay, in *Et liv med geografi: Institutthistorie og selvbiografi – et faglig livsløp knyttet til utviklingen av Geografisk institutt i Trondheim*, by A. Aase. Trondheim: Tapir akademisk forlag, I–XIV.

Jones, M. and Lund, R. (eds) 1983. Sri Lanka: Theme Issue of the Norwegian Journal of Geography. *Norsk Geografisk Tidsskrift*, 37, 137–222.

Khasalamwa-Mwandha, S. 2012. *Spaces of Recovery: An Exploration of the Complexities of Post-War/Disaster Recovery in Uganda and Sri Lanka*. Doctoral theses at NTNU, 2012:356. Department of Geography. Trondheim: Norwegian University of Science and Technology (NTNU).

Kusakabe, K. 2014. Coping Capacity of Small-Scale Border Fish Traders in Cambodia, in *Alternative Development: Unravelling Marginalization, Voicing Change*, edited by C. Brun, P. Blaikie and M. Jones. Farnham: Ashgate, 259–67.

Lein, H. 2014. The Reemergence of Environmental Causation in Migration Studies and its Relevance for Bangladesh, in *Alternative Development: Unravelling Marginalization, Voicing Change*, edited by C. Brun, P. Blaikie and M. Jones. Farnham: Ashgate, 207–18.

Lie, M. and Lund, R. 1991. What is She up to? Changing Identities and Values Among Women Workers in Malaysia, in *Gender and Change in Developing Countries*, edited by K.A. Stølen and M. Vaa. Oslo: Norwegian University Press, 147–64.

Lie, M. and Lund, R. 1994. *Renegotiating Local Values: Working Women and Foreign Industry in Malaysia.* Richmond: Curzon Press/Scandinavian Institute of Asia Studies.

Lie, M. and Lund, R. 2008. Global Enterprises, Local Workers: China as a Meeting Ground, in *Making it in China,* edited by M. Lie and R. Lund. Kristiansand: Høyskoleforlaget, 93–112.

Lund, R. 1977. *NORAD-prosjektenes betydning for kvinnenes stilling: En undersøkelse.* Oslo: NORAD.

Lund, R. 1979. *Prosperity through Mahaweli – Living Conditions in a Settlement Area.* Hovedfagsoppgave i geografi. Bergen: Universitetet i Bergen.

Lund, R. 1981. Women and Development Planning in Sri Lanka. *Geografiska Annaler,* 63B, 95–108.

Lund, R. 1983. The Need for Monitoring and Result Evaluation in a Development Project – Experiences from the Mahaweli Project. *Norsk Geografisk Tidsskrift,* 37, 169–86.

Lund, R. 1986. Development Geography: Development Orientation and Planning Focus, in *Welfare and Environment,* edited by M. Jones. Trondheim: Tapir, 121–43.

Lund, R. (ed.) 1989. Development Geography. *Norsk Geografisk Tidsskrift,* 43, 53–114.

Lund, R. 1992. Welfare, Gender and Social Change, in *Levekår og planlegging,* edited by M. Jones and W. Cramer. Trondheim: Tapir, 240–52.

Lund, R. 1993. *Gender and Place,* Volume 1: *Towards a Geography Sensitive to Gender, Place and Social Change;* Volume 2: *Gender and Place: Examples from Two Case Studies.* Trondheim: Department of Geography, University of Trondheim.

Lund, R. 1998. Kvinner og kjønn – perspektiv og problemstillinger i utviklingsstudier og feministisk geografi, i *Utviklingsgeografi,* edited by J. Hesselberg. Oslo: Tano Aschehoug, 163–92.

Lund, R. (ed.) 2000a. Geographical Research on Forced Migration and Development. *Norsk Geografisk Tidsskrift–Norwegian Journal of Geography,* 54, 89–136.

Lund, R. 2000b. Geographies of Eviction, Expulsion and Marginalization: Stories and Coping Capacities of the Veddhas, Sri Lanka. *Norsk Geografisk Tidsskrift–Norwegian Journal of Geography,* 54, 102–9.

Lund, R. 2008. At the Interface of Development Studies and Child Research: Rethinking the Participating Child, in *Global Childhoods: Globalization, Development and Young People,* edited by S. Aitken, R. Lund and A.T. Kjørholt. London: Routledge, 131–48.

Lund, R. 2012. Researching Crisis – Recognizing the Unsettling Experience of Emotions. *Emotion, Space and Society,* 5, 94–102.

Lund, R. and Agyei-Mensah, S. 2008. Queens as Mothers: The Role of the Traditional Safety Net of Care and Support for HIV/AIDS Orphans and Vulnerable Children in Ghana. *GeoJournal,* 71, 93–106.

Lund, R. and Azmi, F. 2009. Shifting Geographies of House and Home: Female Migrants Making Home in Rural Sri Lanka. *Journal of Geographical Science*, 57, 33–54.

Lund, R. and Blaikie, P. (eds) 2009. The Tsunami of 2004 in Sri Lanka: Impacts and Policies in the Shadow of War. *Norsk Geografisk Tidsskrift–Norwegian Journal of Geography*, 63, 1–96.

Lund, R. and Lie, M. 1989. The Role of Women in the New International Division of Labour – the Case of Malaysian Women in Norwegian Industry. *Norsk Geografisk Tidsskrift*, 43, 95–104.

Lund, R. and Panda, S.M. 2011. New Activism for Political Recognition: Creation and Expansion of Spaces by Tribal Women, Odisha, India. *Gender, Technology and Development*, 15, 75–99.

Lund, R., Agyei-Mensah, S. and Jørgensen, S. (eds) 2008a. New Faces of Poverty in Ghana. *Norsk Geografisk Tidsskrift–Norwegian Journal of Geography*, 62, 135–250.

Lund, R., Dei, L.A, Boakye, K.A. and Opuku-Agyemang, E. 2008b. It is All About Livelihoods: A Study of Women Working in Stone Chip Production in Cape Coast Municipality, Ghana. *Norsk Geografisk Tidsskrift–Norwegian Journal of Geography*, 62, 139–48.

Lund, R., Khasalamwa, S. and Tete, S.Y.A. 2011. Beyond the Knowledge–Action Gap: Challenges of Implementing Humanitarian Policies in Ghana and Uganda. *Norsk Geografisk Tidsskrift–Norwegian Journal of Geography*, 65, 63–74.

Moss, P. (ed.) 2001a. *Placing Autobiography in Geography*. Syracuse, New York: Syracuse University Press.

Moss, P. 2001b. Writing One's Life, in *Placing Autobiography in Geography*, edited by P. Moss. Syracuse, New York: Syracuse University Press, 1–21.

Moss, P. 2001c. Engaging Autobiography, in *Placing Autobiography in Geography*, edited by P. Moss. Syracuse, New York: Syracuse University Press, 188–92.

Nagar, R. 2006. Local and Global, in *Approaches to Human Geography*, edited by S. Aitken and G. Valentine. London – Thousand Oaks – New Delhi: Sage Publications, 211–17.

Nash, H.J. 1994. Women in the Field: Critical Feminist Methodologies and Theoretical Perspectives. *The Professional Geographer*, 46, 54–66.

Owusu, G. 2005. *The Role of District Capitals in Regional Development: Linking Small Towns, Rural–Urban Linkages and Decentralisation in Ghana*. Doctoral theses at NTNU, 2005:118. Department of Geography. Trondheim: Norwegian University of Science and Technology (NTNU).

Owusu, G. 20143. Housing the Urban Poor in Metropolitan Accra, Ghana: What is the Role of the State in the Era of Liberalization and Globalization?, in *Alternative Development: Unravelling Marginalization, Voicing Change*, edited by C. Brun, P. Blaikie and M. Jones. Farnham: Ashgate, 73–85.

Owusu, G. and Lund, R. 2004. Markets and Women's Trade: Exploring their Role in District Development in Ghana. *Norsk Geografisk Tidsskrift–Norwegian Journal of Geography*, 58, 113–24.

Owusu, G., Agyei-Mensah, S. and Lund, R. 2008. Slums of Hope and Slums of Despair: Mobility and Livelihoods in Nima, Accra. *Norsk Geografisk Tidsskrift–Norwegian Journal of Geography*, 62, 180–90.

Palttala, P., Boano, C., Lund, R. and Vos, M. 2012. Communication Gaps in Disaster Management: Perceptions by Experts from Governmental and Non-Governmental Organizations. *Journal of Contingencies and Crisis Management*, 20, 2–12.

Panda, S.M. 2014. Right to Rights: *Adivasi* (Tribal) Women in the Context of a Not-So-Silent Revolution in Odisha, India, in *Alternative Development: Unravelling Marginalization, Voicing Change*, edited by C. Brun, P. Blaikie and M. Jones. Farnham: Ashgate, 191–206.

Perera, B.M.K., Fernando, M.H.J.P., Wicremasinghe, Y.M. and Perera, M.S. 1983. *Chena* Development Experiment – Paindikulama. *Norsk Geografisk Tidsskrift*, 37, 187–96.

Resurrección, B.P. 2014. Discourses That Hide: Gender, Migration and Security in Climate Change, in *Alternative Development: Unravelling Marginalization, Voicing Change*, edited by C. Brun, P. Blaikie and M. Jones. Farnham: Ashgate, 219–40.

Rose, G. 1993. *Feminism and Geography: The Limits of Geographical Knowledge*. Cambridge: Polity Press.

Roth, R. 2001. A Self-Reflective Exploration into Development Research, in *Placing Autobiography in Geography*, edited by P. Moss. Syracuse, New York: Syracuse University Press, 121–37.

Seneviratne, H.M.M.B. 2003. *Settlers of Mahaweli System C and Their Sibling Families at Home Villages*. Dr. polit. thesis. Department of Geography. Trondheim: Norwegian University of Science and Technology (NTNU).

Shanmugaratnam, N., Lund, R. and Stølen, A.K. 2003. *In the Maze of Displacement: Conflict, Migration and Change*. Kristiansand: Høyskoleforlaget – Norwegian Academic Press.

Skotte, H. 2004. *Tents in Concrete: What Internationally Funded Housing Does to Support Recovery in Areas Affected by War; The Case of Bosnia–Herzegovina*. Dr. ing. thesis. Doctoral theses at NTNU, 2004:61. Department of Urban Design and Planning. Trondheim: Norwegian University of Science and Technology (NTNU).

Skotte, H. 2014. Teaching to Learn – Learning to Teach: Learning Experiences from the Reality of an Ever-Changing World, in *Alternative Development: Unravelling Marginalization, Voicing Change*, edited by C. Brun, P. Blaikie and M. Jones. Farnham: Ashgate, 39–53.

Smith, D.E. 1988. *The Everyday World as Problematic: A Feminist Sociology*. Milton Keynes: Open University Press.

Smith, D.E. 1990. *The Conceptual Practices of Power: A Feminist Sociology of Knowledge*. Boston: Northeastern University Press.

Smith, D.E. 2005. *Institutional Ethnography: A Sociology for People.* Lanham
    – New York – Toronto – Oxford: AltaMira Press, Rowman & Littlefield
    Publishers, Inc.
Tete, S.Y.A. 2011. *Protracted Displacement and Solutions to Displacement:
    Listening to Displaced Persons (Refugees and IDPs) in Ghana and Sri Lanka.*
    Doctoral theses at NTNU, 2011:287. Department of Geography. Trondheim:
    Norwegian University of Science and Technology (NTNU).
Tschudi, A.B. 1983. Worlds Apart, in *The Practice of Geography*, edited by A.
    Buttimer. London and New York: Longman, 35–42.
Vågenes, V. 2014. Muted Power – Gender Segregation and Female Power, in
    Alternative Development: Unravelling Marginalization, Voicing Change,
    edited by C. Brun, P. Blaikie and M. Jones. Farnham: Ashgate, 101–15.
Valentine, K.B. 1998. Ethical Issues in the Transcription of Personal Narratives,
    in *The Future of Performance Studies: Visions and Revisions*, edited by S.J.
    Dailey. Washington DC: National Communication Association, 221–5.
Vandsemb, B.H. 2007. *Making a Place and (Re)constructing a Life: The Role of
    Gender in a in Spontaneous Frontier Migration in Tanamalwila, Sri Lanka.*
    Doctoral theses at NTNU, 2007:28. Department of Geography. Trondheim:
    Norwegian University of Science and Technology (NTNU).
Vandsemb, B.H. 2014. Spontaneous Frontier Migration in Sri Lanka: Conflict
    and Cooperation in State–Migrant Relations, in *Alternative Development:
    Unravelling Marginalization, Voicing Change*, edited by C. Brun, P. Blaikie
    and M. Jones. Farnham: Ashgate, 269–85.

Index

References to figures are shown in *italics*. References to tables are shown in **bold**.

ERPs, *see* Economic Recovery
Programmes (ERPs)
Escobar, Arturo 172, 175–8, 185, 187
Esplen, E. 223
ethical issues, and research 319
ethnicity, and climate adaptation 226, 227
ethnocentrism 6, 12
ethnographic research 5, 291
ethnonational identity, and gender 155,
157, 160–61, 165
European Union
*Climate Change and International
Security* 229–30
European Qualifications Framework
for lifelong learning (EU 2008)
39, 43
export processing zones (EPZs)
and gender 135–8
and multinational corporations 137,
148, 149, 150
and trade unions 137–8, 141, 148–9
see also women's agency and human
rights violations (Sri Lanka EPZ)

Fanjingshan Reserve, China, *see* local
values and the Fanjingshan
National Nature Reserve (FNNR)
project (China)
FAO (Food and Agriculture Organization)
221
FDI (Foreign Direct Investment), *see*
Foreign Direct Investment (FDI)
female empowerment, *see* agency;
empowerment; female power
in Hadendowa society (Sudan);
women's agency and human rights
violations (Sri Lanka EPZ)
female power in Hadendowa society (Sudan)
background 101, *102*, 103, *104*
gender segregation 105, 109–11, 112,
113
material capital 106
social capital and marriage bond 106–7
symbolic capital and gender ideals 108–9
women/men segregation and the
marital relationship 109–11
women's muted power and gender
segregation 112–13

women's muted power (*zar* patient
case) 111–12
feminism
and alternative development 4, 6, 11
and autobiographical approaches
307–9
and climate change 223
and climate justice 219–20, 223
criticism of Western feminism 159
and development studies 307
and gender 158–61
and globalization 10
and grassroots women's struggles
195–6
vs instrumentalism 97
and local values focus 171–2, 174
Lund's feminism 172, 175, 187, 318,
322
poststructural feminism 171, 172–7
and social sciences 6
see also empowerment; feminist
geography; feminist politics in Sri
Lanka; gender; women
feminist geography 187, 312, 321, 324
feminist politics in Sri Lanka
conflict, gender and feminism 155–7
context of Sri Lankan civil war
157–8
gender is not enough 158–61, 165
University Teachers for Human Rights,
Jaffna (UTHR(J))
Broken Palmyra, The 157, 161,
162–5
Thiranagama on class and gender
164
Thiranagama's biographical details
155, 156, 157, 161–2
Thiranagama's 'No more tears sister'
(*The Broken Palmyra*) 162–5
see also feminism; feminist geography
Ferguson, J. 232
Few, R. 260
financial crisis (2007–), and Millennium
Development Goals (MDGs) 12
fish traders, *see* coping capacity and female
border fish traders in Cambodia
Flint, C. 290
Fløysand, Arnt 11, 13, 324

Liberian refugees 291
Oromo refugees 291
refugee camps 163, 249, 250–51, 291,
    294
see also displacement; forced migration;
    forced migration and geographical
    research (NTNU); internal
    displacement and women's agency
    in Sri Lanka; internally displaced
    persons (IDPs); migration;
    spontaneous frontier migration in
    Sri Lanka; UNCHR (UN Refugee
    Agency)
remittances 210, 295
Renaud, F. 231
research
    academia vs the real world 40
    academic publishing, Eurocentricism
        of 321–2
    action research 291, 292, 297–9, 321,
        322, 324–5
    actor-oriented perspective 246, 291,
        293, 321
    autobiographical approaches 307–10,
        312–13
    'biographical turn' 309
    collaboration between Global South
        and Global North 12
    critical scholarship 4–5, 298
    'dual imperative' concept 287–9
    ethical issues 319
    histories of knowledge 307–8
    Lund on 40, 318, 319–20, 321–2
    migratory research 292
    participatory research 30, 297, 299
    and political practice 27, 28–9
    and procedural justice 2, 25, 27
    'real time research' (RTR) 298, 319–20
    research funding 28–9
    researchers' patriarchal orientation 198
    as a way of learning 42
    see also alternative development;
        development studies/theory;
        geography; knowledge; learning
        and teaching (urban planning)
Research Group on Forced Migration
    (NTNU) 207, 287, 299, 319

resources, *see* access to resources
Resurrección, B.P. 4, 8, 15, 220, 323
reterritorialization 290
    see also place
right to return, research on 290
right to rights and Odisha tribal (*Adivasi*)
    women (India)
    background: economic reform and
        human rights violations 191–5
    conceptual underpinnings 195–6
    Odisha *Adivasi* women 196–7
    women's resistance movements and
        organizations 198–202
rights-based approach (RBA) 50
Rintala, Sami 47
Ritzer, G. 136
Robbins, P. 5, 26, 29
Robertson, C. 121
Robinson, Sir Ken 40
Rose, Gillian 309
Rostow, W.W. 171
Roth, Robin 309
Rowlands, J. 138–9
RTR, *see* 'real time research' (RTR)
Rural Studios (Auburn University, Alabama)
    50
Ruwanpura, K. 160

Sagmo, T.H. 294
Sahoo, Pravata 202
Sajor, E.E. 78
salmon industry (Chile)
    OECD/CEPAL recommendations 66
    see also Norwegian FDI and Chilean
        salmon industry
Sanderson, D. 45
SAPs, *see* Structural Adjustment
    Programmes (SAPs)
scholarship, *see* knowledge; research
Schön, Donald 42, 49
Schumacher, Ernst, *Small is Beautiful* 171
Schwartz, P. 228
Scott, James C. 32
Scott, Joan W. 227
sea-level changes 212, 230
security, and climate change 219,
    228–32

For Product Safety Concerns and Information please contact our EU representative GPSR@taylorandfrancis.com, Batch 1, Taunus, Verlag GmbH, Kandlerstrasse 25, 81379 München, Germany